THE DESERT BIGHORN

Contributing Authors

Rex W. Allen
Bruce M. Browning
Gerald I. Day
O. V. Deming
William Graf
Hatch Graham
Campbell Grant
Tommy L. Hailey

Charles G. Hansen
Fred L. Jones
Warren E. Kelly
Richard H. Manville
Gale Monson
Norman N. Simmons
Jack C. Turner
Richard A. Weaver

Line Drawings by Patricia Hansen

THE DESERT BIGHORN

ITS LIFE HISTORY, ECOLOGY & MANAGEMENT

Gale Monson and Lowell Sumner,
EDITORS

THE UNIVERSITY OF ARIZONA PRESS
Tucson, Arizona

THE UNIVERSITY OF ARIZONA PRESS
Copyright © 1980
The Arizona Board of Regents
All Rights Reserved
Manufactured in the U.S.A.

Library of Congress Cataloging in Publication Data

The Desert bighorn.

    Bibliography: p.
    Includes index.
    1. Bighorn sheep. 2. Wildlife management.
I. Monson, Gale. II. Sumner, Lowell. III. Allen, Rex W.
QL737.U53D48   599.73'58   80-18889
ISBN 0-8165-0689-2
ISBN 0-8165-0713-9 (pbk.)

Dedication

To Dr. Charles G. Hansen, loved and respected by all who knew him, and long to be remembered for his extensive desert bighorn studies and publications. Prior to Chuck's death in an airplane accident while censusing bighorn in Canyonlands National Park in 1973, he had worked with bighorn with the U.S. Fish and Wildlife Service and, later, the National Park Service. He died doing the thing he liked most, working with the animals he knew so well.

To readers of this book, it will be obvious how much Chuck, as author or co-author of many chapters and contributor to others, had learned about bighorn. His knowledge can now be shared with everyone, as he had hoped. Although all readers did not have the chance to know Chuck, through his writings they may come to feel better acquainted. He was one of the best friends the desert bighorn ever had.

TECHNICAL STAFF
DESERT BIGHORN COUNCIL

Contents

About the Editors & Authors   xv

Preface   xix

Acknowledgments   xxi

1. The Origin and Relationships of American
   Wild Sheep   *Richard H. Manville*   1

2. The Desert Bighorn and Aboriginal Man   *Campbell Grant*   7
   Aboriginal Hunting Methods, 8; The Desert Bighorn in Rock Art, 10; Ceremonial Use of the Desert Bighorn, 28; The Desert Bighorn as an Art Motif, 32; The Desert Bighorn as Artifact Material, 36; The Desert Bighorn as a Food Resource, 39.

3. Distribution and Abundance   *Gale Monson*   40
   Arizona, 41; California, 44; Nevada, 45; New Mexico, 46; Texas, 47; Utah (and Colorado-Wyoming), 48; Baja California Norte and Baja California Sur, 49; Chihuahua, 50; Coahuila, 50; Sonora, 51; Recapitulation of Desert Bighorn Numbers, 51.

4. Physical Characteristics   *Charles G. Hansen*   52
   General Body Characteristics, 52; Hair and Skin Characteristics, 54; Horns, 56; Eyes, 58; Teeth, 59; Exterior Glands, 60; Gaits, 60; Physiological Characteristics, 62; Respiration and Pulse, 63.

[vii]

Contents

5. **Habitat**   *Charles G. Hansen*   64
   Great Basin Desert, 66; Painted Desert, 67; Mohave Desert, 68; Sonoran Desert, 70; Colorado Desert, 71; Baja California Mountain Desert, 72; Chihuahuan Desert, 73; Habitat Requirements of Bighorn, 74; Food, 76; Water, 77; Topography, 77; Climate, 78.

6. **Food**   *Bruce M. Browning & Gale Monson*   80
   Great Basin Desert, 81; Painted Desert, 81; Mohave Desert, 84; Colorado Desert, 88; Sonoran Desert, 89; Baja California Mountain Desert, 93; Chihuahuan Desert, 94; Miscellaneous Food Habits Data, 96; Management Implications, 97.

7. **Water**   *Jack C. Turner & Richard A. Weaver*   100
   Water consumption, 101; Water Shortages, 106; Water Development, 107; Water Balance Physiology, 108.

8. **Senses and Intelligence**   *Charles G. Hansen*   113

9. **Behavior**   *Norman M. Simmons*   124
   Seasonal Activities, 125; Seasonal Movements and Home Ranges, 127; Daily Activities, 133; Daily Movements, 137; The Environment and Daily and Seasonal Activities, 141; Behavior and Topography, 144.

10. **Reproduction**   *Jack C. Turner & Charles G. Hansen*   145
    Puberty, 145; Reproductive Season, 146; Gestation, 146; Lambing, 148; Parturition, 148; Reproductive Potential, 148; Percentage of Productive Ewes, 149; Sex Ratio, 149; Relation to Nutrition, 150.

11. **Growth and Development**   *Charles G. Hansen & O. V. Deming*   152
    Fetal Development, 152; Lambs, 153; General Growth and Development, 156; Growth of the Skull, 163; Growth of the Horns, 165; Age Determination by Horn Growth, 166; Tooth Development, 169; Age Determination by Tooth Replacement, 171; Age Determination by Size and Shape of the Whole Animal, 171; Factors Influencing Growth and Development, 171.

12. **Natural Mortality and Debility**   *Rex W. Allen*   172
    Parasites and Disease, 172; Other Mortality and Debility Factors, 181; Inbreeding, 185.

13. **Predator Relationships**   *Warren E. Kelly*   186
    Wolf, 187; Coyote, 187; Gray Fox, 189; Bobcat, 190; Mountain Lion, 190; Jaguar and Ocelot, 193; Eagles, 193; Impact of Predators on Bighorn Populations, 194.

14. **Competition**   *Fred L. Jones*   197
    Deer, 197; Burros, 202; Domestic Livestock, 207; Other Animals, 214; Within Herds, 215.

15. Population Dynamics   *Charles G. Hansen*   217
       Population Increase, 218; Population Composition, 218; Group Composition, 221; Life Tables, 222; Population Fluctuation, 230; Population Growth, 231; Population Turnover, 235.

16. Sign Reading and Field Identification   *Fred L. Jones*   236

17. Population Survey Methods
                             Norman M. Simmons & Charles G. Hansen   260
       Aerial Surveys, 260; Water-hole Surveys, 265; Surveys by Boat, 269; Surveys on Foot and Horseback, 270.

18. Capturing, Handling, and Transplanting
                Charles G. Hansen, Tommy L. Hailey, & Gerald I. Day   273
       Bighorn Capture, 273; Restocking Former Ranges, 282; Marking Bighorn, 285.

19. The Impact of Modern Man   *Hatch Graham*   288
       Modern Man in the Past, 288; Man Today, 292; The Future Impact of Man, 308.

20. Habitat Protection and Improvement   *William Graf*   310
       Improvement of Forage Conditions, 311; Competitive Utilization and Abuse, 313; Water Improvements and Increases, 315; Steps To Improve and Protect Bighorn Ranges, 318.

21. Habitat Evaluation   *Charles G. Hansen*   320
       Factors for Rating Bighorn Habitat, 320; Bighorn Habitat Index, 335.

22. Hunting   *Warren E. Kelly*   336
       Early History, 336; Harvest Objectives, 338; Considerations Governing the Hunting Seasons, 338; Regulated Hunting, 338; Poaching, 340.

Appendix: Boone and Crockett Club Official Score Sheet   343

Bibliography   345

Index   365

ILLUSTRATIONS

1.1. Approximate distribution of the races of bighorn sheep, *Ovis canadensis,* during period of early exploration of the West.   4

2.1. Movements of the Shoshoneans north and east through the desert bighorn range.   12

2.2. Pecked Shoshonean-type bighorn drawings.   15

2.3. Highly stylized bighorn of the Late Period.   16

2.4. Bighorn panel from the Coso Peak.   16

2.5. (a) Petroglyph representations of the atlatl from the Coso Range, Inyo County, California. (b) Atlatl and dart from Atlatl Rock, Valley of Fire, Nevada. 20
2.6. Bighorn petroglyphs in Coso Range, Inyo County, California. 21
2.7. Non-Shoshonean-type bighorn representations. 23
2.8. Petroglyph representations of "medicine bags." 24
2.9. Ceremonial use of horns. 29
2.10. Rock art representations of men wearing bighorn headdresses. 32
2.11. Rock art figures with bighorn headdresses. 33
2.12. Desert bighorn effigies from the lower Rio Grande region of southern New Mexico. 34
2.13. The bighorn as an art motif. 35
2.14. Head of Hohokam bone hairpin. 37
2.15. Modern basket made by the Coso (or Panamint) Indians of Death Valley in the 1930s. 38
3.1. Distribution and abundance of desert bighorn. 42
4.1. Head conformation of a desert bighorn ewe. 53
4.2. Shedding pattern of desert bighorn. 55
4.3. Rams of some desert bighorn populations exhibit unusually flaring horns. 57
4.4. The bighorn "run," one of the five traveling gaits. 61
5.1. North American desert and its subdivisions, excluding the parts of Mexico where bighorn are absent. 65
5.2. Spring in the Toiyabe Range, Nevada, Great Basin Desert. 67
5.3. On the Desert National Wildlife Range, a typical part of the Mohave Desert. 69
5.4. The Sonoran Desert at the northern boundary of the Cabeza Prieta Wildlife Refuge, Arizona (formerly Cabeza Prieta Game Range). 71
5.5. Colorado Desert habitat. San Ysidro Mountains on the Anza-Borrego Desert State Park, California. 73
5.6. Chihuahuan Desert habitat. The Sierra de los Castro, north of Saltillo, Coahuila, Mexico. 75
6.1. Grass, forbs, or browse? 82

6.2. Bighorn feeding on bud sage *(Artemisia spinescens)*. Blue Notch Canyon, San Juan County, Utah.	99
7.1. A ram kneels to drink, Castle Dome Mountains, Yuma County, Arizona.	103
7.2. Pothole affording water for only a short period before evaporation dries it up. This is the only water hole in the Nopah Range, Inyo County, California.	105
7.3. A "death trap" water hole in the Chocolate Mountains, Imperial County, California. The remains of thirty-four trapped bighorn were found in this steep-sided tank in 1969.	108
7.4. Two bighorn ewes and two lambs nervously scrutinize man-made Eagle Tank before entering to drink. Sierra Pinta, Yuma County, Arizona.	109
9.1. Family group of a matriarch ewe, her yearling ram, and a 4-month-old female lamb. Agua Dulce Mountains, Pima County, Arizona.	126
9.2. A group of sixteen rams in Burro Canyon, Kofa Mountains, Arizona.	128
9.3. Bighorn of both sexes and different ages arranged in a cline.	129
9.4. Bighorn occasionally cross between the Sierra Pinta and the Sierra Cabeza Prieta. Cabeza Prieta National Wildlife Refuge, Arizona.	131
9.5. Two rams about to "clash" at the Red Rock big game research area, Grant County, New Mexico.	138
9.6. Occasionally combat horn-clashing does not result in a head-on collision. River Mountains, Nevada.	139
9.7. A ram lamb exhibits playful behavior by leaping in the air before his keeper at the Arizona-Sonora Desert Museum.	142
9.8. A yearling ram at entrance to cave used for shade and relief from the sun and dessicating winds on hot days. Near Eagle Tank, Sierra Pinta, Yuma County, Arizona.	143
10.1. Variation in breeding and lambing seasons within the ranges of four varieties of bighorn sheep *(Ovis canadensis)*.	147
11.1. Desert bighorn lambs being bottle fed at the University of Nevada Veterinary Medical Laboratory at Reno.	157
11.2. Full figure of day-old lamb.	158
11.3. Full figure of two-month-old ewe.	158

11.4.	Full figure of four-month-old ewe.	158
11.5.	(a) Four-month-old lambs. (b) Six-month-old lambs. (c) One-year-old bighorn.	158
11.6.	"Frisky," a male bighorn lamb born at Corn Creek, Desert National Wildlife Range, Nevada, at the age of 3.5 months.	160
11.7.	Horn ring method of aging desert bighorn.	167
12.1.	"Horny dermatitis" in a ram collected from the Kofa Game Range, Arizona.	175
14.1.	The mule deer, one of the natural competitors of the desert bighorn. Kofa Mountains, Yuma County, Arizona.	198
14.2.	A hybrid 3-year-old ram at Quitobaquito, Pima County, Arizona, in 1922—the result of a bighorn ram and domestic ewe mating.	210
15.1.	Mean bighorn group sizes by group classes and seasons, Cabeza Prieta Wildlife Refuge.	224
15.2.	Mean bighorn group sizes by group classes in two different bighorn refuges.	225
15.3.	Change in group composition by season in Cabeza Prieta Wildlife Refuge.	226
15.4.	Survivorship curves for desert bighorn on Desert National Wildlife Range.	229
15.5.	Natural mortality of desert bighorn by percentage for each yearly age class from 2 years to end of life span.	230
15.6.	Population growth of bighorn released on Hart Mountain, Oregon.	232
15.7.	Population growth and decline of bighorn on Desert National Wildlife Range between 1936 and 1968.	234
16.1.	Bighorn pellets.	237
16.2.	An old ewe positively identified by her broken and dangling left horn, who was followed intermittently in a behavior study for a year. Sierra Pinta, Yuma County, Arizona.	250
17.1.	A large, relatively comfortable blind made of saguaro "ribs." Heart Tank, Sierra Pinta, Yuma County, Arizona.	268
18.1.	Design of a corral trap built around a water hole on Desert National Wildlife Range, Nevada.	275
18.2.	An 8-year-old ram marked on its white rump patch with red dye as it entered the water hole under the electrically operated spray nozzle. Eagle Tank, Sierra Pinta, Yuma County, Arizona.	286

20.1.	Bighorn drinking at water collected from a developed spring on Desert National Wildlife Range, Nevada.	317
21.1.	Land sections classified for desert bighorn sheep use.	332
21.2.	Actual bighorn sheep use of land.	333

TABLES

10.1.	Numbers and Sex Ratios of Bighorn Observed on Desert National Wildlife Range	150
11.1.	Measurements of Individual Bighorn From Kofa Game Range, Arizona	162
11.2.	Percentage of Growth Occurring During Different Age Intervals in Desert Bighorn Ewe Skulls From Desert National Wildlife Range	164
11.3.	Percentage of Growth Occurring During Different Age Intervals in Desert Bighorn Ram Skulls From Desert National Wildlife Range	164
11.4.	Tooth Eruption and Replacement in the Lower Jaw of Bighorn	169
11.5.	Aging Bighorn According to the Replacement of Incisiform Teeth	170
11.6.	Characteristics That Can Be Used To Roughly Distinguish Young From Old Bighorn in the Field	170
12.1.	Parasites Reported From Desert Bighorn	176
15.1.	Theoretical Age Distribution of Surviving Bighorn Based on the Percentage of Natural Mortality of Ewes and Rams on Desert National Wildlife Range for the Years 1953 Through 1967	219
15.2.	Selected Ram : Ewe and Lamb : Ewe Ratios From Desert National Wildlife Range	220
15.3.	Sizes of Ewe, Lamb, and Yearling Bands on Desert National Wildlife Range Between 1947 and 1952	222
15.4.	Sizes of Bighorn Groups in Arizona	223
15.5.	Mean Bighorn Group Sizes in Cabeza Prieta and Kofa-Ajo Mountains	223
15.6.	Life Table for Desert Bighorn Rams	227
15.7.	Life Table for Desert Bighorn Ewes	228
15.8.	Sex and Age Composition of Bighorn From 1951 to 1954 on Wildhorse Island, Montana	228

15.9.	Adjusted Age Composition of Bighorn on Wildhorse Island, Montana, From 1951 to 1954	229
15.10.	Proportion of Bighorn Dying in Relation to Those Available To Die	231
15.11.	Population Growth of Bighorn Released on Hart Mountain, Oregon	232
15.12.	Population Growth and Decline of Bighorn on Desert National Wildlife Range Between 1936 and 1974	233
15.13.	Year of Death for 610 Bighorn Found on Desert National Wildlife Range Between 1936 and 1967	234
15.14.	Mean Mortality Rate of Rams and Ewes on Desert National Wildlife Range	235
16.1.	Sample Bighorn Pellet Group Counts, California	239
21.1.	Categories and Values of the Tools Used To Evaluate Desert Bighorn Habitat	322
21.2.	Classification of Total Scores for One-mile-square Sections of Land Evaluated as Desert Bighorn Habitat, Desert National Wildlife Range, Nevada	324
21.3.	Tool IV : Evaporation	327

About The Editors & Authors

ABOUT THE EDITORS...

GALE MONSON saw his first desert bighorn in 1934 in the Sikort Chuapo Mountains of the Papago Indian Reservation in Arizona while making a range survey on horseback. A few years later he was studying bighorn on the Kofa and Cabeza Prieta Game Ranges, then newly established for the protection and management of desert bighorn under the administration of the U.S. Fish and Wildlife Service. Later he was in charge of Havasu National Wildlife Refuge on the Colorado River for eight years, then became manager of the Kofa and Cabeza Prieta areas and the Imperial National Wildlife Refuge for another eight years. Monson completed his career with the Fish and Wildlife Service as a staff assistant in the Wildlife Refuges Division in Washington, D.C. He then became editor of the *Atlantic Naturalist*. On his return to Arizona he joined the Arizona-Sonora Desert Museum as resident and weekend supervisor. He co-authored with Allan Phillips and Joe Marshall *The Birds of Arizona,* and with Phillips *A Checklist of the Birds of Arizona.*

LOWELL SUMNER began his career in the wildlife profession by studying growth rates in young owls, hawks, and eagles. In 1935 he joined the National Park Service as a biologist to commence a survey of the desert bighorn of

Death Valley, which culminated in the first organized bighorn censuses for that area. He enlarged on this work 20 years later when he helped the California Department of Fish and Game with another census at Death Valley. Sumner studied ecological requirements and presented management recommendations for deer, elk, and bighorn of western National Parks, including Alaska, from 1935 until his retirement. The status and management of desert bighorn in Death Valley and Joshua Tree National Monuments held his special interest and support throughout his career. He published widely in this field and was senior author of *Birds and Mammals of the Sierra Nevada*. At the time of his retirement he was in charge of natural science investigations for the National Park service in Washington, D.C.

ABOUT THE CONTRIBUTING AUTHORS...

REX W. ALLEN for many years was a parasitologist for the Agricultural Research Service, U.S. Department of Agriculture. He has investigated desert bighorn parasites in New Mexico, Arizona, Nevada, and California, participating in field collection programs as well as executing laboratory analyses.

BRUCE M. BROWNING was in charge of the Food Habits Section of the Wildlife Investigations Laboratory of the California Department of Fish and Game. In this capacity he analyzed the stomach contents of many desert bighorn specimens from Arizona, Nevada, and California.

GERALD I. DAY, a veteran researcher for the Arizona Game and Fish Department, was a pioneer in the use of tranquilizers that would make possible the live capture and transportation of big game animals, including bighorn.

OSCAR V. DEMING before retirement was a biologist with U.S. Fish and Wildlife Service wildlife refuges. He served for a number of years at Desert National Wildlife Range, where his main concern was desert bighorn investigations.

WILLIAM GRAF was professor of wildlife management at San Jose State College with a special interest in bighorn management problems. He died on an Alaskan expedition.

HATCH GRAHAM, while a Resource Officer with the U.S. Forest Service, accomplished extensive research into desert bighorn behavior in relation to human activities in the San Bernardino National Forest in California.

CAMPBELL GRANT, well-known archaeologist and author, researched Indian bighorn drawings and artifacts, principally in the Coso Range of California.

TOMMY L. HAILEY was the first custodian of the desert bighorn transplant population in Texas for the state's Parks and Wildlife Department.

CHARLES G. HANSEN for a number of years was a wildlife biologist for the U.S. Fish and Wildlife Service on Desert National Wildlife Range and for the National Park Service in Death Valley. At the time of his death in an airplane accident he was head of the Cooperative National Park Resources Studies Unit at the University of Nevada, Las Vegas.

PATRICIA A. HANSEN, who contributed the line drawings to this volume, lived for a number of years on Desert National Wildlife Range, Nevada, and in Death Valley National Monument, California. At the former place she was foster mother for desert bighorn born in captivity, observing and caring for them from lambhood until they were several months old.

FRED L. JONES, while with the California Department of Fish and Game, carried out extensive field studies of bighorn in the Sierra Nevada, in Death Valley, and in the Santa Rosa Mountains.

WARREN E. KELLY was a desert bighorn investigator for Arizona and Nevada game and fish agencies before being involved in similar work for the San Bernardino National Forest in California and the Humboldt National Forest in Nevada.

RICHARD H. MANVILLE, deceased, was director of the U.S. Fish and Wildlife Service's Bird and Mammal Laboratories, housed at the Smithsonian Institution, Washington, D.C.

NORMAN M. SIMMONS has studied bighorn behavior in Colorado and Arizona, as well as behavior of Dall sheep in Canada's Northwest Territories.

JACK C. TURNER conducted extensive physiological studies of desert bighorn at the Philip L. Boyd Deep Canyon Research Center at Palm Desert, California.

RICHARD A. WEAVER—"Mr. Bighorn" in the California Department of Fish and Game—became famed for his wide experience with bighorn in the far-flung desert mountain ranges of California's Mohave and Colorado deserts.

Preface

Ever since the first white explorers and hunters penetrated the mountains of the Wild West and the Far North, the bighorn sheep has possessed a romantic image, identified with the traditional remoteness and inaccessibility of its craggy, wilderness habitat.

The magnificent curling horns of the mature ram and the exceptionally tasty flesh of younger animals brought about in earlier years the demise of great numbers at the hands of explorers, prospectors, and trophy hunters—and by poachers, even to the present time. But by far the most destructive factor has been the penetration and permanent alteration of the bighorn's habitat by a continually advancing tide of human settlement.

Although there are occasional notable exceptions in large parks and wildlife refuges, the bighorn has for the most part continued to be one of North America's most intractable and unmanageable wilderness animals. Where the wilderness has faded away, so usually has the bighorn.

Most large North American trophy animals were significantly reduced in numbers by the end of the nineteenth century, but with the development of the science of wildlife management and the adoption of more adequate hunting regulations, most species had recovered to a marked degree by the 1920s. Not so the bighorn. Despite many years of total protection from hunting, its numbers continued to diminish, particularly in the Southwest, leading some conservationists to predict early extinction for the various races that are collectively known as the desert bighorn.

In the 1930s, studies to determine the causes and possible remedies for the desert bighorn's alarming decline were undertaken by various state and federal wildlife agencies and some universities, but the animal's elusiveness and the inaccessibility of most of its remaining strongholds severely hampered the research programs. In 1957 the Desert Bighorn Council came into being when a group of wildlife scientists, administrators, and managers gathered for the purpose of pooling the still sparse information obtained up to that time about the desert bighorn. The resulting exchange of findings was recorded as the first number of the *Desert Bighorn Council Transactions,* which has been published annually ever since and continues to be a major source of new information on this animal.

In 1962 the Council proposed an inventory of all existing information on the desert bighorn, a program of cooperative studies with priorities assigned to fill in gaps in existing information, and publication under a multiple authorship of all resulting information in a book designed to appeal to all persons, whether professional or not, who might be interested in any aspect of the desert bighorn, its habits, and requirements for survival. A Technical Staff was appointed, one of whose duties was to plan the organization of the book and review the chapters on completion.

By 1967 volunteers had been secured for the chapters and the editors had been selected. By 1971 all chapters had been completed in preliminary form and, through the generosity of the National Park Service, were duplicated in limited numbers to facilitate a technical review. This came to be known as the "preliminary edition."

The present volume thus is a distillation of 40 years of organized effort to learn about and to rescue one of the most magnificent as well as threatened desert species—one that retains an important part of its original habitat in Arizona and is identified, perhaps more completely than any other native animal, with the wild beauty of Arizona's desert mountain ranges.

THE EDITORS

Acknowledgments

The Desert Bighorn Council is grateful to Carol Beckley, Deene Mills, Shelley Momii, Georgia Razhon, Jeanne Roster, and Martie Sebek for typing the manuscript of the "preliminary edition," and to Mary Mielke for preparation of voluminous correspondence with many prospective contributors during the organizational stages of the project. Thanks go to Lewis E. Carpenter, William H. Carr, Harry L. Gordon, Paul R. Johnson, Seymour H. Levy, Bonnie McGill, Robert P. McQuivey, Carlos Nagel, and E. Linwood Smith for special assistance.

Hatch Graham organized the initial program of uniformly retyping all manuscripts after the first editing and arranged for their subsequent circulation to an extensive panel of technical reviewers. James A. Blaisdell carried to completion the circulation of manuscripts and the resulting feedback of suggestions to authors and arranged for the submission of illustrations; he also performed the bulk of compilation of the Bibliography.

The Council expresses special appreciation to Robert M. Linn for arranging to have the National Park Service duplicate the entire manuscript in its preliminary form as the "preliminary edition" for distribution to reviewers. Special thanks also go to the following reviewers, whose suggestions materially improved the manuscript in many ways:

Rex W. Allen, Bonnar Blong, Donald H. Bolander, Harold T. Coss, Douglas B. Evans, Gerald Gates, Frank W. Groves, Tommy L. Hailey,

Norman V. Hancock, Jack B. Helvie, Rodney T. John, Roger D. Johnson, Cecil A. Kennedy, John H. Kiger, Parry A. Larson, Levon Lee, Jessop B. Low, Wallace G. Macgregor, Robert E. McCarthy, George Merrill, Tom D. Moore, Nick J. Papez, Richard Raught, William H. Rutherford, Edward R. Schneegas, Paul W. Shields, Walter Snyder, George K. Tsukamoto, Ralph E. and Florence B. Welles, George W. Welsh, Wayne W. West, and Lanny O. Wilson.

The Desert Bighorn Council's Technical Staff, which also reviewed all manuscripts and offered important suggestions, comprised the following: James A. Blaisdell, W. Glen Bradley, Gerald I. Day, Ramiro Garcia Perez, Hatch Graham, the late Charles G. Hansen, Al R. Jonez, John P. Russo, Jose Sanchez Samano, Norman M. Simmons, Richard A. Weaver, John E. Wood, and James D. Yoakum. Charles L. Douglas of the National Park Service at Las Vegas, Nevada, was of great help at crucial times when assistance was especially needed.

The reader will note frequent inclusion of contributed, previously unpublished, material throughout the text. The full name of the contributor is given in each case, excluding such phrases as "personal communication" or "in litt."

We give thanks to the following for help in typing the final manuscript: Lydia Berry, Eunice Cooper, Charles L. Douglas, Larry A. Dunkeson, Gerald Gates, Hatch Graham, Fred L. Jones, Warren E. Kelly, Flora Martinez, Cindy Rockenfield, John P. Russo, Norman M. Simmons, LaVerne Taylor, Jennifer Trent, Jack C. Turner, Colleen E. Wilson, and Lanny O. Wilson.

We are grateful to Mario Luis Cossio, former Director General of Wildlife in the Department of Conservation in Mexico, and to Ticul Alvarez S. of the same Department, for their review of our section on the status of bighorn in that country. We wish to acknowledge as well our indebtedness to Pablo Dominguez and his supervisors in the Mexican Department of Conservation for their permission to use Pablo's beautiful photograph as the basis for the cover painting, which was superbly executed by Michelle Colby.

We also thank Valerius Geist for assistance, and for his permission granted jointly with Christian Georgi of Paul Parey Verlag, publisher, for permission to reproduce an illustration from Geist (1968), *Zeitschrift für Tierpsychologie* 25.

The Council considers itself privileged indeed to have, for the enhancement of this book, Patricia Hansen's many inimitable drawings of bighorn in true to life situations.

Finally, we are happy to acknowledge with gratitude the care accorded our manuscript by Marshall Townsend of the University of Arizona Press and his staff, of whom Marie Webner, Assistant Editor, was especially helpful.

DESERT BIGHORN COUNCIL

1. The Origin and Relationships of American Wild Sheep

Richard H. Manville

An afternoon thunderstorm swept across the Sierra Pinacate of northwestern Sonora. A band of bighorn sheep stood against the wind and driving rain, unprotected atop a barren crag. It was November of 1907. A hunter, John M. Phillips, gazed at the impressive spectacle with awe and later described it in his journal:

> The setting sun broke through the clouds behind me, gloriously bringing out all the details. The leader was standing almost broadside to me, his massive head accentuated by the deer-like leanness of his neck and body. The shining sun and the falling rain had formed a rainbow directly back of the pinnacle on which the ram stood. What a wonderful picture it would have made...! That magnificent ram, standing like a statue on the pedestal of red bronze lava, washed by the falling rain and lit up by the setting sun; on one side a head with horns quite as massive as those of the central figure, on the other the heads of two younger rams, and the whole group overarched by a gorgeous rainbow! [Hornaday, 1908].

Such a sight is not soon forgotten. It rouses speculation about where these splendid creatures came from, how they live, what their future is. Such problems are the concern of this volume, devoted to what has been hailed as the noblest of all American game animals and the trophy most coveted by sportsmen.

The buffaloes, cattle, antelopes, sheep, goats, and their kin, which range widely throughout the world, comprise the family Bovidae. They are grazing animals, ruminants, with hollow, unbranched, permanent horns, that usually are present in both sexes. Many of them have been domesticated. The subfamily Ovinae of the family Bovidae includes among others the sheep *(Ovis)* and the goats *(Capra)* of the northern hemisphere. These are stout-bodied animals, with relatively short legs and tails and hairy coats. Both sexes possess horns, but they are much smaller in the females.

Male sheep develop massive horns which diverge, forming a lateral spiral. In male goats the horns may be of great length and spirally twisted, but they curve upward and backward. Sheep have concave foreheads while goats' are convex. They also have glands near the eyes and on the hind feet which goats lack. They do not have glands at the base of the tail as do goats. In sheep the ear is longer than the tail.

The ancestral bighorn stronghold was the wild mountain and desert regions of Eurasia. Here, from the Caucasus Mountains of Asia Minor, across the highlands of Tibet to the deserts of Mongolia and beyond, there evolved the urial or shapu *(Ovis vignei),* the argali *(O. ammon),* the Marco Polo sheep *(O. poli),* the small red or Cyprian sheep *(O. orientalis),* and the bharal or blue sheep *(Pseudois nahura).* Europe had its mouflon *(O. musimon),* now restricted to Sardinia and Corsica, and north Africa the Barbary sheep or aoudad *(Ammotragus lervia).* The Old World sheep are considered in great detail by Sushkin (1925). Somewhere in this vast theater, probably in southwestern Asia and by the Neolithic era, early man was instrumental in developing the progenitors of our modern domestic sheep *(O. aries).*

The Tertiary period was, throughout the world, a time of great geographic change. In the Old World, entire mountain systems—the Alps, Caucasuses, Himalayas, Pyrenees, and others—were being formed; in the New World, the Cordilleran system was developing from Alaska to Cape Horn. Amidst this changing scene, which extended over some 70 million years, a great assemblage of mammals was evolving. Many earlier forms, such as the reptilian dinosaurs, were rapidly declining and passing into extinction. Following the Tertiary came the Pleistocene epoch, or Ice Age, which began about a million years ago. This was a time of alternate stages of hot, dry climates and of cold, wet climates. During the cold periods large parts of the northern half of the North American continent were submerged under tremendous sheets of snow and ice.

When great quantities of water were frozen into the glaciers overlying the land, as is still the case over Greenland and Antarctica, the level of the oceans dropped. Together with a gradual elevation of the land, this resulted in a closer association of the continental masses. Thus there appear to have been several times when a Bering land bridge existed, connecting Siberia and Alaska. Whether this was a bridge of dry land or of ice-covered shallow seas, it furnished a passageway for the movement of land animals both eastward and westward.

By mid-Pleistocene times, immigrants from Asia included such mammals as mammoths, antelopes, and primitive horses. Sheep were relative latecomers, and the first arrivals in North America were perhaps similar to the argali of the Asiatic steppes. Later, primitive men arrived on the American scene, and from them derived the Eskimos and Indians who, many centuries later, welcomed the first white men on the continent.

Probably the early sheep in America arrived not all at one time but gradually, over a stretch of many centuries, and in small bands that represented different racial stocks. Established at first in congenial Alaskan environments, they moved southward as the climates altered. Climatic changes affected the food and cover available, as well as other animal populations competing for these necessities. To all of this the sheep adapted, themselves changing in the process until south of Alaska they evolved into a new and distinct species, our bighorn *(Ovis canadensis)*. Stock and Stokes (1969) discuss some of the possible relationships between American varieties of sheep, past and present.

All this took countless generations, and it still continues. Ultimately the bighorn reached their southern limits in northern Mexico, in the meantime continuing to occupy suitable habitats to the north along their route of travel. The thinhorned Dall sheep *(Ovis dalli)* of Alaska and northwestern Canada may be descended from a later and different influx across the Bering land bridge. Both bighorn and thinhorn have continued to adapt to local situations and have evolved to the point where we now recognize different races or populations within each species.

Quite recently, in terms of geologic time, European men entered on the American scene. They probably found the bighorn dispersed much as they are today, except in much greater numbers and in areas where they are now extinct. Coronado, on reaching Cíbola in 1540, reported to Mendoza, the Governor of Mexico, seeing "some sheep as big as a Horse, with very large horns and little tails" (Winship, 1896).

By 1697, the Jesuit missionaries Juan María de Salvatierra and Francis María Piccolo, working in Baja California southward to La Paz, encountered the mountain sheep. Father Piccolo published a description of this bighorn, regarding it as a sort of deer, "as large as a Calf of one or two years old: Its Head is much like that of a Stag: and its Horns, which are very large, like those of a Ram: Its Tail and Hair are speckled, and shorter than a Stags: But its Hoof is large, round, and cleft as an Oxes." This account served as the authority on the bighorn for 100 years. Father Miguel Venegas repeated the description in his "Noticias de la California" of 1758, and appended an engraving entitled "The Tayé or California Deer," probably the first published picture of the species (Seton, 1929, vol. 3).

In 1800 Duncan MacGillivray, an employee of the North West Fur Company, collected one of a small herd of bighorn on the headwaters of the Bow River, near what is now Banff, Alberta. The specimen was sent to London, where it reached the Royal Society in 1803. Shaw (1804) promptly described it as *Ovis canadensis,* the name which still stands. But the same

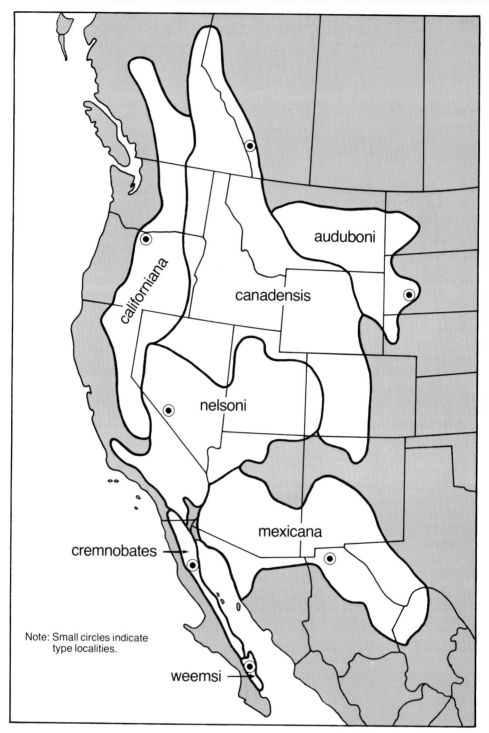

Fig. 1.1. Approximate distribution of the races of bighorn sheep, *Ovis canadensis*, during period of early exploration of the West. The small circles indicate type localities.

specimen was also designated as *O. cervina* by Desmarest, and as *O. montana* by Schreber. Further details of this taxonomic muddle have been summarized by Allen (1912) and Osgood (1913, 1914).

The Lewis and Clark Expedition of 1804–06 noted bighorn at many places, the first entry by Captain William Clark at the mouth of the Cheyenne River, on 1 October 1804, reporting, "On the Mountains great numbers of goat and a kind of anamale with circular horns, this anamale is nearly the size of an Small Elk" (Burroughs, 1961). Various other expeditions, surveys, and collecting parties continued to report on the occurrence of bighorn. Eventually there emerged a picture (Fig. 1.1) of their approximate distribution when white men first explored the West.

Many taxonomists, not aware of the total situation nor of the degree of variation in native sheep, applied distinctive names to local populations. The result was a confusing array of names which overcomplicated the picture. The accumulated materials in various collections were scrutinized by Cowan (1940) and again by Buechner (1960), and following them we now recognize two species, and ten included subspecies or races, of native sheep in North America:

Ovis canadensis: the bighorn sheep or mountain sheep

    O. c. canadensis Shaw, 1804. Rocky Mountain or Canada bighorn. Type from mountains on Bow River, near Exshaw, Alberta.

    O. c. californiana Douglas, 1829. California, lava beds, or rimrock bighorn. Type from near Mount Adams, Yakima County, Washington.

    O. c. auduboni Merriam, 1901. Audubon or badlands bighorn. Type from Upper Missouri; probably the badlands between Cheyenne and White Rivers, South Dakota. Extinct.

    O. c. nelsoni Merriam, 1897. Nelson bighorn. Type from about 5 miles south of Grapevine Peak, on boundary between Inyo County, California, and Esmeralda County, Nevada.

    O. c. mexicana Merriam, 1901. Mexican bighorn. Type from Lago de Santa María, Chihuahua, Mexico.

    O. c. cremnobates Elliot, 1904. Peninsular bighorn. Type from Matomi, Sierra San Pedro Mártir, Baja California, Mexico.

    O. c. weemsi Goldman, 1937. Weems bighorn. Type from Cajón de Tecomaja, Sierra de la Giganta, southern Baja California, Mexico.

Ovis dalli: the thinhorn sheep

    O. d. dalli Nelson, 1884. Dall sheep. Type from mountains south of Fort Yukon on west bank of Yukon River; probably Tanana Hills, Alaska.

    O. d. stonei Allen, 1897. Black or Stone sheep. Type from Che-on-nee Mountains, on headwaters of the Stikine River, British Columbia.

    O. d. kenaiensis Allen, 1902. Kenai sheep. Type from head of Sheep Creek, Kenai Peninsula, Alaska.

Bighorn are found in the highest mountains of the West, as well as in many low desert mountain ranges. They all seem to require steep, rugged terrain with adequate food, and freedom from undue competition from other mammals, including man. Through countless generations they have become adapted to the particular habitats they now occupy. The southern bighorn are generally lighter in weight, color, and pelage than the bighorn found farther north. What we here refer to as the "desert bighorn" is really an *ecologic* entity: any bighorn living under relatively arid desert conditions. We use the term "desert bighorn" to include the four races that typically inhabit dry and comparatively barren desert mountain ranges: *nelsoni, mexicana, cremnobates,* and *weemsi*. Portions of the ranges of *californiana* and *canadensis*, as shown in Fig. 1.1, may also be included under this term; these are areas of intergradation, but they do meet the criteria of "dry and barren."

Probably never as abundant as deer, bighorn were nonetheless well dispersed and in good numbers at the early occupation of the West by white men. Buechner (1960) concurs with Seton's (1929) estimate of 1,500,000 to 2,000,000 bighorn sheep on the continent early in the nineteenth century. Furthermore, they were differently distributed in habitat than they are today. Grinnell (1928) felt that "Until the settlement of the western country the sheep occupied the high plateaus west of the Missouri River, and the mountain country of the Continental Divide west almost to the Pacific, south into Mexico, and north into Alaska.... In old times the wild sheep were not confined to what we call mountains but in many parts of their range lived in a country not very different from that then commonly occupied by the mule deer—that is to say, about the near buttes, rough bad lands or low rocky hills."

But with the westward spread of civilization the herds of bighorn began to diminish. Domestic sheep were introduced near Pikes Peak and elsewhere in Colorado about 1878, and the bighorn began to suffer from scabies and depleted forage (Packard, 1946). The bighorn were essentially gone from the Black Hills of South Dakota about 1887, although a few lingered until 1899. In the Yellowstone region of Wyoming, several thousand bighorn (plus elk, deer, and pronghorns) were slaughtered for their hides from 1870 to 1877.

Severe hunting and heavy snows eliminated bighorn from the Mount Shasta region of California by 1883 (Buechner, 1960). By 1890, in the Gallo and Gallinas mountains of New Mexico, Smith (1890) recorded that "the now almost extinct mountain sheep" was only occasionally seen. Along the Mexican boundary, Mearns (1907) reported that as early as 1894, bighorn were scarce in northwestern Sonora as a result of excessive hunting with firearms by the Papago Indians. These few cases typify the predicament that confronted the bighorn following the advent of European man.

2. The Desert Bighorn and Aboriginal Man

Campbell Grant

Archaeological evidence exists that man began to hunt the desert bighorn many thousands of years ago. Excavations at Ventana Cave in southern Arizona (Haury et al., 1950) have revealed the bones of extinct mammals such as the sloth and horse in association with human artifacts at the lowest levels. These levels have recently been radio carbon dated to about 11,000 years ago (Emil Haury). The first bighorn bones appear at levels dated between 7,000 and 8,000 years.

Bighorn bones are not at any of the known sites of Early Man, though the bones of mammoth, camel, horse, tapir, and dire wolf are relatively abundant. The Ventana Cave evidence gives a clue to the arrival of the bighorn in the Southwest, as the early hunters would surely have taken this superior food animal if it had been there at the time.

It is not hard to imagine the first groups of skin-clad humans with their spears and spearthrowers, moving slowly northeast through the mountains of eastern Siberia toward their momentous entry into the New World. They were probably hunting caribou and bighorn as they entered the ice-free valleys of Alaska, with only the position of the sun to tell them that they were moving south into a new land.

Through the centuries, successive waves of human migrants entered the New World, driven by hunger, pressure from stronger tribes, or sheer wanderlust, and as they moved they often displaced earlier peoples. The story of the desert bighorn and the American Indian mainly concerns the Mongoloid Uto-Aztecans (Shoshoneans) who entered the Great Basin and the Southwest several thousand years ago, and the later Mongoloid migrants, the Athabascans (Navajo and Apache).

The nomads in their southern advance hunted the Dall sheep *(Ovis dalli dalli)* of Alaska, and its relatives, the Kenai sheep *(O. d. kenaiensis)* and the black Stone sheep *(O. d. stonei)*, and finally the Rocky Mountain bighorn *(O. canadensis)*, whose desert subspecies occur deep into northern Mexico.

All that we know about the interrelations between aboriginal man and the desert bighorn comes from three sources: ethnological information, archaeological evidence from shelter caves and ruins, and pictures of bighorn pecked and painted on the rocks.

In the latter part of the nineteenth and the early twentieth centuries, Indian informants (mainly Paiute) gave investigators detailed descriptions of the methods of hunting the bighorn, methods that had been followed from the more distant past until the coming of the whites with their rifles.

There is much archaeological evidence that the aboriginal populations of the Great Basin and the Southwest esteemed the bighorn for more than its succulent meat. There are few dry shelter caves in the region that have not produced artifacts made of the horns, bones, or hides of the bighorn.

The most dramatic evidence of the importance of the bighorn in the lives of the Indian people is represented by the vast number of paintings and petroglyphs of bighorn and the hunting of bighorn. These are found in particular abundance where Shoshonean territories coincided with bighorn ranges.

ABORIGINAL HUNTING METHODS

With their puny weapons, the atlatl or spearthrower and that later invention, the bow and arrow, the Indian hunters had to devise special methods to bring their wily prey within range.

A favored method in the Great Basin took advantage of the habit of bighorn to head for high rocky points when alarmed. On such spots the hunters would wait in stone blinds for other Indians to drive the bighorn past them so that they could be shot. I have seen great numbers of these stone blinds in the Coso Range of mountains in Inyo County, California. One rocky knob, rising no more than 15 m (50 ft) from the sage-covered mesa top, had seventeen of this type of blind. In the construction of these blinds, the natural jumble of shattered rock had been modified by rearranging movable stones and building concealing copings where necessary. Often simple petroglyphs were made in or around the blind by the waiting hunters. Generally in this volcanic region, the blinds are built about 6 m (20 ft) above the floor of the narrow basaltic gorges. These steep-walled canyons were natural bighorn "traps" into which the animals were driven by men and dogs, past the ambushing spearmen or archers.

Wallace (no date) and Hunt (1960) note the same sort of thing in Death Valley, east of the Coso Range.

Unboubtedly there was ambushing at water holes wherever the topography permitted natural concealment of hunting blinds, and this type of hunting could be carried out by a single hunter.

Muir (1901) describes another method of communal hunting for the desert bighorn in Nevada:

> Still larger bands of Indians used to make extensive hunts upon some dominant mountain much frequented by the sheep, such as Mount Grant on the Wassuck Range to the west of Walker Lake. On some particular spot, favorably situated with reference to the well-known trails of the sheep, they build a high walled corral, with long guiding wings diverging from the gateway; and into this enclosure, they sometimes succeeded in driving the noble game. Great numbers of Indians were of course required, more indeed than they could usually muster, counting in squaws, children and all; they were compelled therefore, to build rows of dummy hunters out of stones, along the ridge tops which they wished to prevent the sheep from crossing. And without discrediting the sagacity of the game, these dummies were found effective for with a few live Indians moving about excitedly among them, they could hardly be distinguished at a little distance from men, by anyone not in on the secret. The whole ridgetop then seemed alive with archers.

In my investigations in the Coso Range, I saw several examples of such dummy hunters made of piled rocks on the edges of walled canyons and near spots where the cliffs broke down and the bighorn might seek escape. They were on the north-facing or shady side of the gorges where they would show up in silhouette from below.

This type of entrapment hunting with winged corral was widely practiced in the Great Basin and on the Colorado Plateau. Most of the accounts described for ambush and corral hunting are from Paiute sources. The Shoshonean migrations from California and southwestern Nevada, described later in this chapter in the section on rock art, filled the Great Basin and much of the Colorado Plateau with Shoshonean-speaking people—the Northern and Southern Paiute, Shoshone, and Ute. In speaking of the relationship between the desert bighorn and aboriginal man, we are mainly discussing the bighorn and the Shoshoneans who dominated the original desert bighorn range.

Wetherill (1954) described a trap corral from Arizona:

> On Skeleton Mesa between Kiet Siel Canyon and the left branch of Dogoszhi Bito, and about four miles [6.5 km] from the mesa point above the mouth of Kiet Siel Canyon, are the remains of a wing trap corral built by the Paiutes about 1890. This trap corral abutted against an overhanging

rock shelter forming a V and was built of logs and brush from the pinyon and juniper which grow on the mesa. It was at this place that the last bunch of Mountain Sheep were trapped and killed by the Paiutes.

After hunting horses on this part of the mesa in the fall of 1937, Max Littlesalt (Eshin Elclisto Begay) reported to me that there were five Mountain Sheep skulls and a few fragments of bone left; also traces of the log and brush corral could be seen.

Bandelier (1892) says of the region about the Rito de los Frijoles in New Mexico: "Prior to 1800 game of all kinds, deer, elk, mountain sheep, bears, and turkeys, roamed about the region in numbers, and the brook afforded fish." In a footnote he adds: "All kinds of game mentioned here were abundant around the Rito de los Frijoles in former times, but the communal hunts of the Pueblos, and later on the merciless slaughter of the Apaches, have greatly reduced it."

The volcanic craters of the Sierra Pinacate in northwestern Sonora were favorite hunting sites for the Sand Papago hunters. Each crater has a trail about its rim lying just far enough back from the inner rim to be concealed from the interior. The bighorn in the craters would be flushed out by beaters and shot from the rim with arrows as they fled up and out.

There is little evidence that the Navajos ever made more than a casual thing of bighorn hunting—and why should they? Almost from the beginning of their history in the Southwest, the Navajos had preyed on the Spanish ranches of the Rio Grande Valley, stealing great numbers of domestic sheep and horses; there was rarely a shortage of mutton. However, Judd (1954) reports: "My older Navajo neighbors remarked that they as young men, and their fathers before them, had hunted pronghorns on nearby mesas—and deer, elk, and sheep in the mountainous country north of the Rio San Juan."

Such hunting must have been by young men who had not yet acquired domestic flocks of their own, or by hunters out simply for the sport of bagging a difficult quarry. J. Juan Spillet states that Indians still hunted bighorn on Navajo Mountain, Arizona-Utah, in the 1940s.

There is no doubt that for thousands of years the desert Indians hunted bighorn by the two methods described and continued the techniques at least until the late nineteenth century. It is probable that similar techniques were used by their remote Asiastic ancestors. The evidence of the log and brush trap corrals disappears with time, but many of the rock blinds must have been ancient before the birth of Christ.

THE DESERT BIGHORN IN ROCK ART

Most of the rock drawings in the United States occur west of the Mississippi River. In the previous century, several hundred sites were known in the eastern states. Today a great many of them have been destroyed, mainly through road building and the construction of flood control and hydroelectric dams. In the Great Plains are many scattered sites along the eastern flanks of the Rockies. From the Rockies to the Pacific Coast there are enormous numbers of carved and painted rock art sites, particularly in the canyons created by the big rivers, in wind-eroded caves, and on patinated volcanic rocks.

Of all the designs made on rocks by western Indians, the bighorn is by far the most abundant. Figure 2.1 shows the occurrence of the bighorn motif in the western states. The map can only give an approximation; some dots may represent a single drawing, while other dots mark sites with thousands. The scale of the map also makes the indication of every site impossible. In the Coso region, for instance, only about a third of the sites are shown owing to space limitation.

I have photographed rock drawings of bighorn in many parts of the West, culminating in a survey of the vast concentration of such pictures in the Coso Range. During this survey, carried out in 1966–67, we recorded over 14,000 rock designs in a 94-sq-km (36-sq-mi) area, of which more than half were of bighorn. In the immediately adjoining region we have estimated as many more. As all the problems of interpreting the bighorn drawings exist in the Coso region, I base the present discussion on the studies made there.

The Coso Range makes up the northern half of the huge Naval Weapons Center, an area of roughly 2,850 sq km (1,100 sq mi), with headquarters at China Lake, California. Here, for more than 37 years, the Navy has developed and tested new ideas in missiles, particularly rockets, and the target or impact area closely coincides with the rock art canyons.

The Range lies at the northwestern edge of the Mojave Desert, with the southern end of the Sierra Nevada to the west and Death Valley to the east. From an elevation of about 610 m (2,000 ft) at China Lake, the barren land rises very slowly for 24 km (15 mi) and then rises sharply at the southern edge of the Coso Range to a high point at Coso Peak (2,500 m or 8,160 ft). The only permanent water in the area occurs in the form of springs.

The local rock is basically granitic but is largely overlaid with lava flows, with much evidence of intense ancient volcanic activity. Cinder cones, deposits of obsidian, and many basalt cliffs and gorges formed by block faulting and later erosion give the region a quality that is at once forbidding and fascinating.

The vegetation is typical of the high desert, with large forests of Joshua trees *(Yucca brevifolia)* at the middle elevations and pinyon pine *(Pinus edulis)* and juniper *(Juniperus* sp.*)* at the higher levels. Among the commoner plants are Mormon tea *(Ephedra* sp.*)*, rabbitbrush *(Chrysothamnus* sp.*)*, sagebrush *(Artemisia* sp.*)*, and creosote bush *(Larrea divaricata)*. Herds of feral horses and burros roam the mesas and open canyons. The blacktail jackrabbit *(Lepus californicus)*, desert cottontail *(Sylvilagus auduboni)*, bobcat *(Lynx rufus)*, and coyote *(Canis latrans)* are abundant, while mule deer *(Odocoileus hemionus)* frequent the pinyon belt around Coso Peak. The desert bighorn, portrayed by the thousands on the rocks, are gone, though

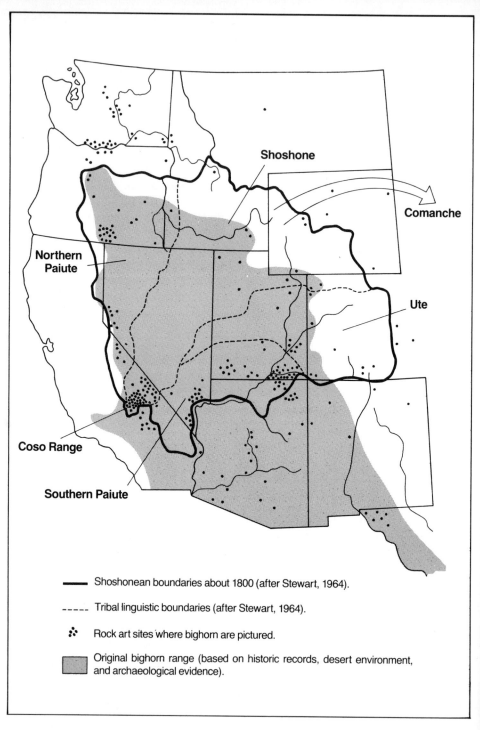

Fig. 2.1. Movements of the Shoshoneans north and east through the desert bighorn range.

a few have recently been reported in the adjacent Argus Range. The evidence of the drawings demonstrates that at some time bighorn occurred in large numbers. It is my opinion that they disappeared from the area a long time ago, and the discussion that follows will develop the reasons for this disappearance.

At the close of the Great Pluvial, which brought the last Ice Age to an end about 10,000 years ago, the Coso area presented a picture vastly different from today. It was surrounded by the huge Pleistocene lakes created by the incessant Pluvial rains and runoff from the melting Sierra snowpack. To the north was Owens Lake, to the south Searles Lake, and to the east Panamint Lake and Lake Manley that filled Death Valley to a depth of 180 m (600 ft).

There is no doubt that man lived in the area when these lakes still held considerable water. The people living near these huge lakes doubtless had a lacustrine culture, with heavy reliance on wildfowl and fish, but as the lakes dwindled and vanished their culture became the typical Desert Culture pattern (Jennings, 1957) that dominated the Great Basin and much of the Southwest for thousands of years. The Desert Culture was mainly dependent on the seasonal harvesting of small hard seeds, with sporadic hunting to vary the diet.

Harrington (1957) excavated a site at Little Lake, on the western edge of the Coso Range, and found great numbers of Pinto Basin points, a type of projectile point used with the atlatl dart. He states that the site was at least 3,000 years old, and possibly a great deal older. In any case, aboriginal man hunted the bighorn in the Coso Range for a very long time.

South of Coso Peak lies Wild Horse Mesa, having an average altitude of 1,525 m (5,000 ft). Cutting through the mesa are two parallel canyons, Renegade and Petroglyph, about 6.5 km (4 mi) apart. There are shallow waterways that wind through the gently rolling mesa, except where the block faulting of the basalt has created many narrow tributary gorges from a few hundred yards to several kilometers long. At the edge of the mesa, the canyons plunge steeply down toward the Coso Basin through gorges up to 150 m (500 ft) deep.

It is in the gorge area on the mesa that the bulk of the rock pictures occur. At the start of the survey, only about 1.5 km (1 mi) of each canyon was known. We explored nearly 48 km (30 mi) of canyons, and in one 13 km (8 mi) long tributary of Petroglyph Canyon we recorded almost 3,000 new designs. In all, sixty-two sites were located and described.

The designs that seem the earliest on the basis of surface patination and erosion, and which I term as in the Early Period, include rather naturalistic bighorn, always drawn with horns to the side and often with an attempt to indicate hoofs. These are almost invariably crudely done. With these are many stylized atlatls, or spearthrowers, with stone weights attached, bighorn horns, solid body anthropomorphs, and many miscellaneous abstract curvilinear and rectilinear patterns. These latter are very common throughout the Great Basin, the Southwest, and adjacent Mexico (Figs. 2.2, 2.3, 2.4).

The drawings I have placed in the Transitional Period feature the introduction of the bow, almost as revolutionary a change as that caused by the later appearance of the rifle. During this period, the atlatl persists, though it is far outnumbered by drawings of men shooting the bow. Bighorn of all shapes are abundant, and many are drawn with care and attention to detail. Many bighorn begin to show the horns-to-front style that will be the dominant type in the Late Period. Many new elements are depicted such as deer (never common), projectile foreshafts, dogs, and human figures with patterned bodies.

Through all periods, the abstract patterns continue to be made, but, with the exception of what looks vaguely like a heraldic shield, the abstract patterns decline in ratio to the naturalistic.

In the Late Period, the atlatl disappears entirely, many bighorn became large and beautifully stylized, and the ceremonial figures attained great importance and are very elaborate and varied. Dogs are often shown attacking bighorn, and long processions of stick figures are common. The shield patterns become complex, and in the many hundreds we recorded no two were alike.

To make a picture on stone, the aboriginal artists employed two basic techniques. The first was by painting: the application to a rock surface of a ground pigment mixed with an oil binder. This oil may be derived from a vegetable or animal source. The Yokuts of the southern San Joaquin Valley used the juice of the milkweed *(Asclepias* sp.*)* mixed with oil derived from the crushed seed of the wild gourd *(Cucurbita* sp.*).* Such paintings (pictographs) are mainly confined to rock shelters, overhanging ledges, and surfaces somewhat protected from the destructive forces of wind and water.

In addition, most paintings occur on light-colored rocks that display the colors to advantage.

The second basic technique (petroglyphs) utilizes cutting or breaking the surface of darkly patinated rock, revealing the lighter original color beneath. The effect is a negative type drawing of light on dark. This can be accomplished in a variety of ways. The two most common techniques are pecking out the design with a hard, pointed stone and abrading the surface with the flat or edge of a stone. This latter technique is used when a large surface is to be covered, and the abrading stone scrapes off the surface patina. A number of pictures in the Coso Range are made by incising with a sharp stone tool and by shallow scratching.

The formation of the black patina on rock surfaces used by the petroglyph makers is an interesting phenomenon. The surface coloring is commonly

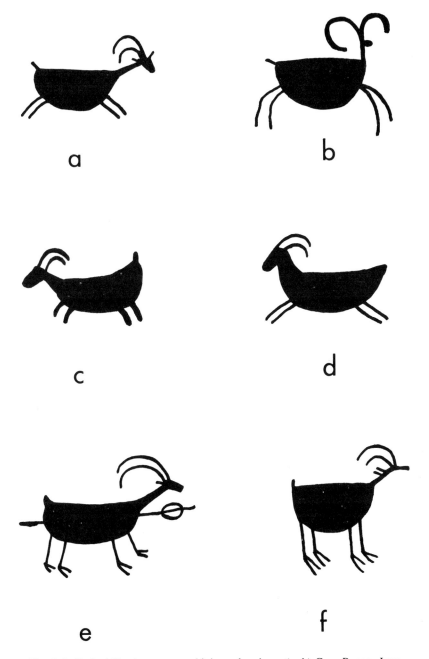

Fig. 2.2. Pecked Shoshonean-type bighorn drawings. (a, b) Coso Range, Inyo County, California (after Grant, 1968). (c) Atlatl Rock, Valley of Fire, southern Nevada. (d) Canyon de Chelly, Arizona. (e) Southeastern Utah. (f) Northeastern Arizona. (c through f after Kidder and Guernsey, 1919).

Fig. 2.3. Highly stylized bighorn of the Late Period.

Fig. 2.4. Bighorn panel from the Coso Peak. Note single deer, and bighorn holding tails in mouths.

called "desert varnish" and its maximum development under favorable circumstances is a blue-black or brown-black color. Often it is quite glossy from the polishing action of flying dust and sand.

Many articles have been written on the formation of desert varnish. Most are in agreement on several points. Water permeates the surface of the rock and, as the stone dries, returns to the surface through capillary attraction, bearing minute amounts of minerals in solution. On reaching the surface, these oxidize to form patina. In addition, heat seems somehow important in creating patina. Patina formation is extremely slow. Many rock-pecked drawings that we know are at least several hundreds of years old look as if they had been made yesterday.

From my studies in the Coso Range and in reading accounts of such patination in other parts of the world, another important factor has emerged. Wherever dark patination has occurred (in southeastern California, southern Utah, South Africa, or Australia), the country was invariably inland desert with high summer temperatures. In addition, each region received the bulk of its rainfall through summer thundershowers. Apparently water and heat must be present *at the same time* for optimum patina formation. A thundershower striking a very hot rock surface passes by and the rock quickly heats up again, accelerating the drying process and the movement to the surface of minerals in solution. This effect of water on hot stone can be seen in northern Arizona, where water has run down over the great overhangs that protect many of the cliff dwellings. The habitual runoff channels over the rock are stained black through patina formation.

When the surface of the rock is broken by the pecking or abrading stone, the lighter original color shows up in sharp contrast to the dark patina. The inner color of the basalt, or sandstone, is usually medium to light but appears much paler owing to the fracturing of many tiny crystals by the artist's tool. Everyone has experienced this effect by striking any rock with another—the impact point shows as a lighter spot. A pane of glass will give the same effect when struck.

As soon as the artist's pecking stone has broken the surface of the rock, the lighter-colored stone that is exposed is ready to start another buildup of patina. Porous sandstone tends to repatinate much faster than dense basalt, but in either case it takes a long time for the design to repatinate to the color of the surrounding surface. This is no help in arriving at exact or even approximate dating but is of great value in relative chronology. If on the same panel there are pecked designs of (1) great contrast, (2) moderate contrast, (3) little or no contrast, we are dealing with drawings made at three different periods with the third type being the oldest. This criterion does not hold with drawings made near canyon floors where excessive sandblasting continues to destroy contrast through surface erosion faster than patina can rebuild.

Relative chronology can also be established by superimposition of one pattern on another. This is quite common throughout the West and is puzzling, as usually there are excellent surfaces nearby with no drawings at all. There are a number of possible explanations for this. Suppose a hunter has drawn a bighorn on a particular rock and that he has success in the ensuing hunt. This surface becomes perhaps a "lucky" rock, and the hunter uses it again and again. Perhaps others utilize the same rock hoping to share in its power. Some rocks are covered solidly with bighorn pictures.

In some instances, another reason must be sought. Often a bighorn is carefully pecked *on top of* an elaborate ceremonial figure and both drawings seem of about the same age (identical patina). Here we may have an example of ritual obliteration, or perhaps the bighorn drawing becomes more effective or "powerful" when superimposed upon the ceremonial figure (shaman?).

As for approximate dating, we have only a few slender clues. Don Martin of Santa Rosa, California, who is a longtime student of rock art and the phenomenon of desert varnish, has made an interesting experiment. He has taken samples of heavily patinated rocks from the Coso Range to his home in Santa Rosa and has exposed these rocks to several seasons of California inner valley weather, where the summers are warm and dry and the winters are cool and wet. Under these conditions, the patina disappeared, thus supporting our theory that patina is *only* formed when heat and moisture coincide. This rules out the possibility that the Coso rocks were patinated during the Great Pluvial when incessant rains were creating the Pleistocene lakes. As the country grew progressively hotter and dryer, conditions favorable to patina formation undoubtedly were created. Between 3,000 and 4,000 years ago, however, the Little Pluvial again brought unusual moisture to the area. It is quite possible that patina formation was arrested and existing patina destroyed during this period. In other words, the desert varnish in the Coso Range utilized by the ancient hunter-artists perhaps may be no older than 3,000 years.

Interpretation of rock drawings is difficult and in many cases impossible. In a very few instances, living Indians who actually made such drawings in their youth have given detailed information to the field workers. Certain clues can be derived from areas in the world where similar drawings are still being made today, for example, the *wondjina* paintings in the Kimberley region of Australia. These mythical creatures are painted in rock shelters, and they represent the ancestral creators of particular clans. They are appealed to for specific things like babies, rain, good hunting, etc. The elaborate sand paintings of the Navajo, used in healing ceremonies, developed from earlier, simpler designs painted on rock.

The aboriginal rock pictures in this country were made for a variety of reasons, though most of them can be classified as ceremonial. Others can be classed as mnemonic or memory aids (tally marks to keep track of numbers) and as design patterns for later reference to decorate pottery, blankets, or sandals. In the Great Plains are panels that are often quite realistic and that obviously record some important event. Some designs have been recognized by living Hopi as clan symbols. There is ethnographic evidence that a few drawings are doodlings, or copying of ancient designs by modern Indians with little knowledge of what they were making.

Lanny Wilson notes an example of rock art apparently used as a location marker: "I found only 11 areas where a man could gain access to Wingate Mesa in southern Utah. At four of the access points, I found the same drawings, a picture of a bighorn and the symbol used for water. From these four access points it is less than 3.2 km (2 mi) to permanent water utilized by the desert bighorn."

Ceremonial pictures were made during rituals held at previously decorated spots. The Indians prayed for things to make their life easier or happier or more complete: they wanted rain (but not too much), health, success in hunting, success in warfare, children, and the like. Important rites were connected with puberty. If the visualization of a supernatural being to be propitiated would aid the singing and dancing recitations, so much the better—Christians have been doing much the same thing for nearly 2,000 years.

It is generally believed that the aboriginal drawings of big game animals are connected with hunting magic. The picturing of the animal to be hunted was an aid in bringing luck to the hunter. This motivation was certainly true of the European Palaeolithic cave hunters who had to kill such animals as the bison, mammoth, bear, and rhinoceros with the feeblest of weapons, from a simple sharpened stick to a stone-pointed spear thrown with a spearthrower. Beyond this, I think that the drawings were mainly of the hard-to-get animals and that where the game was easy to come by it was seldom pictured. The hunter need spend no time drawing pictures of such animals as the rabbit or the antelope, both easily procured in communal drives. Certainly the salmon of the Pacific Northwest, all-important in the economy of the area, was rarely pictured. In the Great Plains, the buffalo seldom figures in rock art.

I believe this theory explains the bighorn petroglyphs. Why was it pictured in such vast numbers in the western states? Perhaps the original reason for picturing the bighorn was because of the difficulty of successfully hunting this wily creature. The earliest drawings of bighorn in the Coso Range were made by atlatl hunters who ambushed them at high points or as they traveled through the narrow gorges to water or feeding grounds.

The Coso bighorn pictures are concentrated in the gorges, at the springs, and on divides between watersheds—all natural hunting areas. There can be no doubt that the motivation for these drawings was to aid the hunter through hunting magic; the picturing of the animal would in some way bring supernatural power to bear on the problem. Heizer and Baumhoff (1962) have noted that in Nevada the rock designs were invariably on game trails or near springs where driving and ambushing were possible.

The atlatl or spearthrower is a very ingenious weapon, in effect adding

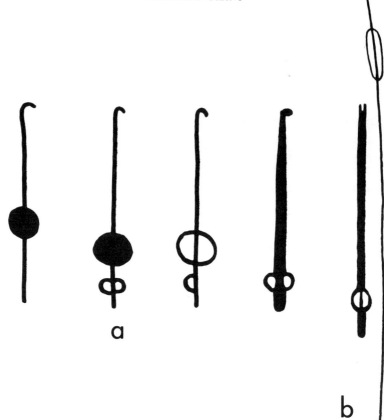

Fig. 2.5. (a) Petroglyph representations of the atlatl from the Coso Range, Inyo County, California. The left three examples are common. Only one of the fourth type is known. (b) Atlatl and dart from Atlatl Rock, Valley of Fire, Nevada.

another joint to the arm (Fig. 2.5). Known examples are between 50 and 60 cm (20 and 24 in.) long. It was a great improvement over the hand-thrown spear, but usually was only effective at close range, perhaps up to 23 m (75 ft). This is not to say that long casts are not possible. In Australia, where the spearthrower is still the favored weapon of many aborigines, casts of up to 120 m (400 ft) have been made. I made an exact copy of a Basketmaker atlatl from a cave burial in northern Arizona. With it, a 1.8-m (6-ft) spear or dart was easily thrown over 60 m (200 ft). With these long throws, however, little accuracy was possible.

A great many drawings in the Coso Range are of the Early or Atlatl Period, and they are mainly of bighorn or weighted atlatls (Fig. 2.6). The Coso atlatls are usually shown as a stick with a hook at one end, finger grips at the other, and one or more round shapes along the weapon to represent stone

Fig. 2.6. Bighorn petroglyphs in Coso Range, Inyo County, California. *Top;* Hunters carrying weighted atlatls, from five sites in the Coso Range. *Bottom:* Hunters and hunting scenes showing the bow and arrow. The lower two scenes show the ultimate in simplification of the bowman.

weights. Such weights have been found with atlatls in Oregon, Nevada, and Arizona. They are believed to have added thrust to the cast and were also "good luck" or fetish stones. The Coso artists had a curious convention in representing the weighted atlatl. The stone or stones (as many as three) are greatly exaggerated in size and in scale to the atlatl shaft and would have weighed several pounds, whereas the specimens recovered average only a few ounces. Perhaps this exaggeration gave the weapon added "power."

During this period, the large numbers of bighorn in the region apparently were able to hold their own against the hunting pressure exerted by the Indians with their atlatls. What was it that brought the Coso rock art to such a peak in the Late Period and why did the tradition of rock drawings die out? I have a theory, based partly on pure speculation and partly on comparisons with other aboriginal areas, to account for both points.

About 2,000 years ago, something revolutionary occurred: the bow and arrow arrived. Whether it passed slowly from region to region or was introduced by nomadic peoples from the north passing through, we shall never know. In the Transitional Period, the bow and the atlatl appear on the rocks, often together or on adjacent panels. The new weapon was a great improvement over the atlatl: it killed at greater range, it was more accurate, and its penetrating power was superior. In addition, arrows could be launched at high velocity while the hunter still crouched in his blind, whereas the atlatl hunter had to stand up to hurl his slower-moving missile.

Some evidence of the effectiveness of the new weapon is shown by the excavation at Danger Cave, Utah (Jennings, 1957). In Level IV (mainly bow and arrow), there is an increase in ungulate bones of 280 percent over Level V (mainly atlatl).

Based on the evidence from Lovelock Cave in Nevada (Grosscup, 1960), the bow arrived in that area about 500 B.C. and completely supplanted the atlatl by A.D. 1. The same 500-year span was probably needed for the weapon shiftover in the Coso region. The Indian was very conservative and took up new ideas slowly. Eventually, however, bow hunting became the accepted method of killing bighorn and, with improved communal hunting methods, the hunting magic motivation became secondary—it was no longer a terribly difficult animal to bag.

With the advent of the bow, the rock pictures became more sophisticated, especially those of bighorn, which are increasingly stylized and occur in many diverse forms (Fig. 2.7). A common form has a boat-shaped body (flat back and bulging belly), with horns shown front view on side view head. With them is a mysterious object (Fig. 2.8) that I have called a medicine bag and that might have contained the paraphernalia of the medicine man.

To understand the relationship between Indians and game animals, it is necessary to know something of primitive religion. Most aboriginal people believed in dynamism or psychic force—that animate and inanimate objects and natural phenomena have powers not perceptible to human senses.

Fig. 2.7. Non-Shoshonean-type bighorn representations. (a) San Francisco Mountains, Baja California; black, painted. After Meighan, 1969. (b) Culbertson County, Texas. After Jackson, 1938. (c) Canyon de Chelly, Arizona; Navajo, white, painted. (d) Hudspeth County, west Texas. After Jackson, 1938. (e, f) Near Three Rivers, New Mexico; pecked.

Fig. 2.8. Petroglyph representations of "medicine bags," from the Coso Range, Inyo County, California.

Some objects are considered to have a special vital force and are regarded as sacred. Belief in the unseen and the supernatural played a large role in most primitive societies, which believed in mysterious omnipresent powers that must be placated and appealed to at every turn. The middleman between primitive man and the supernatural forces that surrounded him was the shaman or medicine man. From man's earliest beginnings, certain individuals have claimed the power of communicating with the unseen world and the ability to exorcise with ceremonies the dark forces that filled the life of aboriginal man with anxiety. There are deep caves in France and Spain where Ice Age man painted the big game animals he hunted 15,000 to 25,000 years ago. With the paintings of such animals as the mammoth, woolly rhinoceros, and giant bison is an occasional figure of a man wearing the skins of animals. This would be the shaman, probably the hunt shaman who directed the ceremonies before the hunt. Such rituals have continued to the present time and can be as simple as singing and dancing around sacred objects of stone, wood, or bone, or as elaborate as the Navajo sand-painting healing ceremonies.

The evidence of the pictured rocks in the Coso area indicates that an actual bighorn hunting cult developed, in which the chief figures were the hunt shamans. Next to the bighorn, the most abundant motif in the Late Period is an elaborately dressed figure (hunt shaman?) with feathered headdress, patterned body, and fringed dance skirt. Such costumes were known in historic times in the Great Basin. These anthropomorphic figures often occupy the highest points in the pecked panels, as though height was connected with power. Often the figures bear a symbolic bow in one hand and a sheaf of three arrows in the other.

Many Indian tribes hold certain creatures in special reverence. The eagle is universally respected as a benevolent deity and enters into rituals and mythology throughout North America. In the Northwest, the salmon was believed to be immortal, ascending the streams to provide food for the people and returning to life the following year to be harvested again. In other areas, the same beliefs were held for deer and especially the buffalo. These important animal deities had to be continually honored and good relations maintained with them, lest they disappear or diminish in numbers. The shamans communicated with these supernatural animal beings through recitations, trances, and ceremonial dancing.

On the basis of the rock drawings, it appears probable that the bighorn became a venerated animal deity in the Coso region and that the people developed rituals in its honor to insure its continued abundance.

It was the duty of the hunt shaman to placate the bighorn in advance, so that it would be willing to be killed without too much difficulty and so that its spirit would return again and again in bighorn form. There is no way of knowing which drawings of bighorn were made in relation to pre-hunt ceremonies and which were made by individual hunters hoping for a little extra luck. Perhaps each hunter pecked a bighorn under the supervision of the hunt shaman—the variations in quality suggest such a possibility.

Smithson and Euler (1964) note the function of the hunt shamans as described by Yuman Havasupai informants:

> Game shamans were also known to the Havasupai. Before extended hunting forays, the shaman would sing special songs calculated to quiet deer and prevent them from running when hunters approached. They were sung at night and repeated four times. Content involved not only information about the habits of the animal, but also that the hunters were sorry for killing the game they needed to assuage their hunger. Locations the hunters intended to visit were also mentioned. Successful hunters frequently prayed to the sun to shine brightly so the deer could be seen. The sun was thought to "own" the animals.
>
> Game shamans could lure only certain species. *Cik paniga*, one of three remembered by our informants, was thought to have power over deer and mountain sheep but not antelope.

Whatever the ritual practices, the great picture galleries from the basaltic gorges of the Coso Range attest to the importance of the bighorn in the economy of the ancient Shoshoneans. But what ended the long tradition of petroglyph making, and where did the people go, and when? These are hard questions, but with some speculation, a few clues, and a little solid evidence, a partial answer is possible.

Geist (1967) studied the habits of the Rocky Mountain bighorn in Canada, and his findings are a clue to the disappearance of the Coso bighorn and the end of the hunting cult. He discusses the inflexible feeding patterns of the bands, patterns acquired from birth as the lambs follow the ewe leader through the routines that they will adhere to all their lives. The migration routes between feeding ground, the salt licks, and the watering places—all are part of a fixed way of life. Nothing can radically change this pattern except unbearable harassment by man. When such a point is reached, the surviving animals abandon the area and do not reoccupy such abandoned territory. With great difficulty they can establish a new feeding area.

Is it possible that at some time during the Late Period the point of overkill or insupportable harassment was reached and the bighorn abandoned the region? Did communal ambush hunts with the bow and arrow along the migration routes and on the ridges, possibly in combination with other adverse environmental factors, eventually prove lethal?

In accordance with this theory, one can postulate an upsurge in rock drawing activity and ceremonialism connected with the bighorn cult, to bring back the vanished bighorn. But the bighorn did not come back. An important food source for what must have been a sizable population at the time vanished, and eventually, when all hope of bringing the bighorn back had gone, the long tradition of rock art in the area ceased. *When* this happened we do not know, but there are a few straws in the wind.

Linguists working on language problems in the Great Basin are in general agreement that the Numic-speaking Shoshoneans (Northern and Southern Paiute, Shoshone, and Ute), who occupied vast areas of the western United States, originated in southeastern California and southern Nevada. Lamb (1958) has postulated about 1,000 years ago for the start of the Numic migrations.

As to what triggered the migrations, one can say with certainty that the Numic people began to move out of a land that would no longer support large numbers of people. In the Coso region, the bighorn had disappeared, and in other areas they were getting warier and scarcer.

Certainly an important factor in forcing the migrations was the drying up of the main Pleistocene lakes and the loss of these great sources of both fish and waterfowl. Antevs (1955) thinks this occurred as long ago as 7,000 years; Davis (1966) believes the lakes reached their present levels about 2,000 years ago. I am inclined to think present levels were not reached until about 1,000 years ago, after having had replenishment during the Little Pluvial of 4,000 to 3,000 years ago.

When the first whites came into the Coso Range in the 1860s looking for silver, they found only a hundred or so Indians living marginally on seed-gathering and a little hunting (there is no mention in the early records of bighorn in the Coso Range, though a few might have survived in remote areas). Some of the Indians' ancestors who had abandoned the area long before had certainly bettered themselves: in Utah and Colorado, they became the Ute; in Nevada and Oregon, Northern Paiute; in Wyoming, the Wind River Shoshone; and in Texas, the Comanche. The Ute separated from the Southern Paiute in historic times and acquired a Plains culture. About the same time, the Comanche became distinct from the Wyoming Shoshone and moved farther south and east. The remnants in the Coso Range spoke a Shoshone-Comanche dialect.

It seems certain that some of the migrating groups of Shoshoneans took their rock-drawing traditions with them as they moved east and northeast.

Where they encountered favorable conditions to hunt the bighorn, they continued to make bighorn pictures; where, because of unfamiliarity with new country or lack of bighorn, the hunting magic formulas were ineffective, the tradition of rock art died out.

In Utah, where the Colorado River is joined by the San Juan River, there are great numbers of bighorn petroglyphs, and here the migrating Shoshoneans struck hunting country comparable to the Coso-Death Valley region. In 1776 Father Silvestre de Escalante set out from Santa Fe in what is now New Mexico in an attempt to find a land route from Santa Fe to Monterey, California. He never made it, but as he turned back toward Santa Fe from northern Utah, he crossed the Colorado with great difficulty in the Glen Canyon region. The Franciscan explorer described what he saw that November day: "Through here wild sheep live in such abundance that their tracks are like those of great bands of domestic sheep. They are larger than the domestic breed, of the same form, but much swifter" (Bolton, 1950).

It appears more than coincidence that all but one of the major concentrations of rock art sites featuring the bighorn are well within the territories of the Shoshoneans. The exception is the middle Columbia River, especially between The Dalles and junction with the Snake River. Here the Sahaptin tribes might well have acquired the tradition of pictured hunting magic from the neighboring Northern Paiute, a short distance to the south. Wherever concentrations of bighorn sites are shown on the map (Fig. 2.1), rock and canyon formations exist to make ambush driving or corral trapping possible.

If, as theorized earlier, the destruction or dispersal of the Coso bighorn bands was a factor in starting the Shoshonean migration, then we have a terminal date for bighorn drawings in the Coso of 1,000 or less years ago. If patina began forming after the Little Pluvial of 3,000 to 4,000 years ago, a tentative dating for the start of the tradition of drawing the desert bighorn would be around 3,000 years ago. By that time, enough patina might have built up on the rocks to attract the picture makers.

As the Shoshonean hunting people moved north and east seeking greener pastures and more bighorn, they left the record of their presence on the rocks. At hundreds of sites in the Great Basin and on the Colorado Plateau, bighorn and ceremonial figures have been recorded. Many are astonishingly like those in the Coso region, the postulated origin of the bighorn hunting cult.

After the Coso concentration of bighorn pictures, the next largest group of sites is in the San Juan area or Four Corners country, where Arizona, Utah, Colorado, and New Mexico come together. Here, after the hand print, the bighorn is the most abundant motif—a few are painted, but the great majority are pecked onto patinated sandstone. Schaafsma (1966) notes that bighorn pictures in Tsegi Canyon, Arizona, are formalized with flat backs and round bellies, as in the Coso Late Period bighorn. The earliest drawings, both painted and pecked, were more naturalistic, often with ears and hoofs. Later, greater stylization was introduced with detail eliminated. Schaafsma suggests that the later type dates from the thirteenth century, and studies in the Glen Canyon region indicate that the realistic style bighorn in that area date between A.D. 1050 and 1250. These dates fit in very well with the Shoshonean migrations into the region from southeastern California and Nevada.

Another major concentration of bighorn drawings occurs in southern Nevada, roughly halfway between the Coso–Death Valley complex and Four Corners. Elsewhere in the Great Basin–Colorado Plateau country, sites are widely scattered, with many in southern Utah. Further south, in what are now the states of Arizona and New Mexico, there are numerous sites, chiefly near pueblo settlements. At A.D. 1000 this was the country of the Kayenta pueblo people around Marsh Pass, the Chaco pueblo people in Chaco Canyon, and the Hohokam and Mimbres peoples in the Gila River drainage. All these peaceful agricultural people were at, or approaching, the peak of their specialized cultures, soon to be shattered by aggressive hunting nomads moving into their world, the Shoshoneans from the north and west and the Athabascans from the north and east.

Far to the south, in central Baja California, there is a mountainous region where the only permanent water occurs at palm oases in deep canyons or in large rock pools or *tinajas*. In this area, particularly in the Sierra de San Francisco, north of San Ignacio, are many large painted panels in dry caves depicting greater than life-sized men and game animals, including the bighorn. On the basis of a wooden artifact dated by radiocarbon, it has been determined that the caves were occupied around 500 years ago. Bighorn still persist in these inhospitable and remote mountains.

Most of the bighorn petroglyphs in the Great Basin and on the Colorado Plateau are so similar in style and technique, with their fat bodies and sticklike legs and horns, that I call them the Shoshonean-type bighorn. Wherever the Shoshoneans hunted, they used the same highly stylized manner of representing bighorn. In other areas of the Southwest, the bighorn is usually drawn far more realistically. The Anasazi in the Canyon de Chelly painted very lifelike bodies but retain the thin horn convention of the Shoshoneans. In the Three Rivers region of central New Mexico, bighorn petroglyphs on scattered basaltic boulders depict the horns very accurately, thick and recurved, other details being subordinated. The Baja California bighorn are shown with thick curving horns, front view, and detailed hoofs.

CEREMONIAL USE OF THE DESERT BIGHORN

At several late sites in the Coso Range are anthropomorphic figures wearing bighorn horn headdresses, some carrying unknown ceremonial objects. In the Gobernador district of northern New Mexico are paintings and petroglyphs of the important Navajo yei, *Ganaskidi*, the Humpback God. Bighorn horns grow from his head (Fig. 2.9). Reichard, in her monumental study (1950), gives some details of the importance of the bighorn in Navajo rituals.

Fig. 2.9. Ceremonial use of horns. (a) *Ganadiski*, the Navajo humpbacked yei; Largo Canyon, New Mexico. (b) *Panwu*, the Hopi bighorn sheep kachina. After Fewkes, 1903. (c) Pecked figures with horn headdresses; Sheep Canyon, Coso Range, California. After Grant, 1968.

The bighorn as a valued game animal plays a major role in Navajo mythology. The Hunchback God shares many characteristics with the bighorn and has supernatural control over it. The hump in the God's back is a feathered bag bearing seeds of all vegetation. All versions of the Night Chant, a famous Navajo healing ceremony, describe the hero lying in ambush for bighorn aided by *Ganaskidi,* the owner and controller of all sheep.

The association of a humpbacked deity with bighorn hunting was another idea borrowed from the Pueblo people by the Navajo. In the Marsh Pass area and at Hagoé in the Monument Valley area, Arizona, there are pecked humpbacked figures playing flutes in association with bighorn. The flute players are almost invariably lying on their backs (Kidder and Guernsey, 1919). In Fewkes Canyon, Mesa Verde, Colorado, painted on the wall of what appears to be a ceremonial room, is a humpbacked figure shooting at a bighorn. The Hopi humpbacked phallic deity *Kokopelli* derives from the concept pictured in these early Pueblo rock pictures.

Ganaskidi looks very like the Hopi mountain sheep kachina *Panwu,* illustrated in Fewkes (1903), and it is likely the Navajo yei was directly adapted from the kachina. During and after the Pueblo revolt of 1680, many Hopis fled to the Navajo country in northwestern New Mexico and lived with them. During this period, the Navajos borrowed heavily from the culturally advanced Hopi both material culture features and religious beliefs. At one time the Acoma Indians are reported to have had a bighorn kachina, *Kac-ko.*

There is evidence from southwestern Arizona and northwestern Sonora that the bones and horns of the bighorn were utilized ritually by the Piman Sand Papagos *(Pápagos Areneros).* These Indians were first described in the area by Father Eusebio Francisco Kino in the late seventeenth century. They were centered near Sonoyta but ranged widely to the Gulf of California by way of the sand dune country of the Gran Desierto, and northwesterly into the present Cabeza Prieta National Wildlife Refuge in Arizona.

Fontana (1962) reports that at the Cabeza Prieta Tanks in the Sierra Cabeza Prieta there were piles of bighorn horns, the remains of which were still visible in 1970. Horns have been seen near Heart Tank in the Sierra Pinta to the east. These were the result of bighorn kills by the Sand Papagos. Apparently they did not use the type of stone blinds reported in Nevada and California (Grant, 1968) but ambushed the bighorn in some way when they came in to water at the tanks.

From very early times, travel through this desolate region has only been possible through knowledge of the available water, and this water almost without exception is found in the rock pools called "tanks" or "tinajas." When Father Kino set out from Sonoyta in 1699 to visit the Indians on the lower Gila River, he went by way of the Sierra Pinta and the Sierra Cabeza Prieta. He called the route "El Camino del Diablo," and noted an abundance of bighorn near the tinajas (Bolton, 1936).

In 1774 Juan Bautista de Anza followed Kino's Camino del Diablo on his successful attempt to find an overland route to the California Missions. On 1 February he reached the Cabeza Prieta Tanks and noted in his diary (Bolton, 1930):

> This watering place is found in the midst of a number of hills whose opening is towards the east. It has six tanks of very good rain water. In the first the animals drink. It is easily filled from the tanks above by climbing up and emptying them through the natural channel in the living rock which connects them all. In them there is a large supply of water, and as much more could be supplied if they were cleared of the rocks and sand with which they abound...
>
> These tanks, the one farther back (Tinaja del Corazón) and another which is ahead of us, are inhabited by Papagos in the height of the dry season and as long as the water gathered in them lasts, through their desire to hunt mountain sheep there. These animals ... live among the cliffs that are highest and most difficult to scale. They are native to dry and sterile regions. Their flesh is better than that of deer. They multiply very slowly and almost never run in the level country because of the impediment of their horns. These horns the Indians are careful not to waste. Indeed, whenever they kill the sheep, they carry the horns to the neighborhood of the water holes, where they go piling them up to prevent the Air from leaving the place. Those who, like ourselves, do not practice or do not know of this superstition, they warn not to take one from its place, because that element would come out to molest everybody and cause them to experience greater troubles.

This Papago belief as reported by De Anza, that the horn piles could control the wind, continued well into the nineteenth century (Fontana, 1962).

Julian Hayden has been doing archaeological work in the Sierra Pinacate (in Sonora below the Sierra Cabeza Prieta) for many years. At Papago Tanks, in the lava area, he reports horn piles similar to those seen by De Anza. This southern branch of the Sand Papagos *(Areneros Pinacateños)* had another curious ritualistic practice in that they cremated portions of the skeletons of bighorn, antelope, and deer. This custom of burning the bones after stripping off the flesh and breaking the marrow bones is very old with the Sand Papagos and persisted until about 1850 when the Pinacateños were decimated by yellow fever. José Juan of the Aliwaipa or main branch of the Sand Papagos in 1936 said that the bones were burnt "to quiet the spirits of the killed animals, so that they would not alarm the sheep living in the area" (Julian Hayden).

There can be no doubt that the bighorn was a most important animal to the tribes living in southern Arizona and New Mexico, particularly among the Piman-speaking people—the Papagos and the Pimas. To these we might add the Hohokam, inasmuch as Gladwin (1957) thinks the Pimas are directly descended from the Hohokam. The only ethnological information we have from the area about the supernatural aspects of the bighorn concerns the horn

Fig. 2.10. Rock art representations of men wearing bighorn headdresses. (a) Petroglyph from Cornudas Mountains, southern New Mexico. From photograph by Kay Sutherland. (b) White painting, Hueco Tanks, Texas. After painting by Forrest Kirkland. Courtesy of University of Texas Press©.

and bone piles of the Sand Papagos, but the many effigies in stone, bone, and shell, the petroglyphs, and representations on pottery suggest that the animal meant much more to the Indians than hide, horns, and meat.

A petroglyph from the Cornudas Mountains, New Mexico, and rock paintings from the Hueco Tanks, Texas (the two locations are only about 40 km [25 mi] apart) depict human figures with bighorn headdresses (Figs. 2.10, 2.11). The art work is thought to be that of the Mescalero Apaches and probably represents one of the supernatural beings known as Mountain Spirits. Masked and costumed men impersonated such beings at important ceremonies (Kirkland and Newcomb, 1967).

THE DESERT BIGHORN AS AN ART MOTIF

Aside from its common use in rock art, the bighorn rarely appears in Indian art. On the walls of a ruined kiva in Chaco Canyon are figures of hunters with bows and arrows, shooting deer or bighorn (Smith, 1952), and wall paintings of bighorn are reported from Zuni by Cushing (1882–83). At

Fig. 2.11. Rock art figures with bighorn headdresses, buff outlined with white, Hueco Tanks, Texas. After painting by Forrest Kirkland. Courtesy of University of Texas Press©.

the destroyed pueblo of Awatovi, the hunter's society had its walls painted with representations of cougar, gray fox, lynx, and coyote in pursuit of elk, bighorn, deer, and rabbit (Smith, 1952).

The only Indians in the Great Basin and the Southwest that used the bighorn as an art motif, apart from rock art hunting magic, were the corn-growing people. They used bighorn casually as food but often employed the dramatic bighorn with its great curving horns to decorate household objects.

A small unfired clay head of a bighorn from Death Valley is noted by Hunt (1960). A number of writers have reported on animal effigies in the Hohokam-Mogollon region of southern Arizona, and Gladwin et al. (1937) picture three effigy vessels featuring bighorn. The finest is a stone vessel carved with a fret design and capped by four bighorn carved in the full round. The other two are fired clay vessels in the form of bighorn, with fret and spiral designs on their bodies. The ingenious designers of animal patterns on the famous Mimbres pottery used the bighorn to great advantage, and modern Acoma potters are copying these designs today.

In the San Andres Mountains of southern New Mexico, numbers of bighorn heads have been found carved in animal bone (Fig. 2.12). In addition, the ancient inhabitants of the region worked with stone and shell to create

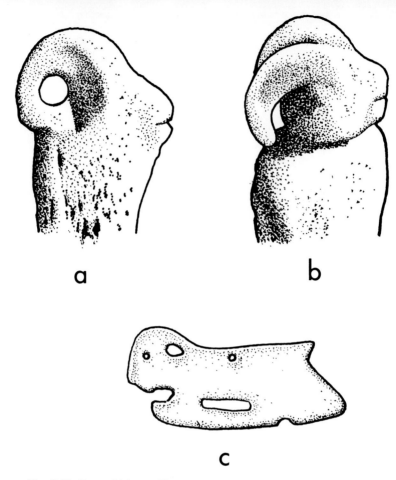

Fig. 2.12. Desert bighorn effigies from the lower Rio Grande region of southern New Mexico. (a, b) Ram heads carved on the ends of animal bones, San Andres Mountains. (c) Shell pendant from the San Andres Mountains. From photographs by Jack E. Gross.

representations of the prized animal for ceremonial or decorative purposes or both. In primitive societies the supernatural or magical significance of a decorated object was of primary importance—if in addition the object was pleasing to look at or to feel, so much the better.

The Hohokam made hair ornaments of bone, shaped like daggers with the wide end often decorated with bighorn heads (Gladwin et al., 1937). Fewkes (1922) illustrates a carved stone bighorn effigy found at Pipe Shrine House, Mesa Verde, Colorado. Some modern Indians in Mexico have created bighorn art objects solely for sale to the white man, with only esthetic appeal. The primitive Seri Indians of Sonora have carved superb animals from native ironwood, and a favorite subject is the head of the bighorn (Fig. 2.13). Mexican Indian potters in the northwestern states often make ceramic bighorn figures.

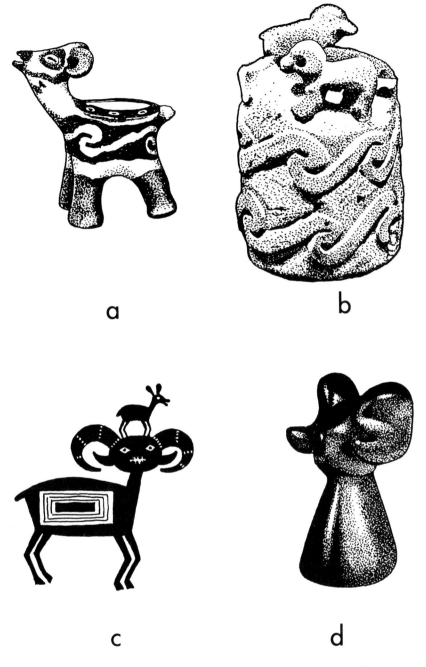

Fig. 2.13. The bighorn as an art motif. (a) Hohokam red-on-buff pottery effigy vessel, Snaketown, Arizona, about A.D. 1000. After Gladwin et al., 1937. (b) Hohokam stone effigy vessel with four bighorn, Snaketown, Arizona, about A.D. 800. After Gladwin et al., 1937. (c) Mimbres black-on-white pottery design, Arizona. After Cosgrove and Cosgrove, 1933. (d) Seri ironwood sheep head, Sonora, Mexico, modern. Author's collection.

THE DESERT BIGHORN AS ARTIFACT MATERIAL

For most of the aboriginal people in the Great Basin and on the Colorado Plateau, life was a daily struggle and every conceivable natural resource—animal, vegetable, and mineral—had to be exploited to the full. This was particularly true in much of the barren and rocky land preferred by the bighorn. So it is not surprising to find the bighorn supplying many needs besides food. Horns, bones, and hide were utilized for many purposes, but the most valuable parts were the great curving horns.

Horn artifacts. The horns of the bighorn are composed of an inner core and an outer, rather thin-walled sheath. This sheath could be softened by prolonged soaking in hot water and steaming, then cut and flattened for the manufacture of a variety of tools, utensils, and ornaments. Blades made of the core have been found at Aztec Ruins National Monument, New Mexico (Richert, 1964).

Twelve examples of sickles made from wild sheep horn have been recovered from archaeological sites in Humboldt and Lovelock caves, Nevada; Barrier Canyon and Glen Canyon, Utah; and Segi Canyon, Arizona. The inner curve had been sharpened, and all examples showed a high polish at this point. Analysis of the silicon on the cutting edge showed that the sickles had been used for cutting grass. All appeared to be of Basketmaker origin and average about 30 cm (12 in.) long (Heizer, 1951). There is ethnographic evidence that the Ute, Southern Paiute, Shivwits Paiute, and northwestern Navajo used such sickles for cutting grass. In the Maze area of southeastern Utah is a rock painting of such a sickle in use (Brodie, 1971).

An unusual use for bighorn horns by the Southern Paiute was reported by Euler (1966): "The Mormon missionaries, such as John D. Lee, saw the Paiutes using bows and arrows in 1854, but it was Carvalho, artist on one of Frémont's expeditions in the same year, who described a compound bow. He observed one made, according to him, of a single mountain sheep horn backed with sinew."

Horn scrapers and projectile flakers have been recovered from sites in the Glen Canyon area (Clark, 1966) and the Kayenta district of northeastern Arizona (Kidder and Guernsey, 1919). Digging-stick tips and blades of horn have been described from Glen Canyon and northeastern Arizona (Clark, 1966). Pendants and hoof rattles are reported from Lovelock Cave (Grosscup, 1960). Spoons made of sections of horns, as well as spatulas, have been excavated at the Lost City site in southern Nevada (Shutler, 1961), and similar spoons have turned up at the Lovelock site (Grosscup, 1960). Horn punches and a spindle whorl were found at Basketmaker sites in the Kayenta region. Horn hoes have been excavated in Chaco Canyon (Judd, 1959).

Kidder and Guernsey (1919) describe a possible arrow-straightener from northeastern Arizona: "In the possession of Mr. John Wetherill, at Kayenta, there is a piece of mountain sheep horn pierced by a series of four round holes

Fig. 2.14. Head of Hohokam bone hairpin (Sedentary Period, about A.D. 1000). From Snaketown, Arizona. Length, 1.6 cm.

The University of Arizona

about three-eighths inch in diameter; their edges are smooth and beveled as if by much use. In our collection are fragments of two similar objects. A guess at the purpose of these specimens is that they served as wrenches to straighten and true up thin, round sticks.''

Grosscup (1960) has recorded a similar arrow straightener from Lovelock Cave.

In addition to the above uses of horn, Clark (1966) lists tubular horn blades, horn seedbeater, gaming pieces, and a horn ornament.

Although somewhat outside the range of the desert bighorn, several other examples of bighorn horn artifacts deserve mention. An atlatl found several years ago in a dry cave east of Condon, Oregon, is unique in several respects. The stone weight was still attached to the atlatl (the first one found in this condition), and the finger grip was made of carved bighorn horn. The workmanship is excellent, but the weapon is so small (38 cm) that it must have been made for a child or as a symbol of authority (Strong, 1969). In the Pacific Northwest, the Tlingit and the Haida made superb horn spoons and ladles, elaborately carved with stylized animals. In addition, the Salish made handsome decorated bowls out of molded mountain sheep horns (Haberland, 1964).

Bone artifacts. Deer and bighorn bones were much utilized by the Indians of the Great Basin and the Southwest, and the most important tool made of bone was the awl, the basic tool of the basketmakers. Awls made of bighorn bone have been identified from Glen Canyon, from Lin Kletson and Pueblo Bonito in Chaco Canyon, New Mexico, at Mesa Verde, Colorado, and at Tse-ta-a in Canyon de Chelly, Arizona. Dagger-like bone hair ornaments are reported from Snaketown, Arizona (Fig. 2.14) and the Mimbres area, New Mexico; they are decorated with carved bighorn heads, but whether the bone itself is bighorn is not known.

Rick Terry, Brooks Institution

Fig. 2.15. Modern basket made by the Coso (or Panamint) Indians of Death Valley in the 1930s, 20 cm (8 in.) in diameter with bighorn design. Note accompanying butterfly motif. Santa Barbara Museum of Natural History.

Basketry. It is possible that Indian basketmakers incorporated the bighorn into their designs at times. An example is a basket made by the last survivors of the Coso (or Panamint) Indians living in Death Valley in the 1930s (Fig. 2.15).

Hide. Tanned bighorn skin was useful for clothing or for the manufacture of bags of all sorts. This type of material is highly perishable, but hide fragments from Glen Canyon have been identified as possible moccasins and children's wrappers (Clark, 1966). In the Kayenta region, skin robes of mountain sheep hide tanned with the hair on have been reported (Kidder and Guernsey, 1919). There are accounts of Paiute women dressing in rabbit skin robes, beneath which they wore skirts of doeskin or mountain sheep skin (Euler, 1966). Sinew has been recovered in both Glen Canyon and the Kayenta area. Bighorn hide must also have been extensively used for thongs and ties of all sorts.

An odd use of the bighorn is noted by Reichard (1950): "A singer said that if the Navajo captured a mountain sheep in a dry season, they butchered it, cut out and cleaned the paunch. Then they slapped it against a stone. In the summer, such an act would bring rain; in the winter, snow."

The use of the bighorn for artifact material was mainly confined to the Basketmakers of the Great Basin and the Colorado Plateau. These people led a foraging and hunting existence for thousands of years before some of them became farmers and developed the Pueblo culture based on the cultivation of corn and the making of pottery. After this transition, the bighorn was of minimal importance to the agriculturalists, who had an assured food supply and who were beginning to develop elaborate cultural patterns with their leisure time.

THE DESERT BIGHORN AS A FOOD RESOURCE

It is obvious that, wherever it was hunted intensively, the bighorn constituted a major food source. In other areas where the main dependence was on agriculture, the animal was a welcome addition when a lucky hunter brought one in. Often the only evidence of this lies in the rock pictures, because in open sites the physical evidence of bones, horn, or hide has vanished. Where there are no rock pictures, the utilization of bighorn as food can be determined only by archaeological evidence from dry caves.

Bighorn food bones were found at the big Hohokam settlement at Snaketown, Arizona (Gladwin et al., 1937), and the historic Pimas, descendants of the ancient Hohokam, are reported to have hunted bighorn (DiPeso, 1956). In a Chaco Canyon ruin, in addition to bighorn bones in the midden, a bighorn skull was found that had been broken for the extraction of brains (Judd, 1954). At archaeological excavations in the lower Glen Canyon, bighorn bones were the major faunal type recovered. Most were long bones and were split and broken as if purposely fractured to obtain marrow (Long, 1966). Woodbury (1965) reports that bighorn were the favorite food all through the Glen Canyon region: "In nearly all sites studied, ungulate bones were found, 158 of which were deer, 1135 of which were bighorn sheep, and 354 of which could not be classified."

Mammal remains at the Awatovi ruin included few bighorn bones, but these Hopi farmers obviously made little effort to hunt them (Lawrence, 1951). An excavated ruin in Tonto National Monument, Arizona, produced a single bighorn mandible (Steen et al., 1962). In the Nevada Humboldt Cave excavation, bighorn bones were in the majority (Heizer et al., 1968).

The seminomadic hunters and food gatherers, particularly the Shoshonean Southern Paiutes, continued to hunt bighorn well into historic times, and bighorn remained an important addition to their diet. In ideal bighorn country like the Colorado–San Juan River region, the animals often were hunted by part-time agriculturalists. However, in areas where farming was done on an intensive scale, as by the Hopi or Hohokam, little effort was made to hunt the animals.

3. Distribution and Abundance
Gale Monson

We look upon desert bighorn as an ecological entity (see Chapter 1), distinguished from other bighorn more on a basis of habitat than morphological differences. Desert bighorn belong chiefly to the warmer desert regions of southwestern North America, where little or no snow falls, and the annual precipitation is 12 to 35 cm (5 to 12 in.).

At the time of that event so fateful to native wildlife—the arrival of European man on American soil—the desert bighorn occupied much of the same territory they do today, but they are extinct or nearly so in extensive areas, notably the Peloncillo Mountains astride the Arizona–New Mexico border, the general region of northwestern New Mexico and extreme northeastern Arizona (where bighorn perhaps never were present), a number of mountain ranges in central and northern Nevada, the trans-Pecos region of Texas, and in most of the canyons and mountains of northern Chihuahua and Coahuila in Mexico.

In some places, perhaps desert bighorn are merely persisting and may not survive. Such areas would include the Big Hatchet Mountains of southwestern New Mexico, at least some of the many small desert ranges of northwestern Sonora in Mexico, and a number of isolated ranges or peaks in the southern parts of Arizona and California.

In measuring desert bighorn abundance we have had very few data until recent decades. Over the greater part of their range they apparently were never common, and in some localities 1978 numbers might be as high as in the exploration period. The accounts of early explorers like James O. Pattie, Col. W. H. Emory, and Lt. J. W. Abert, as well as early settlers, scarcely or seldom mention bighorn.

In times antedating European man by several hundred years, there is evidence, mainly in the form of rock drawings, that bighorn were more numerous in some areas, especially across the northern part of modern desert bighorn range. A reduction in numbers or local extirpation could have taken place owing to climatic and vegetational changes less agreeable to bighorn, or possibly to overhunting by Indians, or both (see Chapter 2).

Figure 3.1 portrays what historical evidence indicates is the range of the desert bighorn, according to categories of past and present abundance. Reintroduction sites are not included.

The object of this chapter is to picture as accurately as possible the early and modern (to 1978) distribution and relative abundance of the animal, on a state-by-state basis, both in the United States and Mexico. Population estimates and distribution are based principally on reports from state wildlife agencies in the United States and the federal wildlife department of Mexico but are not necessarily the official data of these agencies. Desert bighorn are notoriously difficult to census, owing to their scattered numbers in rough, inaccessible country, and the subjoined population data are subject to change from the acquisition of more accurate counts. In some cases at least, conscious efforts were made not to overestimate numbers. It will be noted that geographical reporting units vary from state to state.

ARIZONA

As is the case throughout the sparse accounts of early travelers in the Southwest, bighorn sightings were seldom mentioned by explorers and early settlers in Arizona. The first mention of bighorn in the state (and in North America) is in the writings of Pedro de Castañeda, a private soldier in Coronado's expedition. Castañeda, in the army following Coronado to Cíbola in 1540, mentions sighting a herd of bighorn, apparently in the area of the confluence of the Gila and San Francisco rivers in Greenlee County (Hammond, 1940).

Beginning with J. O. Pattie in 1825 and 1826 (Thwaites, 1905), we find occasional sightings of bighorn recorded in the early Arizona literature, principally from along the Colorado River (including the Grand Canyon). Merriam (1890) saw bighorn on San Francisco Mountain near Flagstaff in 1889, and Cahalane (1939) presents evidence the animals were present in the vicinity of the Chiricahua Mountains about 1900.

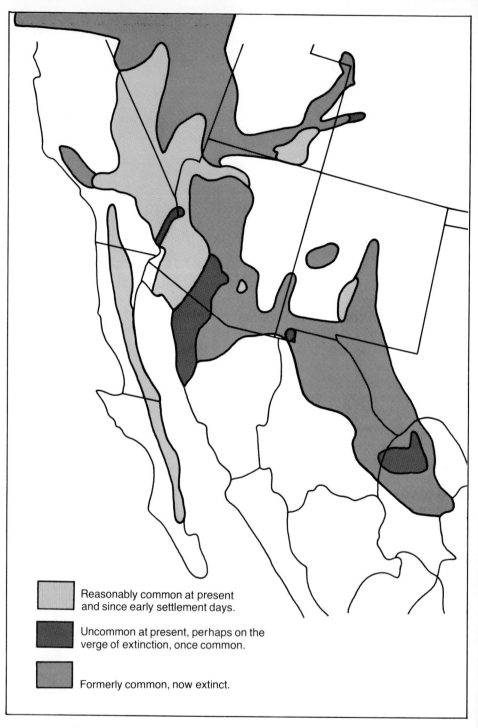

Fig. 3.1. Past and present distribution and abundance of desert bighorn.

Mearns (1907), working with the International Boundary Commission from 1892 to 1894, gives an account of bighorn status in Arizona at that time. He found them along the Mexico border from Nogales westward, as well as satisfactory proof of their presence in the general region of Camp Verde, on Bill Williams Mountain in Coconino County, and in the Santa Rita and Santa Catalina mountains.

Hal Coss of the National Park Service interviewed local ranchers and hunters who testified that bighorn were in the Rincon and Galiuro mountains as late as 1942, and Coss himself in 1974 picked up a set of horns on Tanque Verde Ridge of the Rincon Mountains that he believed could not have been lying there for more than 20 years.

As of 1978, from 2,100 to 2,600 desert bighorn were estimated to live in Arizona, principally in the western one-third to one-quarter of the state. In Coconino County bighorn are found only in the fastnesses of the Grand Canyon. In Mohave County they inhabit the canyon of the Colorado and its tributary canyons, the rougher country along Lake Mead, and the Black Mountains bordering Lake Mohave. To the south, they live in the Mohave Mountains along the Colorado River, but by 1975 they were disappearing from the Aubrey Hills adjoining lower Lake Havasu, owing to various building developments.

Yuma County, southwesternmost in the state, holds the bulk of Arizona's bighorn population. There they are to be found in almost all of the small but rugged mountain ranges, from the Buckskin Mountains just south of the Bill Williams River to the ranges of the Cabeza Prieta National Wildlife Refuge on the Mexican border. They are most abundant in the Kofa Mountains of the Kofa Game Range, and in the adjacent Plomosa and Chocolate mountains. Roughly twenty-five mountain ranges in Yuma County support bighorn.

Fair to good populations of bighorn exist in Maricopa County in the Maricopa, Sand Tank, and Sauceda mountains. The animals in 1976 had disappeared from the White Tank Mountains as a result of pressure from the million-plus people in the Phoenix metropolitan area, which was threatening the small bighorn population of the nearby Sierra Estrella as well.

In Pima County the bighorn are chiefly in the extreme western part, in the Growler and Ajo mountains. The Papago Indian Reservation has numerous mountain ranges suitable for bighorn, but numbers have declined rapidly and it was estimated in 1972 that not more than fifty were left (Brown, 1972). East and north of the Reservation in 1978 a small group was holding out on Ragged Top, an isolated peak within sight of Tucson, while at the west end of the Santa Catalina Mountains, on Pusch Ridge, a band of perhaps fifty was maintaining itself within a few miles of the Tucson city limits. The burgeoning human population of Tucson harasses the bands increasingly as the years pass, and unless something is done to rigorously exclude people from their habitat, the bighorn will inevitably end in extinction.

South and west of Tucson, Seymour and John Levy found the skeleton of a freshly dead bighorn ewe on Gunsight Mountain at the north end of the Sierritas as recently as 1958. In the same year they found fresh tracks and a number of old skulls in the Coyote Mountains at the edge of the Papago Indian Reservation. Several persons have reported to me that they saw a bighorn in the Tucson Mountains in the early 1950s. Bighorn were absent from these locations in 1978.

The easternmost population of desert bighorn is located along the Graham-Pinal county line, in the Aravaipa Canyon country. Here the Game and Fish Department began transplanting bighorn from the Kofa Game Range and other localities in the late 1950s. Initially the animals were held in a fenced pasture, but later the gates were left open and some left the pasture although remaining in the general area. As many as sixty were to be found in and out of the pasture as of 1978.

CALIFORNIA

The bighorn occurs through the desert portions of southern California, from the White Mountains on the north to the Mexican border and east to the Colorado River, including the Chemehuevi Mountains (Robert Ohmart, Susan Woodward). However, there do not seem to be any recent records from those low desert mountain ranges in the western part of the Mohave Desert in eastern Kern County and the westernmost part of San Bernardino County. At the northern end of its California range, the desert bighorn of the White Mountains may rarely come into contact with the "California" bighorn *(Ovis canadensis californiana)* of the Sierra Nevada. There seems to be no intergradation between these two ecologically distinct populations.

Farther to the south, in 1978 there were groupings of about 510 and 110, respectively, in the high San Gabriel and San Bernardino mountains. While these bighorn are found from 800 to over 3,000 m (3,200 to 10,000 ft) in these mountains, and at least for a part of the year occupy a distinctly nondesert biome, in many respects they are similar to the bighorn found not many miles distant in a truer desert setting. Arbitrarily and conveniently we say they are desert bighorn.

Although population data from early days are wanting, it is clear from historical records that desert bighorn once occupied much more of California than was true in 1978. This especially applies to the southwestern mountains, where they formerly ranged as far north as the La Panza Mountains of San Luis Obispo County. Their retreat was likely occasioned mainly by the introduction of domestic livestock and by overshooting. In the 1950s and 1960s the bighorn decline could be attributed to displacement by human recreational use and habitations. Relatively high populations in the San Jacinto and Santa Rosa mountains were being threatened by 1975 by rapidly spreading human developments, mostly roads, homes, and water supply usurpations.

I am indebted to Richard A. Weaver of the California Department of Fish and Game for population estimates that indicate from 3,250 to 3,750 bighorn were found in the state as of 1978. These were distributed about as follows:

San Diego County	390
Eastern Imperial County	160
Central-southern Riverside County	800
Joshua Tree National Monument and vicinity	100
Northeastern Riverside County	15
Southwestern San Bernardino County	15
Southeastern San Bernardino County	200
Northeastern San Bernardino County	390
Northwestern San Bernardino and southern Inyo counties	55
Death Valley and vicinity	600
Northern Inyo and southern Mono counties	155
San Gabriel and San Bernardino mountains	620
TOTAL	3,500

NEVADA

Nevada has many mountain ranges, most of which are below 3,000 m (10,000 ft) and arid. At the advent of European man's settlement of the state, bighorn were found in almost all mountain ranges. One conjectures that those found in the most northern ranges were intergrades with other subspecies—*canadensis* to the north, *californiana* to the west. For our purposes, and using the ecology of the state as a basis for classification, we can say that bighorn, in at least central and southern Nevada, are of the desert type, in this case, *nelsoni*.

Bighorn are gone from northern Nevada, except for introduced *californiana* stock. We know definitely that they formerly occurred in this part of the state in at least the Granite Ranges of Washoe and Elko counties, the Jackson Mountains, the Santa Rosa Range, the Calico Range, the Sonoma Range, the Lake Range, the Simpson Park Range, the Roberts Mountains, the Ruby Mountains, the Schell Creek Range, the Snake Range, and the buffalo Hills. In some of these, particularly the Ruby Mountains, they were common (Barrett, 1965). The last bighorn in the Ruby Mountains were seen in 1921 (Hall, 1946).

The explorer John C. Frémont made numerous sightings of bighorn on the Lake Range east of Pyramid Lake in 1844. Bighorn eventually became extinct in northern Nevada in the 1930s upon disappearing from their last strongholds in northern Washoe and Humboldt counties. In central Nevada they disappeared about the same time from such areas as the Wassuk Range in Mineral County and the Stillwater Range in Churchill County.

In 1978 bighorn were found in parts of central and throughout southern Nevada, with greatest concentrations in southern Lincoln County and in Clark County. Further study may reveal them in places where they are thought to be absent. Some populations are very small and are isolated from other populations by large expanses of low desert or other barriers. Bighorn are occasionally observed in Nevada far from known bighorn habitats. These sightings are usually of rams during the fall and are probably associated with the wanderings of this sex during the rut.

Bighorn in Nevada occur in the following mountain ranges: *Nye County*—Grant Range, Toiyabe Range; *Esmeralda County*—Lone Mountain, Monte Cristo Range, Silver Peak Range; *Lincoln County*—Delamar Mountains, Meadow Valley Range, Mormon Mountains; and *Clark County*—Arrow Canyon Mountains, Black Mountains, Desert Range, East Desert Range, Eldorado Mountains, Highland Range, La Madre Mountains, Las Vegas Range, McCullough Range, Muddy Mountains, Newberry Mountains, New York-Castle-Paiute mountains complex, Pintwater Range, River Mountains, Sheep Range, South Spring Range.

From 1974 to 1976, an inventory of Nevada bighorn numbers was undertaken. The resulting population estimate of 4,270 bighorn in Nevada was based on aerial surveys of each mountain range. An expansion factor was applied to the counts in all ranges, based on the results of fifteen individual surveys on a marked population in the River Mountains, utilizing a modified Lincoln index (McQuivey, 1977).

NEW MEXICO

Desert bighorn never ranged over any considerable parts of New Mexico. They were once common only in the series of low but rugged mountains along the Mexican border east to and including the Guadalupes on the Rio Grande–Pecos River divide, and north to the Sacramento, San Andres, and Peloncillo mountains, as well as to the canyon of the San Francisco River in Catron County. Our chief authority for this information is Bailey (1931), who also mentions authentic records from the great lava flow in Valencia County south of Grants, apparently as late as some time in the 1920s. Aldo Leopold interviewed persons who had definite information that bighorn were present in the Manzano Mountains until the 1920s, and on Yeso Mesa in southeastern Valencia County until about 1900 (George Merrill).

Early explorers seldom mention bighorn in New Mexico in their writings. J. O. Pattie, somewhere along the San Francisco River in January 1825, probably in its lower reaches near the Arizona border, saw "multitudes of mountain sheep" and one was killed by his party (Thwaites, 1905). The word "multitudes" must be taken with a grain of salt, especially since Pattie does not mention seeing bighorn elsewhere in his considerable roamings over southwestern New Mexico and his *Narrative* was constructed much later from memory.

Abert (1848) reports finding mountain sheep "as far south as Valverde" but does not particularize.

"An occasional mountain sheep" was seen in Steins Pass, Peloncillo Mountains, by Cozzens (1967), who does not mention other bighorn seen anywhere in the states of New Mexico and Arizona. The inference from these and other early reports is not one of abundance anywhere in what is now New Mexico.

By the time of Bailey's report, desert bighorn had apparently vanished from all parts of New Mexico except the San Andres-Organ, Big Hatchet-Alamo Hueco, Guadalupe, and Sacramento mountains. In 1941 San Andres National Wildlife Refuge was established specifically to protect desert bighorn. The remnant herd there numbered about 350 animals in 1978, including some ranging off the refuge to the north, and to the south as far as the Organ and Doña Ana mountains. The much smaller group in the Big Hatchet and Alamo Hueco mountains numbered thirty-five to fifty in 1976 (Montoya and Munoz, 1976). The Sacramento Mountains bighorn have not been accounted for since 1942, and the last sighting in the Guadalupe Mountains was in 1946 (Gross, 1960).

An estimate of the total number of desert bighorn in New Mexico in 1978 would be 350 to 390 animals.

Rocky Mountain bighorn, ten from Banff National Park, Alberta, Canada, and eighteen from the descendant flock of introduced bighorn of the same subspecies in the Sandia Mountains, were released on the southwestern side of the Gila National Forest in Catron and Grant counties in 1964 and 1965. The outcome of these releases was at least initially favorable. Surveys conducted in 1974 found bighorn in the release areas, but the number present was not definitely known (George Merrill).

Five desert bighorn ewes were captured in Sonora, Mexico, in early 1972 (Gates, 1972) and transported to a holding pen at Redrock in Grant County. All of the ewes were pregnant when captured. As of 1977, these ewes, plus captives from San Andres Refuge, and their descendants, numbered about forty animals (Montoya and Munoz, 1976; Snyder, 1977).

TEXAS

Desert bighorn originally ranged in the mountains along the drainage divide between the Pecos River and the Rio Grande, and in the Rio Grande canyons of the Big Bend region in trans-Pecos Texas. Apparently widespread and common, they were especially numerous in the more extensive mountain ranges. By the time of Davis and Taylor's report (1939), the total number of bighorn in Texas may still have been as high as 300 head, but they had disappeared from all but the Guadalupe, Apache, Beach, Baylor, Eagle, and Cienega mountains and the Sierra Diablo. In 1961 only twenty-five to thirty were left (Davis, 1961). Some time in the early 1960s they had all disappeared, and the native Texas bighorn was extinct.

From 1957 through 1959, desert bighorn were trapped on Kofa Game Range in Arizona and transplanted to a fenced enclosure on Black Gap Wildlife Management Area south of Alpine in Brewster County. In all, thirteen animals were successfully moved to the 427-acre enclosure, including one ram caught on a ranch near Wickenburg, Arizona, in 1958. By 1974, the increase had resulted in twenty bighorn inside the enclosure and about forty outside. In addition a few Black Gap bighorn were moved in the 1960s to an enclosure in the Sierra Diablo of Hudspeth County. In January 1977 six ewes captured in Baja California, Mexico, were added to the Black Gap population for a net total of about fifty in the two areas (Winkler, 1977).

UTAH (and COLORADO-WYOMING)

In Utah desert bighorn are and have been confined mostly to canyon portions of the Colorado, Green, and San Juan rivers and their tributaries. They apparently are not found on the San Juan River farther east than the Goosenecks and are largely restricted to the general Glen Canyon National Recreation Area–Canyonlands National Park region, much of which is remote, inaccessible, and little known. There are also, on the basis of sight records in the 1960s, small populations of desert bighorn in Capitol Reef National Monument and the San Rafael Swell in Emery County (Norman Hancock). In the 1960s and early 1970s, scattered reports of single animals came in from the western edge of the state, probably of bighorn wandering from mountains in adjacent Nevada.

A small number of bighorn were planted in Zion National Park beginning in 1973, the stock coming principally from Lake Mead National Recreation Area in Nevada. Bighorn were originally present in Zion Park at least until 1953 (McCutchen, 1975).

Wilson (1968) is of the opinion that bighorn were more common prior to the introduction of domestic sheep in the 1880s, especially in the vicinities of Moab and Escalante, where they later vanished. He further feels that bighorn would be more numerous in much of their present range were it not for heavy poaching in the 1940s and 1950s by Indians and uranium ore prospectors and miners.

The Ute Indian tribe released some Rocky Mountain bighorn in the Desolation Canyon section of the Green River, above the town of Green River, in 1968 and 1971. Bighorn formerly inhabiting this canyon, as well as those once found in the area of the Green-Yampa confluence in Colorado and northward along the Green River into Wyoming below the town of Green

River, may have been intergrades between desert and Rocky Mountain bighorn, but we will never know for sure. The 1978 population along the upper Green River was transplanted there by the Colorado Game, Fish and Parks Department and is of Rocky Mountain stock (Norman Hancock).

Desert bighorn range once extended up the Colorado River from Utah into Colorado, where a small herd still existed in 1978 on Battlement Mesa in Mesa and Garfield counties. It has become isolated by human settlement so that now there is approximately a 200-mile gap between this small herd and the nearest desert bighorn in the canyonlands country of Utah (William Rutherford).

As from other states, early-day mention of desert bighorn in Utah is almost lacking in the literature. The writings of the Franciscan missionary Fray Silvestre Velez de Escalante (Bolton, 1950), and the conqueror of the Colorado, Major John Wesley Powell (1874), mention bighorn on a first-hand basis.

The 1978 estimate of desert bighorn numbers in Utah, based principally on aerial surveys, was that 350 to 500 were living in the state (Rodney John, Lanny Wilson, et al.).

BAJA CALIFORNIA NORTE and BAJA CALIFORNIA SUR

Desert bighorn are found in most of the mountainous terrain of the Lower California peninsula, from the Sierra de la Giganta that lies to the north of Bahía La Paz northward to the international boundary. The Sierra de la Giganta constitutes the southernmost extension of range of mountain sheep in the New World. As in other parts of Mexico, we are confronted with a general lack of knowledge of bighorn numbers in Baja California.

Mario Luis Cossio, former chief of the Mexican wildlife department, on the basis of surveys made in 1974 and information obtained earlier in Baja California, considers there are four rather distinct populations on the peninsula, numbering as follows:

1. In the Sierra de la Giganta north to the mountains west of Loreto, 500.

2. In the mountainous area of Las Tres Vírgenes and the Sierra de San Francisco, roughly between Santa Rosalia and El Arco, 500.

3. In the mountains about Bahía de Los Angeles, including the Sierra de Calmallí, the Sierra de San Borja, the Sierra de Calmajue y San Luis, and Cerro Junipero Serra, numerous—1,500.

4. In the mountains from the general area of Bahía de San Luis Gonzaga and El Marmol north to the international boundary, a much larger and more extensive area than any of the first three, 1,500, possibly twice that many.

Cossio believes there is an area of intergradation between *Ovis canadensis weemsi* and *O. c. cremnobates,* represented by population (3) and the southern part of population (4), above.

The four populations ascribed to Baja California by Cossio total 4,000 bighorn. He indicates that this must be regarded as a beginning estimate, and may be considerably in error—too high or too low. Alvarez (1976) estimates, on the basis of surveys made during the hunting season in 1974, that Baja California holds between 4,560 and 7,800 bighorn.

CHIHUAHUA

Bighorn undoubtedly were common at one time in the low mountains of northwestern Chihuahua, as they were in adjacent New Mexico, as well as in the mountains southeastward as far as the western side of Coahuila. In his summary of the species in Coahuila, R. H. Baker (1960) cites evidence of bighorn in the Sierra Mojada in extreme southeastern Chihuahua. In 1974, Angel Salas Cuevas of the Mexican wildlife department saw bighorn and found considerable sign in the Sierra Mojada, the Sierra del Diablo, and the Sierra de San Francisco, a grouping of mountains lying in extreme southeastern Chihuahua from 65 to 130 km (40 to 80 mi) east of Jiminéz. He believes bighorn may be present also in the Sierra de Los Angeles in extreme northeastern Chihuahua. We may, on the basis of the above, suggest there were in 1978 as many as 50 bighorn in Chihuahua.

COAHUILA

Baker (1960) summarizes the status of desert bighorn in Coahuila in the late 1950s as, "Formerly found in arid mountainous areas in the northwestern half of Coahuila, now found in a few localities in the central and northwestern part of the State." He goes on to say: "Bighorn sheep probably occurred formerly in most desert ranges of western and northern Coahuila, at least as far south as the Sierra de la Paila ... From reports received from older residents, bighorns are more at home in the arid rimrock of the foothills and less elevated mountains than in associations of oaks or conifers at the higher elevations. The gradual reduction in number in Coahuila has been the result of both hunting and competition with livestock, chiefly goats. Most of the bighorns now living in Coahuila are in areas either where few people live or where land operators are giving the animals protection from hunting."

Also according to Baker, the only localities in Coahuila where bighorn might still be found in the 1950s were the mountainous area northwestward of San Lazaro (including the Sierra Palo Verde or Sierra de Santa Rosalía, and the Sierra de San Marcos); the Sierra de los Hechiceros in extreme northwestern Coahuila; and the Sierra Mojada (actually in Chihuahua), Sierra del Pino, and Sierra del Rey near the Chihuahua border.

We may presume there were in 1978 possibly fifty bighorn in Coahuila —a sheer guess in view of the lack of new data since 1960.

SONORA

The desert mountains in the northwestern part of Sonora have long been regarded as one of the strongholds of bighorn. They formerly occurred in almost every mountain range as far east as Mexico Highway 15 (from Nogales to Hermosillo) and south to Bahía Kino, where the low mountains inland from the bay represent the southernmost outpost of the animals in Sonora, and where I found them in 1974. In 1978 they were still present in the Sierra Pinacate and other ranges near the Sea of Cortez (the Gulf of California). Since about 1930, it seems there has been some withdrawal or disappearance from the uplands east and north of Mexico Highway 2 (from Sonoyta to Caborca and Santa Ana); at any rate, there were few records from that district in the 1960s and early 1970s except for the Sierra Cubabi near Sonoyta, the Sierra del Humo, and the Sierra del Carrizal (Angel Salas Cuevas).

As to actual numbers in Sonora, the information is scanty. Mendoza (1976) estimates there were no more than 900, an approximation based on a close familiarity with big game in northwestern Sonora. He also states that bighorn were disappearing from the Sierra Pinacate and the Sierra de San Francisco.

In January 1975, twenty bighorn (three rams and seventeen ewes) were transplanted from the Sonora mainland to Tiburón Island, just off the coast of Sonora in the Sea of Cortez (Montoya and Gates, 1975).

RECAPITULATION OF DESERT BIGHORN NUMBERS

From the data included in the foregoing, we can arrive at a figure for a total number of desert bighorn existing in 1978. The data involved in some cases are admittedly sketchy. Nevertheless, we postulate these totals as a point of beginning for the accrual of more accurate data:

STATE	NUMBER
Arizona	2,100– 2,600
California	3,250– 3,750
Nevada	3,700– 4,200
New Mexico	350– 390
Texas (transplants)	50
Utah	350– 500
U.S. TOTAL	9,800–11,490
Baja California Norte and Baja California Sur	4,560– 7,800
Sonora	900
Chihuahua and Coahuila	100
MEXICO TOTAL	5,560– 8,800
GRAND TOTAL	15,360–20,290

4. Physical Characteristics

Charles G. Hansen

GENERAL BODY CHARACTERISTICS

Conformation. Desert bighorn—in contrast to mule deer—are relatively short-legged and generally blocky in appearance, but not as thick-bodied as the northern subspecies. An adult desert bighorn is 76 to 100 cm (30 to 39 in.) tall at the shoulders, and about 152 cm (60 in.) long. The front legs are about 46 cm (18 in.) long, and the chest is about 46 cm (18 in.) deep. The male is normally larger than the female, although some females may be taller and have a larger body than the smallest of the males. The adult male averages about 73 kg (160 lb) live weight, but when in good flesh in early summer may weigh 91 kg (200 lb) or more. An adult ewe averages about 48 kg (105 lb).

Young females are relatively slender until about 2 years old. After they have had their first lambs, generally at the age of 3 years, they thicken in the body but retain a generally sleek appearance. Ewes have much slimmer necks than rams. This is most noticeable in the young ewes, and again as the animals approach old age. Prime ewes are well proportioned animals, with erect head and rounded rump and pelvic region. Healthy ewes do not lose such an appearance in old age as do the rams. See Figure 4.1 for the head conformation of an adult ewe.

Physical Characteristics 53

Charles G. Hansen

Fig. 4.1. Head conformation of a desert bighorn ewe, "Nanny." Desert National Wildlife Range, Nevada.

Prime rams have a thick, blocky appearance, and when in full flesh they are a little too heavy in the body to be considered well proportioned. Old rams often become saggy in appearance; they hold the head lower than do younger animals; their withers are high, and the rump angular. In young rams, the smaller horns and slimmer conformation give them the appearance of prancing when alerted and moving. In contrast, a healthy prime ram with large heavy horns has a regal appearance.

Young lambs appear compact, owing to their short nose and neck. Their legs are long, but the animals are not leggy like a new-born colt or fawn. Lambs quickly fill out and at the age of 2 or 3 months have developed into miniature adult-shaped bighorn, sleek and well proportioned.

Size. Weights and measurements of adults and young animals are presented in Tables 11.1 through 11.8 in Chapter 11.

The respective weights of the head and horns of rams illustrate the comparative bulk of the horns. The average weight of a dry skull is about 1.4 kg (3 lb) and that of dry horns is about 4.5 kg (10 lb). Thus a dry head with horns would weigh about 6 kg (13 lb).

The total weight varies considerably through the year. Rams weighing 91 kg (200 lb) or more in the early summer when they are in good flesh may weigh only 63 kg (140 lb) by winter. Therefore, the time of the year must be considered when the animals are weighed. In Utah mature rams estimated at 77 kg (170 lb) live weight were found to weigh 59 kg (130 lb) field dressed (John, 1968). The loss of weight when a ram is field dressed ranges from 28 to 38 percent and averages 33 percent.

HAIR AND SKIN CHARACTERISTICS

Color. There is considerable variation in the color of the pelage, which changes from dark in the summer and fall to faded or washed out in the spring just before the hair is shed. Young animals tend to have lighter colored coats than prime or old bighorn. Ewes generally are lighter colored than rams.

There is considerable variation in color between geographic areas and within areas. On Desert National Wildlife Range, bighorn are generally medium gray-brown with white on the rump, back of legs, and muzzle. The ewes do not show the wide variation seen in the rams. In some areas of the Wildlife Range the ewes are more gray and less brown, while in other regions the reverse is true. The rams may be quite blackish or even rusty red. The amount of white on the rump, legs, and nose also varies. Some white-spotted bighorn live on the Pintwater Mountains of the Desert Wildlife Range. Their coloration resembles that of a pinto horse, with large splotches of white on their sides or throat and chest (Hansen, 1965d). At least two all-white bighorn have been seen in the Spring Mountains west of Las Vegas, Nevada.

Lanny O. Wilson writes: "The desert bighorn in Utah varies in color from a buff tan, palomino color to a dark slate gray (almost black) on the neck, back, and sides. The brisket tends to be slightly darker than the sides, back, and neck, but becomes increasingly lighter from just back of the front legs. The stomach area is white in many animals. The white rump patch and white band around the nose to the lower jaw are very conspicuous. Rams' horns vary in color from a yellowish-brown to a dark brown."

Molt pattern. On Desert Wildlife Range, the pattern of shed and unshed hair and fleece was used to distinguish individual bighorn during water-hole counts (Hansen, 1964c). These observations showed where on the body and when the winter hair and fleece were shed. The general pattern is illustrated in Fig. 4.2.

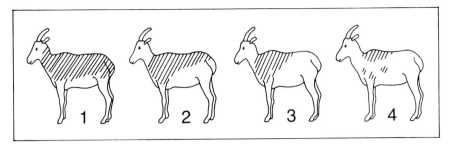

Fig. 4.2. Shedding pattern of desert bighorn.

According to observations made by the author on Desert Wildlife Range, old animals, presumably less vigorous, do not molt as early as the younger animals. Yearlings and lambs shed later than prime bighorn. Ewes having nursing lambs usually retain old hair longer than dry ewes. The first to shed are the prime healthy rams and dry ewes. Bighorn from lower and drier areas molt before those living at higher elevations where it is cooler and more moist.

About one month is required to shed the winter coat. Generally, this occurs from the last of June through July. However, in warm years the first bighorn may complete shedding in May, while during other years this may take place as late as the first of September. The hair and fleece around the chest (Fig. 4.2, drawing 3) is retained the longest in animals that are progressing through a normal molt. This progression may occur so rapidly that a distinct pattern is not easily recognizable. This happens with prime rams, but usually there is a trace of old hair or fleece on the chest region as the last evidence of shedding.

Sick bighorn that retain their coats until late in the season also seem to shed all at one time. Often their coats look patchy by the time they begin to shed. Bighorn that are diseased but not run-down often shed normally. This information about shedding can give some idea as to the number of sick animals in the herd and as to the age and condition of the ewes.

In Utah, Wilson makes the following observations: "By the first of July 1965 most of the bighorn in the White Canyon area had completely shed their winter coats. By the first of July 1966 most of the bighorn were only partly shed, and many were just beginning to shed. The difference in the shedding patterns could probably be attributed to the amount of moisture in the spring and summer. In 1965 it rained almost once a week between April and June 15. In 1966 there was a complete lack of measurable precipitation from April to August. The bighorn were in poor condition in 1966 due to lack of moisture, ample succulent forbs, and green vegetation. These data would suggest that shedding could be closely associated with the physical condition of the animal."

Hair length and thickness. The hair on desert bighorn varies in length from sparse short hairs to a fraction of a centimeter long on the face and at the junction of the legs and body to hairs in the mane that are several centimeters long. The average length of body hair is about 2.5 to 4 cm (1 to 1.5 in.). It grows thick on the sides and top of the body and when combined with the undercoat of fleece is quite dense.

Fleece. The fleece is generally 15 mm (0.5 in.) deep. It is kinky-curly and when stretched out can be 25 to 30 mm (1 in.) long. The fleece is shed with the other hair during the annual early summer molt and is not replaced until autumn. It is rubbed off on trees and rocks where it is sometimes quite noticeable. It appears to be a fine quality wool.

Skin. The skin color of the fetus is black or very dark brown. The rump patch and future white areas show as a lighter color. The color pattern is retained in the adult animals, and the hairless areas are quite black.

The skin of the tongue varies in color. Some desert bighorn have black tongues, and others have pink tongues. Black tongues are more common. Interbreeding of these types produces various combinations of colors. Some combinations show longitudinal stripes; others have the distal part black and the proximal part pink (Hansen, 1970).

Insulation. Desert bighorn are seldom very fat but put on weight in the summer, presumably to prepare for the fall rutting activity and spring lambing. In summer the fat and hair may insulate them from the heat, and the pattern in which these animals molt may reflect this requirement. During the winter most rams are relatively thin; the thick pelage of that season provides insulation that helps in temperature regulation.

HORNS

General. Horns are comprised of a corneous sheath of keratin covering a bony core. They are permanent and elongate from the base. They stop growing in the rut season, at which time a horn ring is formed, leaving a permanent indication of age. Chapter 11 describes much of this in detail. Horns have a relatively smooth appearance in young animals but become rough and scarred in old age. They are used in fighting and feeding and often become worn or "broomed" on the ends. The bony core is covered by a layer of tissue well supplied with blood. The core is honeycombed with chambers or sinuses which reduce the weight of the skull.

Size. Horns of adult rams may measure 76 to 102 cm (30 to 40 in.) along the outside of the curl, while horns of ewes are generally only 25 to 33 cm (10 to 13 in.) long. Horn spread varies from 56 to 66 cm (22 to 26 in.) in rams and from 30 to 41 cm (12 to 16 in.) in ewes. Basal circumference of horns in adult rams averages about 30 cm (12 in.) but may be as great as 37 cm (14.5 in.). In

Jack R. Cooper

Fig. 4.3. Rams of some desert bighorn populations exhibit unusually flaring horns. River Mountains, Nevada.

ewes the basal circumference of horns is from 13 to 15 cm (5 to 6 in.) (Baker, 1967). A ram younger than 6 or 7 years of age generally has a greater tip-to-tip spread than do older rams. Rams of some desert bighorn populations may exhibit unusually tight, or unusually flaring, horns (Fig. 4.3).

Records. Horns of unusually large size are listed in a 1977 publication of the Boone and Crockett Club and the National Rifle Association of America, "North American Big Game." The book shows that a head having a Boone and Crockett score of 205 1/8 points is the largest yet recorded. Two heads picked up in Baja California occupy first and second place.

The Boone and Crockett method of scoring uses the total length and circumference of the horn at four places. (See Appendix 1.) The procedure is: Measure in inches the total length on the outside of the curl. Divide the total length of the longest horn into four equal segments to obtain points D-1, D-2, D-3, and D-4. The distances obtained from the longest horn are used to delimit points D-1, D-2, D-3, and D-4 on the other horn, even though it is shorter. These points are marked on the outside curl of the horns. Then measure the circumference at D-1, D-2, D-3, and D-4 for each horn. These ten measurements are added. The differences between the scores for each horn are then subtracted from the total score for both horns. The resulting number is the Boone and Crockett score.

Record desert bighorn heads in the respective states are:

Arizona	$190^{5}/_{8}$	6th place
California	184	26th place
Nevada	$181^{5}/_{8}$	37th place
New Mexico	$178^{7}/_{8}$	50th place
Baja California, Mexico	$205^{1}/_{8}$	1st place
Sonora, Mexico	$187^{3}/_{8}$	10th place

The record head from Baja California measured:

Length		Base		Third quarter		Spread	
Right	*Left*	*Right*	*Left*	*Right*	*Left*	*Greatest*	*Tip-to-tip*
$43^{5}/_{8}$	$43^{6}/_{8}$	$16^{6}/_{8}$	17	$10^{5}/_{8}$	$10^{6}/_{8}$	$25^{5}/_{8}$	$25^{5}/_{8}$

Seven of the ten largest desert bighorn trophies came from Baja California.

Uses. Horns of the rams are used for fighting and for self-protection. Chapter 9 describes the behavioral use of horns during the rut. They are also used to obtain food and possibly to release nervous tension. Bighorn, especially rams, have learned to obtain food by butting trees or cactus. In some cases the butting dislodges fruit or nuts (see Chapter 8; Russo, 1956); in other cases it breaks spines from cactus to make the succulent flesh available. When bighorn are under extreme stress they have been observed to butt other bighorn, rocks, and posts on Desert National Wildlife Range. The impression is that they use this method to release frustrations.

EYES

"The eyes of young lambs are uniformly dark to the extent that they appear black. The eyelids are somewhat elliptical in shape. A lamb four months of age still retains the dark eyes; a yearling ewe still carries a trace of darkness although the eye was predominately golden, and a two-year-old ewe possessed the typical golden eye coloring of the adults. It is thought that this change of eye color takes place gradually as the animal matures.

"Observations at the Corn Creek headquarters indicated that for a period of a few hours after birth lambs are unable to focus their eyes to the extent that they can distinguish objects with full clarity of vision. This condition appeared to pass during the first 24 hours of life" (Deming, 1953).

The pupil is horizontal and quite dark brown, almost black, in adults. The diameter of the exposed portion of the eye is approximately 3.8 cm (1.5 in.). The lamb's eyes are proportionately larger than those of the adult and seem to bulge most of the time. By the time the animal is 2 months old this large-eyed appearance has been replaced by a more adult aspect.

TEETH

The dental formula ($0/3$ $0/1$ $3/3$ $3/3$) for desert bighorn is the same as for other bovids. A complete set of teeth is acquired by 4 years (Cowan, 1940; Deming, 1952). Tooth replacement is described in Chapter 11.

There is considerable variation from the normal tooth pattern, as well as frequency of damaged and lost teeth, and other irregularities of the tooth row (Allred and Bradley, 1965). Maxillary or upper canines have been reported. Deming (1952) reports upper canines in more lambs than adults. Allred and Bradley (1965) report one adult ram with an upper canine out of 132 rams and 95 ewes examined. Three upper canines were found in lambs and two in rams in a collection of 147 rams, 102 ewes, and 11 lambs from Desert National Wildlife Range. The incidence of upper canines in bighorn from the Wildlife Range was 1.4 percent for rams and none for ewes (Bradley and Allred, 1966). In a sample from California, 3.7 percent of the rams and no ewes had upper canines. In a small sample of skulls from Arizona, 9.1 percent of the rams and one of the two ewe skulls examined had vestigial canines (Bradley and Allred, 1966).

Allred and Bradley (1965) found that approximately 29 percent of the bighorn from the Wildlife Range had one or more missing second premolars. Approximately 12.5 percent of the rams and 54 percent of the ewes had this anomaly. The lower jaw is affected most often. The extreme was found in three ewes, each having two lower and one upper second premolar missing.

A high incidence of necrotic bone of the dental arcade was found in individuals from Desert Wildlife Range (Allred and Bradley, 1966). Osteonecrosis was widespread in ewes 6 years of age or older, and in rams 10 years or older. This condition was not found in lamb skulls, suggesting that it occurs only in adults. A greater incidence of osteolysis and necrosis is associated with teeth in the upper rather than the lower jaw. Allred and Bradley (1966) conclude that damage to the bone surrounding the dental arch was an important factor limiting the life expectancy of bighorn of Desert Wildlife Range.

Bighorn teeth appear to last about 14 years on the limestone soil of the Wildlife Range. The amount of tooth necrosis and wear probably begins to affect the animals' health at about 8 to 10 years of age, because they are unable to masticate food properly. Animals that escape serious osteolysis or necrosis seem to be able to live 5 or 7 years longer than those individuals having such diseases. A captive bighorn ram lived for 17 years on the Wildlife Range on a diet of alfalfa, grain, and a minimum of grass and browse. His teeth were becoming necrotic, although internal tumors were the direct cause of his death. Other bighorn have survived to at least 17 years (Woodgerd, 1964), so that teeth must have the potential to sustain them to at least that age. Wilson (1968) reports that a bighorn in Utah had malocclusion of the premolars and molars with no evidence of any wear. All teeth had extremely sharp spinous processes making mastication practically impossible. This animal was in poor condition and appeared sickly with dull pelage; she was stunted compared with other adult ewes in the area.

EXTERIOR GLANDS

A lacrimal gland is situated below the lower inside corner of each eye. Adjacent skin permits the gland to be exposed or covered. Rams have been seen to rub their lacrimal glands on trees and boulders, an action that probably results in the marking of an individual's territory. Other bighorn have been seen to seek out these signposts.

Small interdigital glands open on top of the foot at the junction of the toes. Occasionally one of these glands does not function properly, and a firm plug of waxy material appears at the orifice. Such waxy material also occurs on the eye gland but is less common. The interdigital gland may function as an organ to deposit scent to mark the passing of an animal (McCann, 1956). Welles and Welles (1961a) describe a ram tracking a ewe, presumably by scent, and I have seen similar although inconclusive behavior.

Desert bighorn have functional sweat glands that help dissipate body heat. They have nearly twice as many per square centimeter but they secrete only a fraction of the amount of sweat produced by man. The cooling process in hot weather appears to result more from the dissipation of heat by panting (Schmidt-Nielson, 1964).

GAITS

The desert bighorn has five gaits in traveling, and these often reflect the mental state of the animal.

The *walk* normally is used by an animal when moving slowly. The body is relaxed, and the neck is at least slightly outstretched. This gait differs little, or not at all, from that of other bovids.

The *pompous* or *alarmed walk* is manifested when the animals are alarmed or are parading before other animals. They hold their heads and necks erect and take slightly higher steps than the normal walk; however, the leg movements are not exaggerated like those of a parade horse.

The stiff-legged bound usually follows an alarm and may develop from any of the other gaits or from a standing start. In this gait the animal bounds off the ground with all four feet and propels itself forward at the same time. This is not usually done as neatly and precisely as by mule deer or antelope and usually is used only for short distances of 15 or 30 m (50 or 100 ft). Animals moving in this manner seem to be deliberately trying to make a lot of noise as they slap their front feet on the ground. This behavior appears only after the animal has been alarmed and may serve several purposes: it may allow the animal to keep its attention on the source of alarm, and it may also

Physical Characteristics 61

Jack R. Cooper

Fig. 4.4. The bighorn "run," one of the five traveling gaits. River Mountains, Nevada.

be the best way to prevent leg injuries when traveling fast through brushy or uneven ground. For example, when all four feet land in the same small area there is less chance of a leg being damaged, since the weight is on four legs instead of one or two as when the animal is trotting or running. Welles and Welles (1961a) state that, "The four-footed, drumming bound ... seems to be responded to by all as a voluntary warning of danger."

The *trot* is similar to that of most other big game animals; it and the walk probably are the most common gaits of bighorn. The trot is often used when the animals are traveling long distances, or are mildly alarmed, or exhibiting a "change of pace." They often trot downhill, apparently through sheer exuberance.

The *run* is the most rapid method bighorn have of covering ground. Usually they do not run unless they are alarmed or are chasing another bighorn. This gait is similar to that used by other big game animals. When seriously frightened, they go into a flat-out run that lowers their body nearly to the ground as they move rapidly (Fig. 4.4).

When two rams are going through the ritual of butting heads, they run at each other for about 9 or 12 m (30 or 40 ft) on their hind legs. This is accomplished in much the way that a human would walk or run on the balls of his feet.

Speed of travel of these animals is solely from estimation. Welles and Welles (1961a) state, "I estimate their speed at 25–30 miles [40–50 km] per hour through terrain that we could scarcely walk across with safety." On flat ground they can probably go faster but presumably for only short distances. During their head-butting ritual it has been estimated they meet head on at a combined speed of about 50 km (30 mi) per hour (Cottam and Williams, 1943).

PHYSIOLOGICAL CHARACTERISTICS

Metabolism is the process by which all living things transform food into energy and living tissue. Heat is a product of metabolism and solar absorption. The body temperature of desert bighorn varies slightly around 40°C (104°F). This temperature was taken of a young animal during cool days, and also on hot days when it was lying in the direct sun and panting heavily (Hansen, 1964b). No record has been made of body temperatures correlated with air temperatures or after heavy exercise. Skin temperatures were taken from two bighorn when the air temperature was 31°C (88°F). On the midback of a 6-month-old ram and an adult ewe the temperature was 35°C (95°F).

Sweat glands scattered over the body aid somewhat in maintaining the body temperature. Hair and wool act as insulation against heat as well as cold. Heat is also lost through respiration, which may be the principal system that regulates body heat in these animals. A difference in temperature readings of 35°C (95°F) on the back where hair was long and 37°C (98°F) by the front leg where hair was sparse could reflect the effect of insulation from the wool and also the effect of sweat glands.

Water retention and cooling are important to desert animals (see Chapter 7). The following characteristics may play an important role in the survival of desert bighorn: (1) The animals retain their thick pelage during the heat of the summer. (2) The rams usually go into the summer with a good amount of fat, presumably in preparation for the rut season. (3) The rams may go for 5 to 7 days without water during dry hot weather. (4) Studies of the shedding pattern show that hair and wool over the chest is the last to be lost. (5) The animals seek shade, higher elevations, cool soil or rocks, and breezes that are cooling. (6) They can drink large amounts of water in a short time. (7) Their internal organs are adapted to withdraw liquids from their foods and conserve body moisture. (8) According to E. L. Fountain, D.V.M., U.S. Army, formerly with the Atomic Energy Commission at Las Vegas, Nevada, three water retention organs in the desert bighorn are enlarged and modified for greater water retention when compared to those same organs in domestic sheep. These are the reticulum in the stomach; the enteric caecum; and the colon. (9) Cooling appears to be achieved mainly by evaporation and by direct conduction to ground and rocks when the animals are in caves or other shade. (10) Desert bighorn can withstand changes in body temperature from 39°C (102°F) to 42°C (107°F) without apparent ill effects. (11) The kidney eliminates excreta with a relatively small loss of water. (12) Bighorn can withstand a certain amount of dehydration, perhaps exceeding the 30 percent weight loss reported in domestic sheep. (13) Domestic sheep exhibit a tolerance to heat and water depletion that exceeds the norm for most animals, while showing a decrease in plasma volume that appears to be extreme (Schmidt-Nielsen, 1964). Desert bighorn may well exhibit an even greater decrease in plasma volume than that of domestic sheep.

RESPIRATION AND PULSE

Average respiration in five desert bighorn on Desert National Wildlife Range was 111.5 per minute for rams, 90.6 for ewes, and 164 for one lamb. Respiration rate ranged from 64 to 192 per minute for rams, 64 to 150 for ewes, and 144 to 193 for the lamb. One ram showed a higher rate while lying or standing in the shade than while in the sun (Logsdon, 1966).

The normal domestic sheep heart beats at a rate of 80 per minute (Logsdon, 1967). The rate in desert bighorn is assumed to be the same.

5. Habitat

Charles G. Hansen

Typical desert bighorn terrain is rough, rocky, and much broken by canyons and washes. Low, rugged mountains of northern Mexico and southwestern Arizona; cliffs along the Colorado River; the sandstone breaks of southeastern Utah; the dissected alluvial fans of Desert National Wildlife Range in southern Nevada; and the broken foothills surrounding a desert mountain range of Death Valley in California—all these are illustrative of bighorn habitat.

The term "desert" is defined qualitatively as dry, brush-covered land with much bare ground exposed; quantitatively as a region receiving less than 25 cm (10 in.) of precipitation annually. Within this large desert biome are a multitude of microhabitats, the result of a variety of plant and animal communities, determined by such factors as variation in soil, topography, temperature, and rainfall.

That area of North America within our definition of a desert embraces about one million sq km (440,000 sq mi) and is illustrated in Figure 5.1. In relation to other deserts of the world, it is fifth in size and is about one-eighth the area of the Sahara. Possibly less than one-third of the entire North American desert biome is presently inhabited by bighorn, although during pristine times possibly the entire region was inhabited (Buechner, 1960).

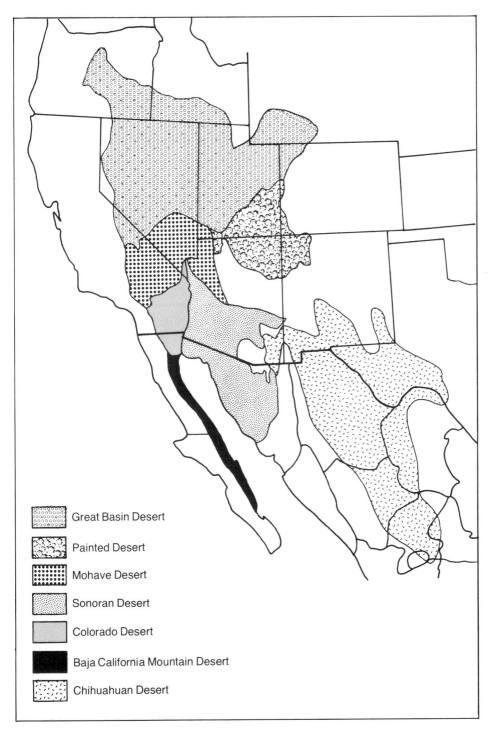

Fig. 5.1. North American desert and its subdivisions, excluding the parts of Mexico where bighorn are absent.

Based on climate and vegetation differences, the North American desert can be divided into seven regional deserts: Great Basin Desert, Painted Desert, Mohave Desert, Colorado Desert, Sonoran Desert, Baja California Desert, and Chihuahuan Desert (Fig. 5.1).

GREAT BASIN DESERT

The largest subdivision of the North American desert is the Great Basin Desert. It occupies most of Nevada and Utah, as well as parts of Oregon, Idaho, Wyoming, and Colorado.

The Great Basin consists of broad valleys and plateaus divided by mountain ranges; elevations are 600 to 2,500 m (2,000 to 8,000 ft) above sea level. Temperature is characterized by cold winters and hot summers. The mean January temperature is between −2° and 4°C (29° and 39°F); the mean July temperature is between 21° and 26°C (70° and 78°F). Mean annual precipitation may range from 14 to 25 cm (6 to 10 in.), depending on elevation and topography. The heaviest rainfall is usually in April or May, and in some areas the rainfall is heavy in autumn also. Winters generally receive moderate snowfall, which is one reason the area is termed a "cold" desert.

In the Great Basin Desert the soft green of the common sagebrush *(Artemisia tridentata)* is the characteristic color. This dominant low-growing shrub extends almost throughout the region, even to the higher elevations. In the lower alkaline plains the smaller, short-branched shadscale *(Atriplex confertifolia)* is common, as well as other drought-resistant plants like hop sage *(Grayia spinosa)* and winter fat *(Eurotia lanata)*. At least three types of Mormon tea *(Ephedra* spp.*)* occur, as well as greasewood *(Sarcobatus vermiculatus)*, rabbitbrush *(Chrysothamnus* spp.*)*, and snakeweed *(Gutierrezia* spp.*)*. Cacti are scarce.

The mountains support stands of pine *(Pinus* spp.*)* and juniper *(Juniperus* spp.*)*. In the south part of the Great Basin Desert the pinyon-juniper woodland consists of singleleaf pinyon *(Pinus monophylla)* in Nevada and common pinyon *(P. edulis)* to the east, mixed with either of three species of juniper *(Juniperus osteosperma, J. scopulorum,* and *J. occidentalis)*. In Oregon and Idaho pinyons drop out and junipers prevail. Often mixed with these woodlands are species of mountain mahogany, the most common being *Cercocarpus ledifolius* and *C. montanus.* Above the woodland, if the elevation is sufficient, ponderosa pine *(Pinus ponderosa)* may occur, with firs *(Abies* spp.*)* or Douglas fir *(Pseudotsuga menziesii)*.

Grasses in much of the bighorn range make up as much as 25 percent of the vegetative cover. Bluegrass *(Poa* spp.*)*, fescue *(Festuca* spp.*)*, wheatgrass *(Agropyron* spp.*)*, bentgrass *(Agrostis* spp.*)*, brome *(Bromus* spp.*)*, and squirreltail *(Sitanion* spp.*)* are more prevalent in the north, while galleta *(Hilaria*

Fig. 5.2. Spring in the Toiyabe Range, Nevada. Great Basin Desert.

spp.), needlegrass *(Stipa* spp.*)*, and grama *(Bouteloua* spp.*)* are common in the south. Indian ricegrass *(Oryzopsis hymenoides)* is much in evidence throughout.

Typical bighorn habitat in this desert is open, rugged terrain, with common sagebrush or shadscale usually constituting the ground cover. Open woodlands or forests are used in summer (Fig. 5.2), but deep snow in winter forces the animals to lower elevations unless sufficient windswept ridges are kept bare in the higher mountains. Desert bighorn in the Great Basin Desert are found only on its southern fringe in south-central Nevada.

PAINTED DESERT

This desert encompasses much of the brightly colored sandstone mesas and canyonlands of Utah, Arizona, Colorado, and New Mexico (Jaeger, 1957; Sutton and Sutton, 1966). This scenic and colorful land has hot, dry summers;

winters are cold with occasional snow. January temperatures average below freezing; in July they are just above 21°C (70°F). Most precipitation occurs from August through February, averaging about 13 cm (5 in.) annually in the valleys and around 23 cm (9 in.) on the mesas. Elevations range from 1,000 to 2,100 m (3,500 to 7,000 ft).

Characteristic vegetational communities are comprised mainly of common sagebrush, shadscale, Mormon tea, four-wing salt bush *(Atriplex canescens)*, hop sage, winter fat, green molly *(Kochia americana)*, bitterbrush *(Purshia tridentata)*, snakeweed, rabbitbrush *(Chrysothamnus viscidiflorus)*, bud sage *(Artemisia spinescens)*, and horsebrush *(Tetradymia spinosa)*. At lower elevations blackbrush *(Coleogyne ramosissima)* predominates. At higher elevations pinyon-juniper woodland is typified by common pinyon, Utah juniper *(Juniperus osteosperma)*, cliff rose *(Cowania stansburiana)*, mountain mahogany *(Cercocarpus* spp.*)*, and *Yucca* spp. (Wilson, 1968). The major grass species are wild ryegrass *(Elymus* spp.*)* and Indian ricegrass.

The bighorn range comprises the canyons, cliffs, and mesas of the canyonlands of southeastern Utah and the Grand Canyon of Arizona. An important bighorn habitat in southeastern Utah is afforded by the rugged Wingate cliffs and the rocky talus slopes immediately below them.

MOHAVE DESERT

The Mohave Desert consists of a series of mountain-separated undrained basins in southeastern California, southern Nevada, and northwestern Arizona (Fig. 5.3). It is the home of the Joshua tree *(Yucca brevifolia)*, which delimits the margins of this desert. The dominant shrub is creosote bush *(Larrea tridentata)*, whose lacy, deep yellow-green appearance replaces the soft gray-green of the sagebrush of the Great Basin Desert.

This also is a high desert, with elevations mostly between 600 and 2,000 m (2,000 and 6,500 ft). However, in Death Valley about 1,425 sq km (550 sq mi) lie below sea level, and nearby Telescope Peak rises above 3,350 m (11,000 ft). January temperatures average about 7°C (45°F); those in July, about 30°C (85°F). The rainfall averages between 2.5 and 13 cm (1 and 5 in.) a year, most of which falls in the winter and spring. Snow blankets the higher peaks in winter; scattered summer cloudbursts of great intensity occur almost every year. Except along the Colorado River, water is generally scarce, and most drainages have only intermittent flows that depend on local weather conditions.

A number of plant communities, all of them used by bighorn, are found in the Mohave Desert (Bradley, 1964). In addition to Joshua tree and creosote bush, the primary shrub species include green Mormon tea *(Ephedra viridis)*, Mohave yucca *(Yucca schidigera)*, banana yucca *(Y. baccata)*, shadscale,

David B. Marshall

Fig. 5.3. On the Desert National Wildlife Range, a typical part of the Mohave Desert.

four-wing salt bush, hop sage, winter fat, green molly, blackbrush, and burro bush *(Ambrosia dumosa)*. Desert holly *(Atriplex hymenelytra)*, sweet bush *(Bebbia juncea)*, and stickweed *(Stephanomeria* sp.*)* are also common in much of the desert.

In the Mohave Desert, pinyon-juniper woodlands are present above 1,070 m (3,500 ft) and may extend to around 2,000 m (6,500 ft), where ponderosa pine forests begin to be evident (though in some cases ponderosa pine is absent). There is often a belt of chaparral on these mountain ranges between 1,370 and 1,500 m (4,500 and 5,000 ft). Summer range of bighorn on some mountains may include subalpine conditions above 2,700 m (9,000 ft).

The most common grasses in the Mohave Desert, where this class of plant is generally scarce, include brome *(Bromus* spp.*)*, salt grass *(Distichlis stricta)*, fluffgrass *(Tridens pulchellus)*, wheatgrass *(Agropyron* spp.*)*, muhly *(Muhlenbergia* spp.*)*, dropseed *(Sporobolus* spp.*)*, big galleta *(Hilaria rigida)*, and bluestem *(Andropogon* spp.*)*.

Bighorn occur throughout the mountains and their foothills. On Desert Wildlife Range, where they are relatively abundant, their escape terrain is steep, rough, and rocky; the animals penetrate the rolling hills and gently sloping alluvial fans, but use of this type of habitat is generally confined to feeding areas within 1.6 km (1 mi) of steep terrain.

On the southwestern edge of the Mohave Desert rise the San Bernardino and San Gabriel mountains of California, a part of the Transverse Ranges that form a barrier separating the Mohave Desert from the Pacific Coast. These mountains contain considerable amounts of bighorn habitat that differ vegetatively from those of the rest of the Mohave Desert. The mountains receive the bulk of their precipitation, which amounts to from 50 to 75 cm (20 to 30 in.) annually, from marine storms.

These mountains rise to 3,000 m (10,000 ft). The streams that dissect them form an intricate arrangement of deep canyons and sharp, narrow ridges. The major "cover" types are browse, cliff, and barren alpine. During winter the bighorn are in the lowland browse type, where there is an abundance of herbaceous vegetation together with shrubs such as redberry *(Rhamnus crocea),* chaparral whitethorn *(Ceanothus leucodermis),* and desert and curl-leaf mountain mahogany *(Cercocarpus betuloides* and *C. ledifolius).* As spring approaches, the bighorn gradually move up to their summer range, where the principal forage plants are buckwheat *(Eriogonum* spp.*),* cream bush *(Holodiscus* spp.*),* mountain whitethorn *(Ceanothus cordulatus),* and a few perennial grasses and forbs (Light et al., 1967).

SONORAN DESERT

The Sonoran Desert is the most varied of the North American deserts. It comprises northwestern Sonora in Mexico and south-central and southwestern Arizona. This desert consists of low, arid plains separated by barren, often detached mountains (Fig. 5.4). Sometimes it is referred to as a tree desert because of the sizable trees and treelike cacti. Mean temperatures range from about 10°C (50°F) in January to about 30°C (85°F) in July. The rainfall averages from a mere trace in the west to 25 cm (10 in.) in the east. Elevations range from sea level to nearly 1500 m (5000 ft).

Characteristic plants are palo verdes *(Cercidium floridum* and *C. microphyllum),* ironwood *(Olneya tesota),* mesquite *(Prosopis juliflora),* and the saguaro or giant cactus *(Carnegiea gigantea).* Other dominant plants include creosote bush, burro bush, barrel cactus *(Echinocactus wislizeni),* whitethorn *(Acacia constricta),* brittle bush *(Encelia farinosa),* ocotillo *(Fouquieria splendens),* jojoba *(Simmondsia chinensis),* desert agave *(Agave deserti),* beargrass *(Nolina microcarpa),* and elephant tree *(Bursera microphylla).*

Many grasses occur in this area, usually not abundantly, including grama *(Bouteloua* spp.*),* galleta *(Hilaria rigida),* and sacaton *(Sporobolus* spp.*).*

Norman M. Simmons

Fig. 5.4. The Sonoran Desert is typified by low but rugged mountains separated by wide alluvial valleys. Northern boundary of the Cabeza Prieta Wildlife Refuge, Arizona (formerly Cabeza Prieta Game Range).

The terrain comprises low, rocky, rugged mountain ranges, often isolated by broad tree-filled valleys or plains. The bighorn live on these mountain ranges and occasionally traverse the broad valleys. The volcanic and granitic mountains contain scattered potholes or "tinajas" that hold runoff water for long periods. Bighorn rely upon these for water.

COLORADO DESERT

The Colorado Desert lies to the west of the Colorado River in southeastern California and northeastern Baja California. Outside of the heavily cultivated regions of the Imperial and Coachella valleys and the Colorado River delta area, the terrain is sand and rock desert. The dominant plants are creosote bush, burro bush, and brittle bush. Smoke tree *(Dalea spinosa)*, littleleaf palo verde *(Cercidium microphyllum)*, and desert willow *(Chilopsis linearis)* occur in the broad washes cutting the alluvial fans at the foot of the steeply rising mountains.

Bordering the Imperial Valley to the west are the Laguna Mountains, a part of the Peninsular Range that stretches southward into Baja California. The Santa Rosa and San Bernardino mountains lie to the north. To the east of the Imperial Valley, many lesser mountain ranges complete the basin's edge. Elevations range from below sea level to 2,100 m (7,000 ft).

The Colorado Desert is very dry and hot. Precipitation averages less than 13 cm (5 in.) annually, and snow is almost unknown. Rain falls principally in summer, mostly in cloudbursts. Temperatures average about 35°C (95°F) in July, and about 15°C (60°F) in January. The surrounding mountains are mainly creosote bush-covered, but those on the western side are high enough for pinyon-juniper woodland and ponderosa pine forests. Plants in these ranges important to bighorn are brittle bush, barrel cactus *(Echinocactus acanthodes)*, ironwood, catclaw *(Acacia greggii)*, ocotillo, galleta, needlegrass *(Stipa* spp.*)*, jojoba, spurge *(Euphorbia platysperma)*, and sweet bush.

The mountain ranges within the Colorado Desert are primarily of granitic and volcanic origin. This type of rock produces potholes or "tinajas" that hold runoff water. Since the rainfall in most of these mountains averages less than 10 cm (4 in.) annually, many of these tinajas are important sources of water to bighorn. Progressively more precipitation occurs in the mountains over 900 m (3,000 ft) high, but in the taller mountains (Santa Rosa and San Jacinto) the bighorn range is generally below 1,200 m (4,000 ft) (Weaver et al., 1968; Weaver and Mensch, 1969; Weaver and Mensch, 1970b).

BAJA CALIFORNIA MOUNTAIN DESERT

We here treat all of the 1,000-km (650-mi) long portion of the Baja California Peninsula occupied by desert bighorn as the Baja California Desert. The bighorn habitat consists of an irregular, sometimes discontinuous escarpment running north and south the length of the Peninsula, falling off on the eastern side. The varied topography includes mountain ridges, hills and tablelands, canyons and outwash plains, largely granitic or lava. Bighorn are found chiefly on the eastern side of this long elevated tract, from the Sierra de Juarez at the international boundary and the high (over 3,000 m, or 10,000 ft) Sierra de San Pedro Mártir on the north of the Sierra de la Giganta (over 1,700 m, or 5,700 ft) in the south.

Rain falls mainly during the summer and early fall, amounting to between 9 and 12 cm (4 and 5 in.) annually. Tremendous downpours occur occasionally. The temperature is very warm, the mean annual temperature being about 23°C (74°F).

The vegetation is generally scanty, consisting of various large cacti, including the cardón *(Pachycereus pringlei)*; various prickly pears and chollas *(Opuntia* spp. and *Machaerocereus gummosus)*; the tree yucca *(Yucca valida)*; one of the magueys *(Agave shawii)*; the famous boojum or cirio

Paul R. Johnson

Fig. 5.5. Colorado Desert habitat. San Ysidro Mountains on the Anza-Borrego Desert State Park, California. Plants in the foreground include brittle bush, cholla, ocotillo, and creosote bush.

(Idria columnaris), which is confined to the central part of the desert; elephant tree *(Pachycormus discolor)*; palo verdes *(Cercidium microphyllum* and *Parkinsonia aculeata)*; ocotillos *(Fouquieria splendens* and *F. diguetii)*; *Bursera* spp.; creosote bush; brittle bush; mesquites *(Prosopis* spp.); limber bushes *(Jatropha* spp.); and bur sages *(Ambrosia* spp.). Live oaks *(Quercus* spp.) occur in the Sierra de la Giganta. Grasses are scarce and poorly represented.

CHIHUAHUAN DESERT

The large Chihuahuan Desert extends farthest south of the North American deserts. Ninety percent of it lies in north-central Mexico. It includes much of the state of Chihuahua and parts of Coahuila, Durango, Zacatecas, and San Luis Potosí, as well as portions of western Texas and southern New Mexico. It has many inland basins where ephemeral shallow lakes form snow-white salt flats after the rainy season. The area is broken by isolated or parallel chains of steep-walled, massive mountains with a north-south orientation, rising above broad brush- and grass-covered plains. Many of the mountains are composed of limestone, whereas volcanic activity is evident mainly in the central and western portions of Chihuahua (Jaeger, 1957).

Like the Great Basin, the Chihuahuan Desert is high, being bordered on the west by the Sierra Madre Occidental and on the east by the Sierra Madre Oriental; nearly half of it is over 1,200 m (4,000 ft), and its upper limits are well above 1,800 m (6,000 ft) (Shreve, 1942; Lowe [ed.], 1964). The major rainfall occurs in the summer and is from about 8 to 25 cm (3 to 10 in.) annually.

The area is devoid of large vegetation, and in the basins and plains the growth consists of grasses or low desert shrubs. The major plants are creosote bush, lechuguilla *(Agave lechuguilla)*, sotol *(Dasylirion wheeleri)*, barrel cactus *(Echinocactus wislizeni)*, allthorn *(Koeberlinia spinosa)*, varnish bush *(Flourensia cernua)*, ocotillo *(Foquieria splendens)*, and mesquite. Sotol is so conspicuous that this desert is sometimes called the Sotol Desert. In the Mexican portion of the Chihuahuan Desert, two characteristic plants are candelilla *(Euphorbia antisyphilitica)* and guayule *(Parthenium argentatum)*. On the upper slopes of the desert is a yucca belt dominated by *Yucca carnerosana;* in the lower valleys occurs another yucca *(Y. filifera),* similar to the Joshua tree but with the blossoms hanging down instead of being upright. Burro grass *(Scleropogon brevifolius)* is conspicuous in the valley floors (Jaeger, 1957).

In the mountains there are grasses, accompanied in some places by oak *(Quercus* spp.*)*, pine, Douglas fir, and maple *(Acer* spp.*)*. Principal plants found in areas inhabited by bighorn, other than those previously mentioned, are black grama *(Bouteloua eriopoda)*, prickly pear *(Opuntia englemanni)*, skunk bush *(Rhus trilobata)*, mountain mahogany *(Cercocarpus montanus)*, mock orange *(Philadelphus microphyllus)*, Mexican tea *(Ephedra trifurca)*, whitethorn, catclaw, and Palmer agave *(Agave palmeri)*.

Almost all of the isolated mountain ranges found in this desert are desert bighorn habitat. However, bighorn have disappeared from nearly all of them because of overshooting and too much competition from cattle, domestic sheep, and goats.

HABITAT REQUIREMENTS OF BIGHORN

Certain basic habitat factors are required by bighorn for survival. These are food, water, and escape terrain. A fourth factor, not always considered in evaluating habitat, is space. This could be described as a lack of crowding. In the case of southwest desert regions, it is hard to believe that there may not be enough space for the limited number of bighorn present there to survive. Although these animals are gregarious, they seem unable to stand much togetherness when they are frustrated by external stimuli or internal drives.

Harry L. Gordon

Fig. 5.6. Chihuahuan Desert habitat. The Sierra de los Castro, north of Saltillo, Coahuila, Mexico. Foreground plants are *Yucca filifera;* other plants include creosote bush and other representative shrubs and grasses.

Harassment by other bighorn, other large mammals, or man can cause individual bighorn to become run down physically, perhaps from an improper diet due to nervous tension. If possible, they move away from a situation where they cannot relax most of the time.

On Desert Wildlife Range are two distinct strains of desert bighorn, distinguishable by tongue color. One strain is nervous and high-strung, while the other is comparatively mild and relaxed in disposition (Hansen, 1970). It is quite conceivable that a population of high-strung bighorn might leave an area because of what, for other bighorn, would be only a minor harassment.

J. T. Light gives a possible example of how and where the differences in space and bighorn behavior may occur in two distinct areas in southern California: "In comparing mountain ranges, the Santa Rosa bighorn are more curious than highstrung. In the San Gabriel and San Gorgonio ranges the bighorn are highstrung by comparison with those in the Santa Rosa area. This may be due to the confined nature of their range which is fully stocked and penetrated by man."

Desert bighorn have survived under conditions that have been too extreme for other large mammals. In part this has been accomplished by their ability to follow the habits and behavior of their parents (Geist, 1967). But should these ancestral patterns be broken the population may not be able to survive. The patterns may be unintentionally broken down by human activities, such as those of hikers or photographers "trying to get close enough for a good look."

Bighorn need space to carry on their daily activities in a relaxed manner. They will accept man and his activities if they are allowed to come to man in their own manner but will retreat in front of man's advances. All the food and water they can use is of little value to bighorn if they cannot have enough space to satisfy these requirements in their own way.

FOOD

Amount. A determination of food requirements will have to be made for each area where bighorn are found. I am not aware of any research on the amounts of food required by bighorn in their natural habitat. Tests on Desert Wildlife Range show that approximately 1.5 kg (3 lb) of dry alfalfa are consumed by penned bighorn daily. This amount increased to about 1.8 kg (4 lb) in late fall after the rut season. By late January and early February, the amount again averaged about 1.5 kg.

Availability and distribution. The species of native foods utilized are discussed in Chapter 6. To maintain thrifty animals, these foods must be available whenever needed. Since many plant species produce food only during certain years in a desert climate, a wide range of habitat is needed by the bighorn if they are to find their required food.

The condition of food plants also is quite important. For example, on Desert Wildlife Range, Joshua trees may be utilized only once every 5 to 10 years, depending upon whether many are blooming. In Death Valley when the food supply is too far from the currently used water supply, the animals appear to seek out new water sources (Welles and Welles, 1961a). But such movements are advantageous only if there is sufficient water at the new source.

Optimum food supply in desert bighorn habitat should consist of a wide variety of plant species and should be situated within a radius of about 9.7 km (6 mi) from permanent water (Hansen, 1967b). On Desert Wildlife Range, and very likely on other areas, there is an abundance of food for bighorn when a whole mountain range, including the foothills, is considered. However, it may not be available because of weather, water supply, disturbance by humans or their livestock, or seasonal growing conditions.

In a desert climate, seasonal weather variations may or may not bring about expected changes in the vegetation, for 5 to 10 years may elapse without sufficient precipitation to allow plants to produce required amounts of biomass. Even 2 years of drought may reduce available forage to such a level that only a starvation diet can be obtained, forcing bighorn to the fringes of their normal range to obtain sufficient food.

WATER

The large subject of bighorn-water relationships is detailed in Chapter 7. However, some comments regarding the part played by water as a component of desert bighorn habitat are given here.

Water availability. Water must be available at all times to maintain a healthy herd. The presence of some disturbing factor may interfere with the use of water, even if water is abundant and permanent.

Lanny O. Wilson, speaking of southern Utah, and J. T. Light, speaking of the Colorado Desert mountains, say that water is the greatest limiting factor for desert bighorn. With the construction of adequate water holes the area utilized by desert bighorn in these areas in the early spring, summer, and late fall could be almost tripled.

Distribution of water in most of the desert bighorn range limits their population size, according to Welles and Welles (1961a), Russo (1956), and Hansen (1965b). In some areas water is not limited; bighorn habitats in San Andres National Wildlife Refuge, New Mexico, along the Colorado River in Arizona, and in the White Mountains of California have abundant water.

Because of their characteristic wariness around water, bighorn prefer open space around their drinking place so that they can see anything suspicious. A spring closely surrounded by dense brush or with a corral or other man-made obstacles around it often will discourage bighorn from using water. A water hole situated away from their normally rough, rocky terrain usually is shunned. They may travel a few hundred meters from rough, broken terrain to get water in flat or rolling desert, but generally they will not venture far from their preferred escape terrain.

TOPOGRAPHY

The preference of desert bighorn for rough, rocky and broken terrain, cut up by washes or other open areas, has been mentioned. Caves and the shelter of trees are used during inclement weather and to escape eagles and aircraft. Bighorn do not normally run long distances but depend upon their climbing and hiding ability to escape their enemies. Without the proper topographic features, bighorn will be absent.

Simmons (1969b) reports that there are no elevational zonations of vegetation in the low, narrow mountain ranges of Cabeza Prieta National Wildlife Refuge in Arizona, and that the effects of slope so noticeable in more massive mountains are hard to recognize there. During relatively cool months or when rainstorms left pothole water, the bighorn seemed to prefer feeding on alluvial fans at canyon mouths where vegetation was denser. However, another effect of topography on bighorn distribution in Cabeza Prieta Refuge was notable. Largely because of the physical discontinuity of the habitat, bighorn tended to live in isolated groups. As a consequence, they developed local behavior patterns suited to the islandlike habitats they occupied.

On Desert Wildlife Range, lambs usually are born in roughest terrain. Such terrain generally has caves or overhanging rocks that offer lambs protection from predators and weather. Although ewes do not always seek out rough areas and may have their lambs on open desert slopes (Simmons, 1969b), rough terrain with caves usually is sought.

Daytime beds are often selected more or less haphazardly, but many nighttime bed grounds on Desert Wildlife Range seem to be chosen with great care. These may even be ancestral locations, as evidenced by the number and depth of the beds and the amount of fecal material. The location is often near the top of a ridge or along a spur from which much territory can be seen. Such a location enables bighorn to escape within a matter of seconds, either by running out of sight over a ridge or by moving rapidly down the mountain. In some areas suitable bed ground may be hard to find, and this could conceivably be a limiting factor in the distribution of bighorn, especially when seasonal changes in bed ground are required.

CLIMATE

Precipitation. Climate in desert bighorn habitat varies from alpinelike conditions on the tops of high peaks to desert conditions below sea level in Death Valley, but of course it is mainly of the desert type. The range of annual precipitation may be from less than 2.5 cm (1 in.) in the Mohave and Sonoran deserts to over 50 cm (20 in.) in the White Mountains of California. Bighorn consume snow, dew, and free water from runoff or springs. The amount of precipitation in an area is important since it affects the vegetation and the amount of free water available. In Baja California where dew is sometimes heavy, it apparently provides sufficient moisture to maintain bighorn in good condition. In some areas dew may collect in small potholes in sufficient quantity to sustain bighorn (Ives, 1962). In areas where sufficient snow falls, it is often used by bighorn.

Temperature. Simmons (1969b) found that on Cabeza Prieta Refuge when wet bulb temperatures were 18°C (65°F) or higher, most bighorn bedded in the shade. When bighorn did not bed in caves, they most often bedded on light-colored rock or on light-colored, coarse soil. The surface temperatures of white- to tan-colored igneous and tan metamorphic rock on which bighorn bedded averaged 4°C (14°F) lower in the shade than in the sun. Tan-colored silt, which seemed to be avoided by bighorn on hot days, reacted rapidly to solar radiation and fluctuated in surface temperature more than rock. Temperatures of silt in the sun averaged 7°C (26°F) higher than rock temperatures in the sun when both were approximately the same color.

One of the most obvious effects of temperature on bighorn activities was noted during June water-hole surveys on Cabeza Prieta Refuge. Most bighorn responded to the earlier warming of ambient temperatures on an east slope by watering sooner in the morning than bighorn on the west slope. Bighorn on the east slope began moving around earlier in the evening than those on the west slope, and those that watered in the evening on the east slope began watering about 2 hours earlier. The peak of the watering activity on the east slope was reached in the morning; the peak on the west slope was reached in the evening.

Temperature inversion. With respect to bighorn habitat in southern California, Hatch Graham states, "In the San Gabriel Front we find layers of warmer air forming at high elevations making living conditions more favorable for bighorn than might be expected. We don't know much about them...." Light et al. (1967) reported that adverse weather conditions direct the movement of bighorn to warmer and drier exposures or slopes. Such local climatic conditions vary considerably with elevation and exposure in the San Gabriel Unit of Angeles National Forest.

Evaporation. Evaporation is affected by wind, temperature, and humidity. Generally desert bighorn live in areas of high evaporation rates. In areas of limited water distribution, sparse vegetation, and a high evaporation rate, bighorn often seek protection from the climate by not moving in the heat of the day, lying in the shade or out of the wind, and not moving too far from water. Evaporation not only dries out the animal but also dries out its food. Consequently, the longer and drier the dry period, the less distance the bighorn will or can travel from their supply of water. This sequence of events can cause overgrazing in the vicinity of a water hole, resulting in poor conditions for survival.

Drought. Desert bighorn can withstand a certain amount of drought (Welles and Welles, 1961a; Simmons, 1969b), but it is unlikely that a herd could maintain itself during a prolonged period without precipitation or free water, although this does seem to be the case in Sonora (Mendoza, 1976). If these animals could live (not just exist) without free moisture of some kind, then there would be bighorn on many mountain ranges of southern Nevada and California presently unoccupied by them. Drought is usually reflected in the condition of the vegetation and the quantity of free water present.

6. Food

Bruce M. Browning and Gale Monson

Information on the food habits of the desert bighorn comes principally from three sources: direct sight observations, rumen content analyses, and inferred observations, i.e., observed utilization of food plants and knowledge that bighorn are in the vicinity. Empirical data abound; quantitative data are less plentiful. The results of 169 stomach analyses are reported. A possible fourth source, that of fecal pellet analysis by microscopic techniques (Sparks and Malechek, 1968; Williams, 1969), has not been developed and utilized in the case of desert bighorn except for some preliminary sampling done in southwestern Arizona in 1978 (Smith et al., 1978).

Many similarities are found in the feeding habits of the bighorn throughout much of its range. However, habitat (vegetational) differences are sufficient to warrant a discussion of food habits on the basis of the seven southwestern deserts described in Chapter 5. The present chapter closes with a summary and a consideration of food habits knowledge as it affects bighorn management.

The relation of bighorn food habits to vegetative classes (grasses, browse, and forbs) is important in any evaluation of bighorn habitat. It is generally held that the wild sheep of the world are primarily grass eaters, or grazers.

Whether this is true of desert bighorn is a controversial matter. Considerable amounts of browse and forbs are found in stomach contents, and bighorn are often observed feeding on them, even where grasses are plentiful. The body of data indicates that desert bighorn are to some degree opportunistic and adaptable feeders; that is, they eat whatever is at hand and appeals to their appetite in whatever locality or season they are in. Generalities about the food habits of desert bighorn over their entire range are difficult to make because of the tremendous variety in both composition and density of the flora in the expanse of range they inhabit.

GREAT BASIN DESERT

The Great Basin Desert forms only the northern rim of the desert bighorn's range. Data on food habits from this desert are lacking.

PAINTED DESERT

Bighorn are found in the Painted Desert mainly in that portion that lies in Utah. Most of the available food habits information in Utah has been reported by Wilson (1968) and Irvine (1969b). Wilson's data were obtained by direct observation and recording of minutes of feeding per food plant in the White Canyon area of San Juan County. Irvine's report is concerned with limited direct and inferred observations, and analyses of eight stomachs obtained from hunter-killed rams in November 1967 and October 1968. He worked for the most part in Red Canyon, not far from Wilson's White Canyon.

Wilson found that bighorn used grasses 38.0 percent of feeding time that totaled 3,865 minutes of observation. Browse comprised 45.0 percent, and forbs another 17.0 percent. His results show, therefore, equal dependence on grass and browse. He qualifies his results as being "somewhat biased toward browse" because of the ease with which bighorn can be seen feeding on shrubs and larger forbs as compared with the low-growing grasses (Fig. 6.1).

The eight stomachs reported on by Irvine averaged, by weight of contents, 12 percent grass, 36 percent browse, and 52 percent unknown. Irvine felt that samples were biased because grasses are digested faster and are washed more readily through the screens used to run the samples. Furthermore, the samples were taken only in the fall (two in September, six in November) and of course made up hardly more than a token number.

Important Food Plants

Grasses. Little galleta *(Hilaria jamesii)* is the grass preferred by desert bighorn in the Painted Desert region during spring, summer, and fall, according to Wilson and Irvine. Galleta made up a quarter of the total diet from

James D. Yoakum

Fig. 6.1. Grass, forbs, or browse? Can you tell, even at this distance?

March to November and was over half of the diet in October 1966, according to Wilson's observations; Indian ricegrass *(Oryzopsis hymenoides)* accounted for 10.8 percent of the feeding minutes, salina wild rye *(Elymus salina)* for 2.3 percent, downy chess *(Bromus tectorum)* for 2.1 percent, and desert needlegrass *(Stipa speciosa)* for 0.3 percent, in the White Canyon area. Wilson mentions that downy chess may be more important than recorded because of the difficulty in observing its use, since it commonly grows closely with galleta. The same is true of Nevada bluegrass *(Poa nevadensis),* which grows in close association with salina wild rye.

Browse. Wilson observed 1,729 feeding-minutes on browse in White Canyon. In that feeding time only four species contributed more than 1 percent of the total diet. Blackbrush *(Coleogyne ramosissima)* is the most important of these, making up a quarter of the diet and being used year round. It was also the most used browse throughout the year in Irvine's study area.

The three other significant browse species observed by Wilson to be eaten by bighorn were singleleaf ash *(Fraxinus anomala),* about 10 percent of the total diet; snowberry *(Symphoricarpos longiflorus),* 3.5 percent; and Mormon tea *(Ephedra* sp.), 1.2 percent. As indicated earlier, browse comprised 45.0 percent of the total diet.

Forbs. Only three forbs contributed over 1 percent of the diet as determined by Wilson's feeding-minutes technique, although sixteen species are listed. Russian thistle *(Salsola kali)* comprised 9.4 percent of the total diet. (Both Wilson and Irvine report that the bighorn sought its succulent green growth during July and August where the ground had been disturbed by the many uranium mine roads throughout the study areas.) Fivehook bassia *(Bassia hyssopifolia),* another roadside weed, made up 4.5 percent of the diet. Rubber weed *(Hymenoxys richardsonii)* was the only other forb contributing over 1 percent to the total diet.

Wilson states that his results may not necessarily show preference; he cites the use of dead, dry seed stalks of sego lily *(Calochortus nuttallii)* which receive low use (less than 0.1 percent) but which are eagerly sought, even in a dry state, although uncommon in the plant association. He further reports that many forbs are eaten when completely cured and that adult ewes were observed digging for food, particularly after long periods of drought and extremely high temperatures, when he believed they were seeking the bulbs of weakstem mariposa *(Calochortus flexuosus).*

Seasonal Influences

Seasonal data on Utah bighorn are sketchy. However, Wilson noted some significant changes between summer and winter diets. He cites heavy use of grass, notably Indian ricegrass, after heavy rain and snow in early November. Browse, principally blackbrush, constituted most of the November diet, about 14 percent higher than the average percentage consumed through the spring and summer. In October, grass, principally little galleta, was about two-thirds of the diet. Hence, a few differences in the seasonal diet are evident, but a clear seasonal pattern is not discernible. Irvine in his Red Canyon study found that "Dominant plant species change during the year.... In early spring *Bromus tectorum* was readily available and sought by the sheep. Following maturity of *Bromus,* the forbs flowered and dominated the formerly open spaces. The browse species were dominant in late summer and during the winter."

MOHAVE DESERT

Desert bighorn range through a variety of plant communities on the Mohave Desert, from as high as the ponderosa pine belt or Transition Zone down through pinyon-juniper woodland, sagebrush, and Joshua tree associations to the low-level browse communities near or even below sea level in Death Valley. We also treat here, since they are adjacent to the Mohave Desert, the plant communities of the high San Bernardino, San Gorgonio, and San Gabriel mountains in the Transverse Ranges of southwestern California. In this last connection, see Chapter 5 on Habitat.

The Mohave Desert bighorn populations have been studied more than those of other deserts, since within its boundaries are government-administered areas where desert bighorn are found, like Desert National Wildlife Range in the southern tip of Nevada and Death Valley and Joshua Tree national monuments in vast arid southeastern California. Nevertheless, rumen analyses are few, none being available for California and only eighty-two for Nevada. The bulk of our knowledge of bighorn food habits perforce comes from direct sight observations of the feeding animals.

Of the Nevada stomachs analyzed, all but twelve are from Clark County at the southern tip of the state, and the bulk of the latter are from Desert Wildlife Range. They were taken chiefly in the winter. The Clark County bighorn, according to these analyses, had eaten 71 percent grass, 13 percent browse, and 5 percent forbs, the remaining 11 percent consisting of unidentified items. The analyses were undertaken by Barrett (1964), Yoakum (1966), and by the Nevada Department of Game and Fish; the laboratory work in all cases was performed by the California Department of Fish and Game's Wildlife Investigation Laboratory.

The other twelve stomachs came from the Silver Peak Range in Esmeralda County, partly in September and partly in December, and were analyzed by Yoakum (1964) with these results: grass, 59.5 percent; forbs, 32 percent; and browse, 8.5 percent. Range analyses in the Silver Peak Range reveal that 22 percent of the vegetative cover was grass; 74 percent browse; and 4 percent forbs. Yoakum concludes that grass was the most significant forage class for bighorn in the Silver Peak Range and a preferred food during the fall and winter months.

A locality-by-locality analysis of the contents of the eighty-two stomachs plus those of an additional ninety-five stomachs collected since 1966 is given by Brown et al. (1977).

To summarize the many field observations of bighorn food habits in the Mohave Desert, McLean reports (1930) that grass "seemed" to be the staple food of bighorn inhabiting the Inyo Mountains of Inyo County, California. Welles (1965), working in Joshua Tree National Monument, observes that grasses also headed the list of important foods.

On the other hand, after several years' observation of bighorn feeding in Death Valley National Monument, Welles and Welles (1961a) conclude that "present evidence indicated a preponderance of shrub utilization in the diet." They point to the general scarcity of grasses throughout the entire region as a good reason for bighorn dependence on browsing rather than grazing.

Important Food Plants

Grasses. Big and little galleta *(Hilaria rigida* and *H. jamesii)* and needlegrass *(Stipa* spp.*)*, principally desert needlegrass, are the most important grass species for bighorn almost throughout the Mohave Desert. Other less important grasses are fluffgrass *(Tridens pulchella)*, gramas *(Bouteloua* spp.*)*, alkali sacaton *(Sporobolus airoides)*, bluegrasses *(Poa* spp.*)*, beardgrass *(Andropogon* spp.*)*, downy chess, fescue *(Festuca* spp.*)*, desert saltgrass *(Distichlis stricta)*, and Indian ricegrass (McCullough and Schneegas, 1966; Welles and Welles, 1961a; Welles, 1965; Lewis, 1960; Light et al., 1967; Yoakum, 1964; Buechner, 1960).

Browse. Bighorn feed on a large variety of browse plants in the Mohave Desert, with no one or two plants preferred. In the Nevada part of the desert, the most used browse species appear to be sagebrush *(Artemisia tridentata)*, mountain mahogany *(Cercocarpus intricatus)*, Mormon teas *(Ephedra* spp.*)*, winter fat *(Eurotia lanata)*, salt bushes *(Atriplex* spp.*)*, cliffrose *(Cowania stansburiana)*, and turpentine broom *(Thamnosma montana)*.

In Death Valley, Welles and Welles (1961a) report that no one plant could be considered the most important. However, they considered desert holly *(Atriplex hymenelytra)* to be "probably more nearly in general year-round use than any other plant." Sweet bush *(Bebbia juncea)* and desert milk-aster *(Stephanomeria runcinata)* are important Death Valley bighorn foods, especially in drought years, and in washes which, under those conditions, are practically the only areas that get enough moisture to produce green feed. These two plants are acceptable forage in their dormant state, also. The Welleses stress that bighorn sometimes use plants only after they are in a dried-out or cured condition. At high elevations in Death Valley, Mormon tea, four-winged salt bush *(Atriplex canescens)*, and hop sage *(Grayia spinosa)* are used by the desert bighorn. Shadscale *(Atriplex confertifolia)*, another desert salt bush, is reported as used by bighorn (Dixon and Sumner, 1939).

Some of the browse "high on the list" of bighorn foods in Joshua Tree National Monument are blackbrush, scrub oak *(Quercus dumosa)*, California buckwheat *(Eriogonum fasciculatum)*, and skunk bush *(Rhus trilobata)* (Welles, 1965).

The flowering buds of the Joshua tree *(Yucca brevifolia)*, Mohave yucca *(Yucca shidigera)*, and banana yucca *(Yucca baccata)* are described by Deming (1964) as eaten in April and May in Nevada. The green seed pods, commonly called "apples," are eaten also. Deming states that bighorn often stand on their hind legs, with front legs braced against the trunk of the Joshua tree, to reach the succulent buds and pods. The long fruiting stems of the Utah agave *(Agave utahensis)* are nipped in two, and the buds and blossoms eaten from the ground.

Charles G. Hansen has observed bighorn taking the dry fibers of yucca leaves that had been dead for several years. He considers this "roughage" and also has observed bighorn stripping the bark from trees, as well as dead spines from cactus. Hansen once observed a band of ewes, lambs, and young rams consuming the berries and green foliage of junipers *(Juniperus* sp.*)* on Desert Wildlife Range.

Although Welles (1965) considers blackbrush an important food plant on Joshua Tree National Monument, its importance on the Nevada mountain ranges is in question. There, although use of blackbrush leaves and stems is constant, large amounts are seldom taken. Blackbrush occurred in two-thirds of the stomachs examined from Desert Wildlife Range but made up only 2 percent by volume.

Some of the preferred and staple foods of bighorn in the San Gabriel Mountains of California are desert mahogany *(Cercocarpus betuloides)*, coffeeberry *(Rhamnus californica)*, silktassel *(Garrya* sp.*)*, and spiny grease bush *(Glossopetalon* sp.*)* (Kennedy, 1963). Kennedy acknowledges that "it is difficult to assign these plants to sheep or deer use." Light et al. (1966), studying the San Gorgonio Mountains to the east of the San Gabriels, record heavy use of hollyleaf cherry *(Prunus ilicifolia)*, Mormon tea *(Ephedra* sp.*)*, redberry *(Rhamnus crocea)*, desert mahogany, cream bush *(Holodiscus microphyllus)*, snowberry *(Symphoricarpos* sp.*)*, horsemint *(Monardella linoides)*, and chaparral whitethorn *(Ceanothus leucodermis)*. It must be recalled that these particular California habitats are not typical of the Mohave Desert but are included here because they are adjacent to the Mohave Desert, and the desert bighorn inhabiting them are considered taxonomically the same as those of the Mohave Desert.

Forbs. Forbs are undoubtedly an important component of bighorn diets in the Mohave Desert, particularly on the lower and more southern parts. In Nevada, filaree, mostly the red-stemmed filaree *(Erodium cicutarium)*, an exotic plant, is considered an important early spring food. Deming (1964) reports that bighorn "paw up and eat the entire plant." Buckwheats *(Eriogonum* spp.*)* are important on both Nevada and California bighorn ranges, being rated highly in Joshua Tree National Monument (Welles, 1965). In Death Valley, the common desert trumpet *(Eriogonum inflatum)* receives heavy use (Welles and Welles, 1961a).

Other forbs important to desert bighorn are honeysweet *(Tidestromia oblongifolia)* and Mojave thistle *(Cirsium mohavense)* which were used "voraciously" in Death Valley, while pebble pincushion *(Chaenactis fremontii)* and several of the desert evening primroses *(Oenethera* spp.*)* received "heavy use" (Welles and Welles, 1961a).

In the San Gabriel Mountains, some of the forbs used by bighorn were penstemons *(Penstemon* spp.*)* and giant hyssop *(Agastache* sp.*)* (Robinson and Cronemiller, 1959).

Ober (1931) records bighorn in Death Valley digging with their feet for bulbs, and they are "apt in securing and eating roots yielding moisture," including roots of the desert poppy *(Arctomecon* sp.*)*. Lewis (1960) reports that San Gabriel bighorn seek out wild parsley (Umbelliferae) roots.

Seasonal Influences

Data on use by season are not substantial. Barrett's summary (1964) shows a noticeable increase in the use of forbs in spring on some Nevada ranges. On the basis of the few stomach samples available, and these only for Nevada, grass use is erratic, there being no clear-cut seasonal pattern of use.

Deming (1964) states there are two vegetative seasons in the Nevada portion of the Mohave Desert. The same is true throughout the southwestern deserts. The first growing season may begin as early as December, or as late as March or April, depending on winter rainfall and temperature conditions. The second occurs in response to summer rains. Between these two growing seasons are vegetative dry seasons. Some of the grasses and forbs are the first plants to produce growth in the year, rather than browse. Deming states, on the basis of inferred data, that early green feed is eaten eagerly by adult bighorn. At this time they abandon their winter diet of browse, dry grass, and cured forbs. As the new spring growth of forbs and grasses dries up, the bighorn revert to "dry rations" at lower ranges, or, in some areas, move to higher elevations to find green feed. He says that this latter may account for the excellent lamb survival in the Sheep Range of mountains on Desert Wildlife Range, since the high protein content of green plants is a necessity in the diet of fast growing lambs.

Water, including seasonal precipitation, apparently is the key to the seasonal food habits of the desert sheep in the Death Valley region. Welles (1957) says that bighorn there "use certain plants at certain stages of growth, according to systemic demands, as free water substitutes, and a dozen other suggested reasons."

Welles and Welles (1961a) state, "It would appear difficult if not impossible to discuss bighorn food by the season without discussing water and rainfall and their effect on vegetation." They consider that the relative importance of ephemeral plants as forage is not yet defined, for they never observed abandonment of an area in Death Valley due to exhaustion of ephemeral food supply.

The classic study by Welles and Welles documents, with many examples, that bighorn feed on certain plants for weeks and then drop them from use completely. This irregular feeding behavior further complicates any attempted analysis of seasonal use of certain plants. And, in extremely arid portions of bighorn range such as Death Valley, dry periods are not limited to definite seasons of the year but actually may last for several years at a time!

Jerome Light reports that upward movement of bighorn in the San Gabriel Mountains from winter to summer range appeared to coincide with plant development, or greening.

COLORADO DESERT

This desert lies in California and Baja California between the lower Colorado River–Gulf of California and the higher mountains of southern California and the upper portion of Baja California, grading off into the Mohave Desert to the north. Desert bighorn are most numerous on the east- or desert-facing slopes of the Coastal Ranges of California that extend into Baja California as the Sierra de Juarez and the Sierra San Pedro Mártir. These bighorn have been studied mainly in the Santa Rosa Mountains and Anza-Borrego Desert State Park of California. However, little work has been done on food habits, and data from the handful of stomach analyses are slight. What data are presented below are inconclusive.

Likewise, the relative importance of browse, grass, and forbs in the diet of these bighorn is unclear. What evidence there is points to all three classes of food as being important.

Important Food Plants

Grasses. Although grasses are probably an important element of desert bighorn food on the Colorado Desert, almost no data are available as to the extent of use, either as a forage class or by species.

Browse. Jojoba *(Simmondsia chinensis)* appears to be a key food plant. The leaves occurred in five of seventeen stomachs analyzed from California, making up 2.7 percent of the total food in the stomachs (Bruce Browning). This probably does not reflect the true picture. Jones et al. (1957) list jojoba as an important food in their report on the Santa Rosa Mountains, together with white ratany *(Krameria canescens)* and bee sage *(Hyptis emoryi).* It was also listed as one of the browse species taken by bighorn in Anza-Borrego State Park (Jorgenson and Turner, 1972). Other browse eaten in the State Park included Mormon tea *(Ephedra* sp.*),* desert agave *(Agave deserti),* scrub oak, desert mistletoe *(Phoradendron californicum),* California buckwheat, desert apricot *(Prunus fremontii),* catclaw *(Acacia greggii),* mesquite *(Prosopis juliflora),* ratany *(Krameria grayi),* and laurel-leaf sumac *(Rhus laurina).*

Forbs. From what we know of bighorn food usage in adjacent deserts, we can presume that forbs make up a substantial amount of the diet in the Colorado Desert. In fifteen stomachs analyzed (by Bruce Browning), little identifiable forb material was isolated, with one exception: 7 percent of the identifiable food consisted of the small desert spineflower *(Chorizanthe brevicorna)*.

SONORAN DESERT

Vegetation in the Sonoran Desert reflects varied habitat conditions. Although shrub and cactus species tend to be much the same throughout, and thus individual species are usually available to bighorn, grasses are a different matter. Grasses are fairly abundant in the more northern, eastern, and higher reaches of the desert, but, when we examine the low, less extensive, and hotter and drier mountains of extreme southwestern Arizona and northwestern Sonora, we find that grasses drop out as a major component. Water is in the same category—there is less rainfall and there are fewer water holes and almost no springs in the lower and hotter parts of the Sonoran Desert. Here is a situation where one can truly examine water-food relationships.

Some browse species, especially jojoba and mesquite, are lacking over considerable portions of desert bighorn territory as one proceeds southward and lower in the Sonoran Desert toward the Gulf of California. This situation is exemplified by Kofa Game Range and Cabeza Prieta National Wildlife Refuge in Arizona, areas less than 100 km (60 mi) apart in their most proximal portions, with the latter largely devoid of jojoba and mesquite, although littleleaf palo verde *(Cercidium microphyllum)*, ironwood *(Olneya tesota)*, and catclaw *(Acacia greggii)* are just as common as on the Kofa.

Fluctuations in the annual rainfall, ranging from almost none in some years to more than 250 mm (10 in.) in others and incorporating "dry" and "wet" periods or cycles lasting 2 to 3 years, are a characteristic of the Sonoran Desert and have profound effects on the amount and quality of food available. This is particularly true of the ephemeral forbs and "six weeks" summer grasses, the former of which when cured may last 2 or 3 seasons and, according to field observations, provide an important bighorn food source (Gale Monson). It is speculated that bighorn populations have adjusted to the more severe periods or cycles, when food and water conditions are minimal with respect to their welfare.

Bighorn stomach collections offer clues to bighorn diet, and we have some data for the Arizona portion of the Sonoran Desert—none from the Sonora portion. Charles R. Hungerford of the University of Arizona contributes data from ten stomachs taken in winter from Pima and Mohave counties, which are in the higher and cooler part of the Sonoran Desert. The contents showed bighorn food to be 34.4 percent grass, 35.1 percent browse, and 21.5 percent forbs (9 percent unidentified). A June 1956 stomach examined from Kofa Game Range in Yuma County was 100 percent browse, including 15 percent cactus.

Fifteen bighorn stomachs collected principally in January and December from southwestern Arizona, and analyzed by the California Department of Fish and Game's Laboratory, show that grass comprised 26.4 percent of the diet and browse 21 percent; 48.3 percent was forbs and 4.3 percent was cactus. Another twelve desert bighorn stomachs, one for each month, were examined in a bighorn-deer competition study on Kofa Game Range. In these, grass was 33 percent of the food; browse 62.5 percent; forbs 3.0 percent; and cactus 1.5 percent (Russo, 1956).

Overall, of forty-three stomachs collected in Arizona, almost equal amounts of grass, browse, and forbs appeared. Small amounts of cactus (hereinafter included with browse) were utilized. Sight feeding observations generally confirm the equal ratios.

Important Food Plants

Grasses. Russo's summary (1956) of plants utilized by bighorn reports slim tridens *(Tridens muticus)* as "heavily used" and fluffgrass *(Tridens pulchella)* as "moderately used." He lists three species each of three-awn *(Aristida)* and galleta *(Hilaria)* and adds two species of grama *(Bouteloua)* as "utilized." Halloran and Crandell (1953) list two gramas and consider one, side-oats grama *(Bouteloua curtipendula),* as preferred in the summer and fall months on Kofa Game Range. These observers also consider big galleta as preferred food year long, and they add that bush muhly *(Muhlenbergia porteri)* is a preferred summer food.

Browse. Workers express a consensus within the limited data available as to which browse plants are important to bighorn in the Sonoran Desert. Several plants are "preferred" or "heavily used," or occur as a significant percentage of the volume of food in stomachs examined. These are catclaw, jojoba, brittle bush *(Encelia farinosa),* littleleaf palo verde, ironwood, mesquite, Mormon tea *(Ephedra* spp.), fairy duster *(Calliandra eriophylla),* silver bush *(Ditaxis lanceolata),* ocotillo *(Fouquieria splendens),* burrobush *(Hymenoclea salsola),* desert thorn *(Lycium andersonii),* ratany, and the agaves *(Agave* spp.).

Jojoba, where it occurs, is, as Buechner (1960) reports, "singularly the most abundant and sustaining browse for bighorn." Russo (1956) calls it "one of the most important year-around forage plants" and finds that it was 38.1 percent of the food eaten by twelve bighorn taken one per month on the Kofa Game Range. It is a preeminent food plant.

Leguminous browse, represented by catclaw, whitethorn *(Acacia constricta),* ironwood, palo verde, mesquite, and fairy duster, contributes significantly to bighorn food. Although leaves and twigs are the principal parts of the plant eaten, the flowers are utilized also. Bighorn have been observed feeding on the wind-drifted fallen flowers of ironwood on the floor of canyon washes, almost in vacuum-cleaner fashion, in the Buck Mountains of southwestern Arizona (Robert Yoder).

The widespread Mormon teas are still another of the more important bighorn food plants in the Sonoran Desert, some stomach samples being composed of as high as 70 to 90 percent of the stems of these plants. Brittle bush, a composite, does not often occur in stomachs collected but still appears to be used extensively by bighorn. Gale Monson reports that in February and March on Cabeza Prieta Wildlife Refuge brittle bush flowers and buds are favorite food items with the bighorn, who come down to feed on them on the canyon floors where the plant grows best.

Silver bush is a browse worthy of note. Buechner (1960) calls it "the choicest browse of the desert bighorn sheep." Russo (1956) reports it is heavily used locally. Gale Monson's observations on this plant on Kofa Game Range and Cabeza Prieta Wildlife Refuge lead him to believe that much of the reported bighorn feeding on silver bush is actually done by the large vegetarian lizard, the chuckwalla *(Sauromelas obesus)* or the desert iguana *(Dipsosaurus dorsalis),* or both. Monson's conclusion is based on observed chuckwalla feedings, on the cleanness of the cutting of plant stalks, and on his observations that bighorn tracks are seldom found where the cutting has occurred. He further states that the name "silver bush" is somewhat of a misnomer, since the plant, although a perennial, is more a forb in growth form than it is a bush.

Desert mistletoe is eaten avidly by bighorn year round on Cabeza Prieta Wildlife Refuge, according to Gale Monson. They reach the mistletoe, which grows in large clumps on littleleaf palo verde and ironwood in particular, by standing on their hind legs and resting their forefeet on the host plants, goat-fashion.

Several cacti are recorded as significant food items of the bighorn in the Sonoran Desert. Russo (1956) relates a personal observation of saguaro use in the close vicinity of free water, and hence concludes "the saguaro as well as other plants of the cactus family serve as more than an emergency water supply." Simmons (1963), who observed the eating of succulents such as cacti, speculates on the possibility of their being used for their moisture.

Gale Monson's observations on Cabeza Prieta Refuge indicate to him that the bighorn are actually more fond of the dry scar tissue that forms over the "bites" taken out of the saguaro than they are of the green cactus. He notes that blacktail jackrabbits *(Lepus californicus)* also feed on the saguaro and that their work should not be confused with that of bighorn. Bighorn butt the outer layer of a saguaro to get at the inner layer, which they consume. Saguaro fruit also seems to be attractive to them. This is true also of both prickly pear and cholla, of which the fruits and new growth are utilized.

Surprisingly enough, pincushion cactus *(Mammillaria* spp.*)* occurred in twelve of the fifteen Arizona stomachs examined by the California Department of Fish and Game's Laboratory, making up 4.3 percent of the volume. Small amounts of this same cactus occurred in three of the twelve stomachs Russo (1956) examined. Barrel cactus *(Echinocactus* sp.*)* occurred in several of the stomachs examined by Charles R. Hungerford and Russo.

Forbs. Of the many forbs used by bighorn in the Sonoran Desert, three are preeminent. These are Indian wheat *(Plantago* spp.*)*, filaree *(Erodium* spp.*)*, and globe mallow *(Sphaeralcea* spp.*)*. Russo (1956) lists two species of Indian wheat, *P. insularis (= P. fastigiata)* and *P. purshii.* The first of these is abundant on Arizona ranges, especially after a wet fall and winter. It is heavily used by bighorn. Halloran and Crandell (1953) list the same two species as preferred during the winter months on Kofa Game Range. Filaree (mainly *E. cicutarium*) is not as widespread on bighorn ranges as Indian wheat but is heavily used where it occurs. Species of globe mallow, especially *S. ambigua,* are important foods. Halloran and Crandell (1953) list them as preferred on Kofa Game Range. Russo (1956) lists three species and one subspecies of globe mallow as being used in the spring.

Russo (1956) quotes an unpublished Arizona Department of Game and Fish report by P. Welles who observed that rock moss *(Selaginella rupicola)* was one of the principal food plants in the Santa Catalina and Tucson mountains. This is the only report of rock moss as bighorn food.

Seasonal Influences

The paucity of data makes it difficult to correlate the above-mentioned forage preferences into a seasonal pattern. Four stomachs collected by Russo (1956) in August through October showed forbs were 75 percent of the plant material eaten. The percentage of forbs in winter stomachs (six in January, four in December) varied from 2.5 to 68.5; grass was 57 percent of the December contents, only 18.8 percent in January.

Russo (1956) states that seasonal growth is important in determining preferences and palatability, key plants, and forage requirements, but he adds that "ephemeral plant-forms utilized by sheep represent a very small part of their diet." He reasons that because many of the ephemeral plants grow only under suitable climatic conditions, the "benefits derived from the temporary use of these plants as sheep forage is thought to be negligible." Gale Monson believes that dried annual growth is important in the diet of the bighorn, on the basis of observing the animals often feeding on it, even when plenty of browse was available nearby. Even in dry years substantial amounts of such growth are carried over from a more favorable season.

At Cabeza Prieta Wildlife Refuge, Monson (1964) observed that bighorn quickly respond to new growth caused by summer rains on some key browse species that grow along washes.

BAJA CALIFORNIA
MOUNTAIN DESERT

A report by Sanchez (1976) on the analyses of stomachs from bighorn taken during hunting seasons in the two states of Baja California gives us the only information available on bighorn food habits in the Baja California Desert. The stomachs were collected as follows: in Baja California Norte, twenty-four (six in spring and three in fall in Arroyo Grande, on the east side of the Sierra San Pedro Mártir; three in spring near Matomi; one in spring near El Marmol; and eleven in fall in the Sierra de La Asamblea); and in Baja California Sur, five (two in fall in Las Vírgenes, three in fall near Loreto).

The breakdown of data from these stomachs shows that in Baja California Norte, the bighorn in question consumed approximately 43 percent grass, 33 percent browse (including cactus), and 24 percent forbs. In Baja California Sur, the results were approximately 53 percent grass, 23 percent browse, 17 percent forbs, and 7 percent unidentified.

Because of the relatively low number of samples, and the fact that nineteen out of twenty-nine were taken in the fall, and ten of twenty-nine in the spring, the report cannot be considered reflective of actual year-round bighorn food consumption. Nevertheless, it does indicate that bighorn eat a surprising amount of grass in this extremely arid country.

The analyses of these stomach contents were performed at the headquarters of the Department of Conservation and Propagation of Wildlife in México, D.F.

Important Food Plants

Grasses. No attempt was made to break the grasses down according to species. However, the genera *Muhlenbergia, Panicum,* and *Aristida* are thought to prevail, on the basis of identified seeds.

Browse. According to the percentage of occurrence in stomach contents, the following species would be considered important: *Acalypha californica* (a spurge), found in eight of twenty-nine stomachs; *Acer* sp. (maple), in six stomachs; *Ambrosia* sp., in five stomachs; palo verde *(Cercidium peninsulare),* fairy duster *(Calliandra* sp.*),* barrel cactus *(Ferocactus* sp.*),* coral-vine *(Antigonon leptopus),* jojoba, and brittle bush *(Encelia* sp.*),* in three stomachs; and ironwood, salt bush *(Atriplex* sp.*),* and *Salvia* sp., in two stomachs. Five additional browse genera were each found in only one of the twenty-nine stomachs.

Forbs. A total of twenty-two species of forbs was identified in the twenty-nine stomachs. Of them, the following occurred in five or more stomachs: *Euphorbia* sp.; *Amaranthus* sp.; filaree *(Erodium cicutarium);* spiderling *(Boerhaavia erecta);* and peppergrass *(Lepidium* sp.*).*

CHIHUAHUAN DESERT

Almost all our data on bighorn foods in the Chihuahuan Desert are provided by workers in New Mexico. A 2-year Texas study in the Black Gap Wildlife Management Area, located on the east side of Big Bend National Park, is restricted in significance because all observations were made inside a 173-hectare (427-acre) enclosure holding bighorn that were not native to this desert (Tommy L. Hailey). However, it is the only source of information from Texas that is based on more than casual observation or the cursory examination of the contents of a few stomachs.

In San Andres National Wildlife Refuge, New Mexico, documented feeding observations demonstrate the ability of desert bighorn to subsist on the three forage types (Halloran and Kennedy, 1949). Four grass species, twenty-three browse species, and six forb species are identified as used, with the comment, "It is believed that if every plant eaten by the bighorns was listed it might ultimately include almost every plant on the range."

Eleven bighorn stomachs were collected in the Big Hatchet Mountains of Hidalgo County, New Mexico, in January 1954 (Gordon, 1956). Of the identifiable portion of these samples, 21.9 percent was grass; 76.8 percent, browse; and 1.3 percent, forbs (32.6 percent of the browse was cactus). Caution must be exercised in interpreting these data, however, for "a period of extended drought and severe competition with cattle and deer" occurred prior to the collection, and perhaps the food items taken at that time were determined largely by necessity rather than preference (Parry Larsen). Six stomachs taken on the San Andres Refuge in November 1968 and 1970 contained about 50 percent grass and 50 percent browse (John Kiger).

The 2-year Texas study referred to above, based on a feeding-minutes technique of field observation, provided data that indicate the captive bighorns' overall year-round diet consisted of 52 percent grass, 37 percent browse, and 11 percent forbs. The limited amount of forbs taken may have been due to lack of availability because of the type of terrain.

Important Food Plants

Grasses. Grass remains in stomachs collected in the New Mexico portion of the Chihuahuan Desert were not identified as to species. Needlegrass *(Stipa* sp.*)* was used in spring and cottontop *(Tricachne californica)* and gramas *(Bouteloua* spp.*)* were used in August on San Andres Refuge (Halloran and Kennedy, 1949). Bigelow bluegrass *(Poa biglovii)* was "eagerly sought" by bighorn in the oak-blue grama association at higher elevations in the Big Hatchet Mountains of New Mexico (Buechner, 1960).

The study in the Black Gap Management Area of Texas showed that chinograss *(Bouteloua breviseta)* was primarily utilized (145 times out of 497 feeding observations, or about 30 percent); other grasses eaten on the area were sideoats grama, *Muhlenbergia* sp., tanglehead *(Heteropogon contortus),* fluffgrass, and slim tridens.

Browse. Despite the limited data, key New Mexico browse plants are indicated. Silktassel *(Garrya* spp.*)*, mountain mahogany *(Cercocarpus* spp.*)*, fendler bush *(Fendlera rupicola),* winter fat, and Mormon tea *(Ephedra* spp.*)* are listed as important forage plants. Not comprising much of the vegetative cover, but conspicuous on the range and in the bighorn diet in New Mexico, are sotol *(Dasylirion wheeleri),* agaves *(Agave* spp.*),* sacahuista *(Nolina microcarpa),* and soapweed *(Yucca* spp.*)* (Halloran and Kennedy, 1949; Buechner, 1960; Larsen, 1962a). Silktassel and mountain mahogany are considered as "key plants" on San Andres Refuge, receiving year-round use by bighorn (Halloran and Kennedy, 1949). Fendler bush is considered important in New Mexico, also. It begins its growth in February and March, and the tender twigs are valuable food at that time. Its growth stays green despite drought conditions (Halloran and Kennedy, 1949; Buechner, 1960).

Prickly pear *(Opuntia engelmannii)* is, according to reports, much used by desert bighorn on New Mexico ranges, being noted in stomach analyses especially (Gordon, 1957). During the November 1968 San Andres bighorn hunt, much evidence of prickly pear use was noted; the horns and muzzles of the harvested rams were stained purple from prickly pear fruit being butted and eaten (John Kiger, Parry Larsen).

Forbs. The use of forbs in New Mexico, by the evidence, is light. The only species mentioned as dominant and as important on San Andres Refuge is big mallow *(Sphaeralcea incana),* which appears in the spring and summer diets (Halloran and Kennedy, 1949). Of unusual occurrence is a fern (Polypodiaceae), which was found in eight of eleven stomachs examined from the Big Hatchet Mountains by Gordon (1956).

On Texas' Black Gap Area, forbs contribute significantly to bighorn diet and increase slightly in use in summer and fall. Spurge *(Euphorbia* sp.*)* was used by bighorn in over half of the observations made of forb use (Tommy L. Hailey).

Seasonal Influences

Seasonal data are not available from New Mexico, except for scant references regarding individual plant or forage type use at certain times of the year. For instance, the eager use by bighorn of grasses that are green in response to summer rain is described (Halloran and Kennedy, 1949).

Quarterly summaries of the Black Gap Area in Texas show a seasonal pattern. Grass was used steadily throughout the 2 years. Browse use showed a slight tendency to increase in the winter period of December through February, and forbs exhibit a higher use in the summer and fall months, probably the result of seasonal rains (Tommy L. Hailey).

MISCELLANEOUS FOOD HABITS DATA

Food habits of young bighorn. Welles and Welles (1961a) in many observations of lambs feeding in Death Valley find that lambs begin learning during the first week of life to eat things adults eat and that, by the time they are 2 weeks old, to a limited degree they are actually doing it. Three of fifteen lamb stomachs analyzed from California showed the same choice of food items as the adults'.

Deming (1964) states that lambs turn to green plants as their first solid food. According to Hansen (1960), lambs begin feeding on vegetation early and generally are weaned at about 6 months.

The earliest that Wilson (1968) observed small lambs feeding in the White Canyon area of Utah was 7 July, when lambs were estimated to be 3 to 4 weeks old. They were feeding on the same plants as adults, with one noticeable exception: the older bighorn were digging for the roots or bulbs of some plant. Food preferences of lambs in White Canyon were 51.8 percent grass, 34.6 percent browse, and 13.6 percent forbs. Wilson thinks that the data are biased to browse because of the difficulty in observing the small lambs among the rocks, brush, grass clumps, and adult bighorn. Galleta and blackbrush were used most.

A seasonal trend appeared in the lambs' diet in White Canyon. As they progressed in age from July to September, the amount of grass in the diet decreased while the amount of browse and forbs increased.

Mineral requirements. Little is know regarding mineral requirements of desert bighorn. Wilson (1968) on several occasions observed bighorn eating large quantities of clay in the White Canyon area of southeastern Utah. Phosphorus was the principal mineral found in areas from which the clay samples were taken. He speculates that lack of phosphorus or calcium, or some other mineral, could be one factor causing low lamb survival in White Canyon and other desert areas.

Honess and Frost (1942), studying the Rocky Mountain bighorn in Wyoming, also observed bighorn eating clay. Phosphorus again was the principal mineral in the clay from the natural salt lick being used. These authors state that lack of certain minerals may cause pernicious results, especially if ewes carrying lambs are deficient in lime (calcium) and phosphorus. They point out that trace elements, such as cobalt, copper, iron, manganese, and iodine also are important for normal growth and maintenance.

Russo (1956), studying desert sheep in Arizona, found no natural licks and did not detect geophagy in the examination of pellets. Because of the low reproductive rate of the bighorn, however, Russo felt that lack of essential minerals during pregnancy was a subject for investigation. Experiments in Arizona with plain, iodized, and phosphorus salt blocks gave negative results; after 2 years there was no indication that the salt had ever been used. It is possible that the minimum mineral requirements are supplied in the wide variety of desert vegetation consumed.

Charles G. Hansen reports there are no known salt licks on Desert National Wildlife Range. Bighorns penned at headquarters did use salt, plain and mineralized, but did not consume it in large quantities. Wild sheep used plain salt placed at water holes. The use was slight the first year, but the following years the blocks were used more.

Welles and Welles (1961a) observed a month-old lamb and an adult ewe eating ordinary gravel in two different locations in Death Valley. They state this was the only indication they had that supplementary minerals may be added to the diet in a separate form.

Observations and experiments made during other work on Rocky Mountain bighorn (Smith, 1954; Moser, 1962) show positive use of salt and other minerals. Whether or not this craving or preference for salt or a particular mineral indicates a physiological need is an unanswered question, however.

MANAGEMENT IMPLICATIONS

Forage types. Desert bighorn have the ability to subsist on a diet consisting of all of the major forage classes: grass, browse (includes cacti), and forbs. It is difficult to conclude which forage type the bighorn prefer on the basis of available data. Geographical and elevational differences in availability of food plants, as well as bias involved with the different methods by which the data are gathered (by direct observation, "inferred" observation, or stomach analyses), make data difficult to evaluate.

Through the northern part of their range and at higher elevations desert bighorn consume much grass. In the lower, southern, and more arid parts of their desert range, bighorn rely more upon browse and forbs, owing possibly to lack of available grass. When conditions are adverse, the desert sheep use forage in dead, dried-out condition.

Interpretation of forage preference is influenced by the method of observation. Data obtained by "inferred" observation invariably run high to browse. Data obtained by careful direct observation of feeding bighorn sometimes are qualified as "biased" by the observer because of the difficulty of identifying the forage class. Stomach analyses usually result in a higher preference rating for grass, at least in areas where grass is a common component of the natural vegetation.

The inference from much food habits data is that desert bighorn prefer grass when it is available. However, when grass is not available, the desert sheep are able to use browse and forbs, even where some of these plants are in a dessicated and decadent condition. This applies to adult bighorn only; whether lambs in all populations can survive on a milk diet without the mother ewes eating grass is not known.

Habitat types. Because of the wide geographical and elevational distribution of bighorn throughout the deserts of the southwestern United States and northern Mexico, a variety of habitat types is used. In the lower elevations of the western portion of the bighorn ranges in the Mohave Desert, blackbrush-needlegrass and blackbrush-galleta vegetative associations are used extensively. At higher elevations in the Mohave and Painted deserts juniper-pinyon habitat types supply forage. In some areas, the association of big sagebrush and juniper is used. California buckwheat and needlegrass form another important type of vegetation for desert bighorn inhabiting the mountains of the Mohave Desert. At very low elevations, such as Death Valley, associations dominated by burrobrush, creosote bush, and salt bush are used at certain times of the year.

In the Sonoran Desert, associations of browse plants (jojoba especially) with big galleta are key bighorn habitat.

In the eastern extension of the desert bighorn range, in the Chihuahuan Desert, communities of grama grasses are characteristic. Gramas in association with dominant and codominant browse such as silktassel, mountain mahogany, and sotol are important habitat types for feeding desert sheep.

The attraction of canyon, or desert wash, plant communities for bighorn is well documented.

Seasonal food habits. Partly because stomach analysis results are biased by being based on hunter-killed fall samples, quantitative data on seasonal food habits are sketchy, at best. Personal observations give some insight, and what data there are reveal seasonal differences but no clear seasonal patterns (Fig. 6.2).

As has been pointed out, there really are only two vegetative seasons throughout most of the desert bighorn range. A first growing season begins as early as December or as late as March. The second arrives in response to summer rains. Indications are that the bighorn's diet reflects the seasonal changes in vegetation, showing an increase in grass and forbs during the spring growing season and after the summer rains. Browse use increases in the fall and winter.

Conflicting reports and opinions are given regarding the significance of ephemeral vegetation in the bighorn diet.

Food supplies. The consensus is that, despite a history of drought and overgrazing by domestic animals, no shortage of food exists where desert sheep are now present. No reports imply that bighorn numbers are declining solely because of lack of food.

Management implications. The most significant management implication derived from food habits studies concerns bighorn forage requirements. To transplant desert sheep into new habitat in order to extend their range, knowledge is needed of the forage type they prefer in the locality from which they are moved. Sites should be selected in areas where competition with domestic stock is under firm control.

Charles A. Irvine

Fig. 6.2. Bighorn feeding on bud sage *(Artemisia spinescens)*, the first browse to green up in the spring. Blue Notch Canyon, San Juan County, Utah.

If the primary objective of a bighorn plan is to maintain, improve, or extend bighorn numbers through adequate management of the habitat, and if, indeed, desert sheep do prefer a particular class of natural food, then conversion of the present plant community to a preferred one becomes an obvious management tool, and control of competition imperative. Buechner's (1960) conclusion regarding the status of the desert ranges of the southwestern United States may be particularly significant. He states that despite a long history of drought and competition with livestock, "forage supplies appear sufficient for the number of bighorn present." Little management of forage then would be needed to preserve the "status quo."

In either case, it is imperative to have factual data on the food habits of the bighorn. The present status of desert bighorn food habits studies over much of its range is insufficient as a basis for sophisticated field management and transplant programs. The facts should be obtained by means of seasonal stomach collections and analyses and accurate field observations of feeding bighorn, followed by careful correlation of the results with valid, quantitative data obtained by surveys of range vegetation.

7. Water

Jack C. Turner and Richard A. Weaver

Reflecting evolutionary origins, terrestrial life is an aqueous phenomenon. One-fifth of the earth's terrestrial surface is desert. The low-latitude deserts are among the most hostile of terrestrial environments, epitomizing the extremes of aridity, temperature, and inaccessibility.

The desert's absence of free water poses the most intractable problem for any living organism, despite the evolution of complex regulatory mechanisms. The major problem encountered by bighorn is not one of continual water loss but of a discontinuous water intake. Indeed, water availability, or rather the lack of it, is the single most limiting factor of desert bighorn populations.

The maintenance of an adequate water balance and a suitable body temperature are major interrelated physiological problems which confront desert bighorn. The high air temperature of summer, irregular rainfall, and high radiant solar loads which characterize the bighorn's environment bring on a problem of increased water expenditure for temperature regulation. The heat load of an animal exposed to hot daytime desert conditions primarily arises from the incidence of direct solar radiation, high air temperatures, and reflected radiation from the ground and sky (Lee, 1968). Maintenance of a constant body temperature in an environment from which an animal is gaining heat requires the expenditure of energy, which further contributes to the heat load.

In common with other animals, desert sheep must regulate the composition and specific gravity of their body fluids within certain limits. The physiological set of these limits dictates the bighorn's propensity for survival. The influx of water must equal the water lost through urine, feces, respiration, and evaporation, if an animal is to remain in water balance. When this balance is not maintained by sufficient water, dehydration results. The relation of water and electrolyte metabolism (relative salt concentrations on either side of cell membranes) in conjunction with dehydration has been studied in a few desert ungulates but is not completely understood (English, 1966a, 1966b; Macfarlane et al., 1961).

WATER CONSUMPTION

The amount of water required by bighorn is dependent on many factors, including body size, activity, forage moisture content, ambient air temperature, and humidity. Water is available to animals from three sources: (1) metabolic water formed as a result of oxidative metabolism; (2) preformed water found in the food; and (3) surface water, e.g. streams, water holes, dew.

Metabolic water. No diurnal mammal is known to be capable of subsisting solely on metabolic water. Metabolic water is formed by the union of hydrogen atoms (produced by burning of proteins, carbohydrates, and fats) with oxygen atoms (from respiration). The metabolic rate of the camel (Schmidt-Nielsen et al., 1967) and merino sheep (Alexander et al., 1962) in thermal neutral surroundings is not different from that predicted on the basis of body weight (Brody, 1945). Production of metabolic water in bighorn probably shows the same relation to metabolism and body weight as it does in other desert-dwelling ungulates.

Preformed water. The relation of preformed water in the diet of desert animals to the total water balance has been sparsely investigated. The oryx, eland (Taylor, 1969a), and dik-dik (Tinley, 1969) are independent of surface sources of water, surviving on early morning dew present on dry browse and forage. Dry food, which at midday contains only 1 percent free water, can contain as much as 43 percent water in the early morning (Taylor, 1968). This source of preformed water could be of significance to bighorn only if nocturnal feeding were confirmed and the presence of dew verified.

Quantitative data are not available, but bighorn typically water in the early morning, although they will come to water at almost any daylight hour. Generally, the hotter the weather the earlier the bighorn will water. Since dew rarely forms during the dry season in the low southwestern deserts, little or no net gain of water is obtained from early morning watering. However, early morning watering reduces unnecessary heat gain, therefore reducing the evaporative water loss. Nocturnal watering on moonlight nights was observed on Cabeza Prieta Wildlife Refuge by Simmons (1969b), but this would appear to be an exception to bighorn behavior observed in the remainder of the Southwest by Richard E. Weaver as well as by Tommie L. Hailey and Gale Monson; and see Turner (1973).

Food plants of desert bighorn have been identified by various persons (Chapter 6). A singular feature is the wide diversity of plants eaten. Most of the published data relate to field observations; few stomach analyses are available. Relative quantities of each plant eaten, and water and electrolyte concentrations of these plants, have not been determined. Although considerable quantities of water can be obtained through the diet, not all intake of water via the food can be considered a net gain. There is a portion that is obligatorily committed to the voidance of salts ingested with the food.

Cactus has been reputed to act as an emergency store of water for the stranded desert traveler. Cactus has also been postulated to explain how bighorn survive in areas of severe drought. Indeed, fully hydrated cactus could serve as a source of water. Under drought conditions, however, cacti become dehydrates, as do bighorn. Investigations by Turner (1973) of osmotic concentration and electrolyte composition in several cacti reportedly used by bighorn, including barrel cactus *(Echinocactus acanthodes),* hedgehog cactus *(Echinocereus engelmannii),* and pencil cholla *(Opuntia ramosissima),* have shown that when these cacti become dehydrated, the water which is obligatorily committed on the part of any bighorn eating them for the voidance of the ingested electrolytes and dry matter is greater than the water obtained by eating the cactus. Therefore, under conditions of water stress, some cacti are not available to bighorn as a source of water, for their ingestion could create added problems for a bighorn facing dehydration. The same is without doubt true of other plants eaten by desert bighorn.

Drinking. Most ungulates, including the camel, are dependent upon surface sources of water to maintain water balance over prolonged periods of time. Camels, guanacos, donkeys, and domestic sheep are capable of consuming 18 to 23 percent of their body weight in water within brief periods of time (Adolph and Dill, 1938; Macfarlane et al., 1956; Schmidt-Nielsen et al., 1956; Rosenmann and Morrison, 1963). These animals drink to restore lost body water; however, overdrinking may function as a reserve mechanism for maintaining adequate water for evaporative cooling. The rapidity with which these animals can consume water would convey the biological advantage of having to spend less time exposed to the high predator pressures around a water hole.

Bighorn rams drink as often as ewes, and lactating ewes water more frequently than dry ewes and young animals (Charles G. Hansen; Simmons, 1969b; Turner, 1970, 1973; Welles and Welles, 1961a). Ewes are more likely

Water 103

U.S. Fish and Wildlife Service

Fig. 7.1. A ram kneels to drink, Castle Dome Mountains, Yuma County, Arizona. Bighorn may drink for as long as 3 minutes at a time.

to begin using water earlier in summer than rams. This probably relates to lactating ewes requiring more water. Lambs will drink water when they are 2 or 3 days old.

Bighorn are capable of consuming large volumes of water, although the amount of water consumed by different individuals is quite variable. This variation is probably due to two factors: small animals drink less per unit weight than large ones, and the longer the bighorn remain away from water, the more they consume on their next visit. Although rams generally consume the largest volumes and weight-relative percent of water (Fig. 7.1), large amounts of water in relation to body weight also are consumed by the smaller ewes and yearlings (Turner, 1970, 1973).

Amounts of water consumed by bighorn around a water hole in midsummer were measured. A ram consumed the largest amount of water (18.7 liters) taken at one visit to the water hole. This was also the greatest weight-relative amount (23 percent). However, large amounts in relation to body weight were not confined to older rams. A ewe, a ram lamb, and a 4-year-old ram all consumed more than 20 percent of their estimated body weight when drinking.

Extended absence from the water hole correlates well with consumption of large quantities of water. There is no apparent difference in water consumption amount between rams and ewes. If it is assumed that the animals drink to replace lost body water, bighorn apparently lose water at a decreasing rate the longer they remain away from water. The curve becomes asymptotic at 4 percent body weight per day. This implies that bighorn must drink a minimum of 4 percent body weight per day to maintain water balance during the summer.

Drinking rates are variable. Rates as high as 2.8 liters per minute have been observed. No correlation exists between amounts of water consumed and length of time spent drinking (Charles G. Hansen, 1971; Turner, 1970).

Desert bighorn of both sexes and all ages require free water regularly during hot dry periods. Some desert ranges have no permanent springs; however, most mountain ranges do have numerous potholes or small rock depressions that hold water for a short time after rains (Fig. 7.2). In some mountain ranges, whose geological structure is favorable, large collection tanks (tinajas) have been formed. These hold large amounts of rainwater for long periods of time, sometimes never drying up. Tinajas and potholes often occur in ranges where perennial sources of water are available. As soon as rain falls, relief on the perennial sources is afforded, and reliance on them is temporarily eliminated. The importance of these potholes and tinajas is evidenced by trails and observed heavy use.

Bighorn occasionally water at residences along the base of the San Jacinto Mountains, within the city limits of Palm Springs, California, although their normal range is adequately watered. However, such occurrences are highly unusual anywhere in the overall desert bighorn range. They almost never come down to a water source, e.g. windmills, that are located even a short distance from rocky escape terrain.

Seasonal water use. The hypothesis that bighorn can exist without a permanent source of water is not widely accepted. The frequency with which bighorn must drink is dependent upon many factors, but two general trends are evident. During hot, dry summer periods, bighorn water daily when possible, but have remained independent of free water for periods of 5 to 8 days (Blong and Pollard, 1968; Charles G. Hansen; Simmons, 1969b; Turner, 1970, 1973; Welles and Welles, 1961a, 1966). Monson (1958) reported that bighorn on Cabeza Prieta Wildlife Refuge did not drink for a period extending from July to December (6 months) or possibly longer, Mendoza (1976) reports that bighorn in some Sonora ranges go without drinking. Although such observations are not uncommon, investigators have concluded bighorn in such areas either utilize alternate, temporary, or unknown sources of water or leave the range. Diligent, but futile, aerial and ground searches for unknown water sources and the persistence of fresh sign or bighorn in these desert

Richard A. Weaver

Fig. 7.2. Pothole affording water for only a short period before evaporation dries it up. This is the only water hole in the Nopah Range, Inyo County, California. (Nopah is Indian for "no water.")

ranges argue against the various alternatives. The question of whether or not bighorn can do without free water for periods of several months (including the hottest of the year), or even longer, in such areas as northwestern Sonora remains unanswered.

During winter months, bighorn may water daily if their feed is dry and water is available (Lanny O. Wilson). Death Valley bighorn water, on an annual average, every 10 to 14 days (Welles and Welles, 1961a). Aldous (1957) observed bighorn eating ice and similarly rare observations of the animals eating snow have been recorded. Simmons (1964) reports little or no use of perennial water holes on the Cabeza Prieta Wildlife Refuge during December through February.

Bighorn in southern California deserts seldom utilize free water before June (Richard A. Weaver). Independence of water holes which during summer months are frequented daily is attributed to reductions in the bighorn's avenues of water loss due to reduced environmental heat load and a complementary increase in water available through the food. If the water contentent of the diet is sufficiently high, bighorn probably can obtain all their exogenous water preformed in their food and thereby become completely independent of water holes.

Apropos of seasonal use is McQuivey's (1978) comment from Nevada, "It is significant that, because of water distribution, several major use areas and some entire mountain ranges, such as Arrow Canyon Range, Desert Range, Southern Spring Range, and the North McCullough Range, support bighorn during the fall-spring months but do not have sheep use during the summer period."

WATER SHORTAGES

Lack of water is the single most limiting factor for bighorn herds in the desert. Bighorn will reluctantly move away from an area with a dried water source and attempt to reestablish themselves around a different water hole. Once a spring has been dry and abandoned by bighorn for several years, they may not return when water is restored. The overall effect is a loss of bighorn habitat and probably a loss of that portion of the bighorn population affected. It is a remote chance that displaced bighorn will successfully find another water source. If a new source is chanced upon, the influx of bighorn is likely to exceed the carrying capacity of the available resources, resulting in sufficient stress that only a small percentage of the population will survive (Hansen, 1971).

When there is a water shortage, dominant animals take the available water first; often the younger animals go without or receive inadequate water. Jones et al. (1957) observed ewes with lambs searching for water in dry canyons, severely taxing their lambs. Inadequate water could contribute to low lamb survival in some areas. In situations such as these, bighorn do not abandon the area, but productivity, herd survivorship, and eventually the total population is reduced. Sumner (1957) accurately predicted a decline in the bighorn population of Joshua Tree National Monument on the basis of a failing water resource.

Bighorn carcasses and remains are more frequently found near water than elsewhere in their range. This apparently relates to a diseased or debilitated animal's increased demands for water, as well as decreased mobility. Data from animals captured around a water hole indicate a significantly higher frequency of disease in animals first to return to water after prolonged absence due to cool weather or rain (Jack C. Turner). Old rams have been observed to disregard their normal fear of civilization (i.e., dogs, people, vehicles) to drink, only to move a short distance to bed down and die. Predation and scavenging of diseased animals around water holes would also contribute to the observed mortality.

Much potential bighorn habitat is used infrequently or seasonally because of the absence of water. Water shortages result from unfavorable distribution of water sources as well as lack of permanence. Little is known about home range or how it may fluctuate seasonally, but it is generally concluded that availability of free water is a key factor in determining home range and distribution. The distance that bighorn range from water either in summer or winter depends upon prevailing weather and forage. Wilson (1971) reports summer and winter home ranges of 6.7 and 29.8 sq km (2.5 to 11.5 sq mi) respectively for Utah bighorn. Although rams had a larger home range, ewes on Cabeza Prieta Wildlife Refuge during spring-summer dry periods had an average home range of only 0.8 sq km (0.3 sq mi) (Simmons, 1969a). Bighorn leaving wintering areas at the beginning of summer on Desert National Wildlife Range in Nevada have moved 32 km (20 mi) to water (Charles G. Hansen). Water sources should be not more than 10 km or 6 mi air line apart for minimum use of the range and should be evenly distributed in suitable habitat at 2-km or 1.2-mi intervals to maximize bighorn utilization of the habitat.

A "death trap" tinaja discovered in the Chocolate Mountains, Imperial County, California (Mensch, 1969), indicates bighorn preference for certain water sources and their reluctance to leave good habitat. The remains of 34 bighorn were found in the 4.5-m (15-ft) diameter tinaja (Fig. 7.3). Bighorn could drink when the tank was nearly full. However, when water receded beyond their reach, bighorn would fall or be crowded into the pit. The 3-m (10-ft) vertical sides did not provide sufficient purchase to allow trapped bighorn to escape. Many animals died in this trap, but the population continued to use the water source and remain in the area despite the Colorado River being only 5 km (3 mi) away. Although the area near the river is rolling terrain, bighorn were not using the river. The salt cedar *(Tamarix* sp.*)* barrier, which bighorn seemingly are reluctant to penetrate, and competition with burros were apparently insurmountable obstacles.

WATER DEVELOPMENT

Important to the management concerns of bighorn is the development of water. Many factors influence the success and speed with which new watering sources are accepted by bighorn. Almost any source (mine tunnels, leaking water pipes, swimming pools, etc.) will eventually be used if they are present in the immediate range of bighorn. The immediate acceptance of new sources depends on many habitat-related variables. Bighorn prefer to drink from sources not surrounded by brush or other obstructions, where their vision is not obscured for 100 m (300 ft) in all directions. Bighorn are wary of trees, rocks, or cliffs which overhang a water source (Fig. 7.4). Any new sources should be established close to existing trail systems and within bighorn habitat

J. L. Mensch

Fig. 7.3. A "death trap" water hole in the Chocolate Mountains, Imperial County, California. The remains of thirty-four trapped bighorn were found in this steep-sided tank in 1969. Blasting away rock to provide a safe entry ramp has removed this danger to wildlife.

and with access to escape terrains. Water holes with livestock corrals or fences are usually shunned by bighorn. Water sources in steep narrow gorges are not the most acceptable to bighorn owing to their vulnerability to predators. Where bighorn use water in these situations, they will remain on a prominent point for a long period of time, come quickly to drink, and leave. Because frequently such water holes are the only source of dry-season water, they are of great importance to bighorn and ordinarily receive heavy use by them, the predator problem notwithstanding. See Chapter 20 for further information on water development.

WATER BALANCE PHYSIOLOGY

It is not disputed that bighorn require water in some form. Although desert bighorn require comparatively small quantities of water relative to their body size, activity, and preferred habitat, even this amount is not always available. Despite the total absence of free water in many desert ranges, especially in northwestern Sonora, the bighorn persist. The mechanisms that allow bighorn to exist in these areas are not clearly understood.

Body fluid distribution. The total water content of desert ungulates is typical of vertebrates, ranging from 70 to 80 percent of the body weight (Till and Downes, 1962; Panaretto, 1963; Macfarlane et al., 1961). The water in

Norman M. Simmons

Fig. 7.4. Two bighorn ewes and two lambs nervously scrutinize man-made Eagle Tank before entering to drink. Sierra Pinta, Yuma County, Arizona.

the rumen of sheep and other ruminants constitutes from 10 to 25 percent of the total body water of the animal and occasionally may reach more than 30 percent (Dukes, 1955). Merino sheep, during the first 3 days of water deprivation, draw upon rumen water to compensate for the unavoidable loss of fluid through urine, feces, and respiration (Hecker et al., 1964). During periods of water deprivation, rumen water probably contributes to the fundamental water homeostasis in all ruminants.

Evaporative water loss. Evaporative cooling is the primary mechanism for heat dissipation for desert ungulates (Brook and Short, 1960a, 1960b; Schmidt-Nielson et al., 1956; Taylor, 1969a). Domestic sheep and wildebeest depend primarily upon panting for heat dissipation, in contrast to cattle and the camel, which depend upon sweating (Schmidt-Nielsen, 1964; Taylor, 1969b). This is reflected in the increased respiratory rates of domestic sheep and wildebeest (up to 300 and 200 respirations per minute, respectively) when subjected to ambient temperatures greater than 45°C (Blight, 1959; Macfarlane et al., 1958; Taylor, 1969a). The camel and domestic cattle, however, show very little increase in respiration rate.

Since a major part of the desert ungulate's heat load arises from solar radiation, thermal panting may have a possible advantage over sweating through gaining less heat from the environment. The surface where evaporation takes place must be cooler than body temperature and have a rapid blood flow; however, the thermal gradient would tend to cause the animal to gain heat. On the other hand, if the respiratory tract is the evaporative surface, the skin temperature can exceed the body temperature and cutaneous blood flow can be low. This would decrease the heat gained from a hot environment. But an advantage would only be gained if the resulting low environmental heat load would offset any increased metabolic heat produced by panting.

The introduction of a barrier (e.g., hair) between the skin and the surrounding air has the immediate effect of reducing the radiation exchange. If the skin were previously gaining heat, the insulation will continue to do the same. The immediate effect of the insulation will soon dissipate as the insulation gains heat from the environment and transfers this energy gain to the underlying skin.

The heat gain or loss of the insulation will be determined by the amount of energy re-radiated and reflected from the surface of the insulation and by the rate of transfer of absorbed energy to the skin by the insulation. Glossy pelage of many animals is responsible for reflecting as much as 40 percent of the direct solar radiation (Lee, 1950). Forced convectional cooling at the surface and re-radiation significantly reduce the heat load (Macfarlane, 1964). The amount of heat transferred through the insulative layer is related to the specific heat and mass of the insulation. In warm environments, the physical rejection of heat at the surface of the insulation would reduce the heat load subject to absorption and, therefore, the total heat load of the animal. Although the addition of more insulation would greatly retard the exchange of heat from the environment to the animal (and vice versa), it would create additional problems. The necessity for locomotion limits the amount of insulation of many animals. Often this problem is circumvented by localization of insulation (Morrison, 1966; Macfarlane, 1964). The retardation of evaporative water loss, and the associated problem of the loss of metabolic heat, must also be considered.

The efficiency of evaporation is reduced to zero in animals with thick pelage. Cooling obtained by evaporation is most effective at the level of the skin and decreases with the inverse of the distance from the skin to the point of evaporation. It is advantageous, therefore, to keep interference with evaporation minimal, or to rely upon mechanisms other than cutaneous evaporation for cooling.

Respiratory evaporative water loss appears to be directly related to ambient temperature, whereas cutaneous water loss is allied to humidity (Brook and Short, 1960b). Lee and Phillips (1948) have studied the adaptability of livestock to climatic stress. The presence of sweat glands in cattle and their importance in evaporative cooling is pointed out by Dowling (1958). However, the greater tolerance of Brahman cattle does not appear to be correlated with efficiency of sweating but with their greater surface area per unit weight, which in turn results in greater heat dissipation by convection, radiation, and vaporization (Dukes, 1955). Their lower basal metabolism may also be an important factor in increased heat tolerance.

Although domestic sheep possess approximately twice as many sweat glands per unit as man and one-third as many as domestic cattle, their sweat production is 1.6 and 5.4 percent of that volume produced under similar conditions by man or cattle, respectively. The contribution of sweat glands to temperature regulation of wild or domestic sheep is not known.

Evaporation accounts for the major expenditure of water in animals exposed to hot, arid conditions (Schmidt-Nielson et al., 1956). Mechanisms for conserving water would produce their greatest effect if directed toward a reduction of evaporative water loss. Controlled daytime hyperthermia has been demonstrated to be of adaptive significance to the camel (Schmidt-Nielsen et al., 1957), donkey (Schmidt-Nielsen et al., 1957; Adolph and Dill, 1938; Dill, 1938), and some breeds of domestic sheep (Lee and Robinson, 1941; Macfarlane et al., 1956, 1958). The obvious effect of body size undoubtedly underlies the degree to which controlled hyperthermia can function.

Urine and fecal water loss. Urine volume and concentration are fundamental considerations in the use of water for excretion. When water is freely available to the normal animal, that which is consumed and not required for the maintenance of equilibrium is excreted via the kidneys. Ruminants in general show a decrease in urinary and fecal water loss when subjected to water deprivation. This can be attributed in part to the accompanying decrease in food intake as a result of restricted water, a concurrent fall in metabolic rate, and an increased reabsorption of water from the distal tubules of the kidney under the influence of an antidiuretic hormone (ADH) (Blaxter et al., 1959; English, 1966b).

Food intake does not show a decrease in the camel or the eland when either is experiencing water deprivation. However, significant changes in water lost through urinary and fecal routes are apparent (Taylor and Lyman, 1967). The fecal water loss of the eland and domestic cattle experiencing water deprivation is very similar when expressed in liters per 100 kg of body weight per day. The difference is a 25 percent higher food consumption per 100 kg of body weight by the eland, and a higher digestibility coefficient.

Kidney function. The concentrating ability of the kidney determines to a great extent the water that is obligatorily committed for the voidance of electrolytes. Urine flow has been determined for several desert ungulates. Although urine flow is dependent upon degree of dehydration, a dehydrated camel loses 0.4 percent body weight per day via urine when stressed 8 to 10 days without drinking water (Macfarlane et al., 1959). The corresponding urine-plasma ratio for total concentration was 7.6, indicating an efficient kidney for conserving water (Macfarlane et al., 1956, 1958).

Urinary concentration of sodium is much lower than potassium in ruminants not experiencing water deprivation (English, 1966a). This is incidental to their herbivorous diet. Potassium appears to be actively secreted by the renal tubules (Macfarlane et al., 1956). Urinary chloride concentrations in domestic sheep are much higher than sodium concentrations (English, 1966a). There is no parallelism in urinary excretion of sodium and chloride by the kidneys of healthy domestic sheep. Chloride anions, along with bicarbonate and phosphate anions, probably contribute to balancing the potassium cation in the urine.

Domestic sheep experiencing dehydration show urinary potassium/sodium ratios less than unity and as high as five upon rehydration (Macfarlane et al., 1961). Sodium is selectively retained with rehydration to maintain physiological levels of sodium within the body. Similar conditions are found in the urinary sodium and potassium concentrations of dehydrated and rehydrated domestic cattle (Taylor and Lyman, 1967). Probably this situation exists in all ruminants.

Loss of body fluids. The influx of water must equal the water lost if an animal is to remain in water balance. When this balance is not maintained, water is lost from the intracellular and extracellular spaces. Many mammals cannot tolerate a water loss greater than 10 percent of their initial body weight. A camel deprived of drinking water for 8 days lost 26 percent of its total body water and 17 percent of the initial body weight. Although the extracellular fluid volume showed the greatest proportional loss in water, the intracellular fluid volume lost a greater total amount. The smallest relative water loss occurred in the plasma portion of the extracellular fluid. Merino sheep that had been dehydrated for a similar period of time lost 23 percent of their body weight. The plasma volume was reduced to half the initial volume, and the total extracellular fluid was reduced by a similar proportion (Macfarlane et al., 1956). This is in sharp contrast to the camel, which suffered little loss of plasma volume in dehydration.

Reductions in blood volume have been regarded as the main cause of explosive heat rise and death in the desert. It has been suggested that one of the reasons for the camel's tolerance to dehydration is its ability to maintain its plasma volume. Some desert breeds of domestic sheep are able to approach and exceed the camel's tolerance to dehydration, but they also show a tolerance to decreased plasma volume (Macfarlane et al., 1962). This tolerance may relate to differences in the mode of thermal regulation. Panting is the main mechanism of heat dissipation in domestic sheep, whereas sweating is the mechanism employed by the camel. The reduced blood volume apparently does not significantly affect heart rates of dehydrated domestic sheep, although it does in many other mammals (Lee and Robinson, 1941; Eyal, 1954).

8. Senses and Intelligence

Charles G. Hansen

When bighorn are observed under field conditions, it is often difficult to determine which of their sense organs is receiving the strongest stimulus, or to which stimulus the animal is responding. When a man observes a bighorn raise its head and point its ears and nose toward him, which sense organ has perceived him? What stimulus is the animal reacting to when it bounds off, or when it puts its head down and continues to feed? If the animal continues to feed and is killed by a hunter, does this mean that the animal has not received any stimuli? Or does it mean that the ram did not have the intelligence to react to the stimuli? If the animal bounded away, did it do so because it was acting through instinct or through learning?

It is very difficult to know when a stimulus is received simply by watching for an animal's reaction to it. When a bighorn reacts, we do not know if it is a learned or an instinctive response. If it is learned, how do we determine the degree of intelligence? Only through objective studies of bighorn behavior will we be able to determine the extent of their intelligence.

This chapter draws upon the observations and impressions of many persons who have studied desert bighorn under various circumstances.

SENSES

Vision. Desert bighorn appear to depend primarily, though not exclusively, upon their sense of sight to detect danger. On several occasions I have seen bighorn running away from me when I was still about 1.6 km (1 mi) from them. In most of these cases the wind was such that it was doubtful that the animals heard or smelled me.

Welles and Welles (1961a) state, "The fact that they have good vision is too well known to warrant general discussion. We have seen ample proof of this in distant vision and in close vision associated with the incredible surefootedness of the species. There is some question, however, involving their intermediate vision suggested by a seeming inability to recognize well known objects within certain limits."

Desert bighorn apparently do not have as good vision at night as they do by day. At Corn Creek Field Station of Desert National Wildlife Range, Nevada, the hand-raised lamb "Tag-along" followed people outside at night, where she occasionally walked into wire fences and other objects that also were difficult for humans to see. Her actions left the impression that she had no better night vision than a human.

Hearing. Bighorn are able to hear relatively well, and they continually move their ears to pick up sounds. When they are alert, they usually have their ears turned to the side or to the back, presumably to pick up sounds.

Welles and Welles (1961a), in discussing the hearing of bighorn in Death Valley, state, "What little evidence we have ... indicates a healthy normal condition at least equal in sensitivity to ours. We have no evidence in Death Valley of hearing impairment by ticks as has been suggested by observers in other areas."

The ears of adult and especially old bighorn on Desert Wildlife Range and on Kofa Game Range in Arizona often become plugged with ticks and wax or scablike debris (Norman Simmons). It is not unusual to find animals with the inner surfaces of their ears infected from the irritation caused by ticks and mites. The exudate may be so excessive that the pinna is completely plugged and full, which would suggest an almost complete loss of hearing. Simmons tested an afflicted ewe's response to noise and found a definite loss of hearing. However, bighorn with plugged ears do not seem to be unduly hampered by such reduction in hearing ability.

Smell. Bighorn use their sense of smell to pick up scent from other bighorn as well as from other animals and objects. Their scent stations are established by rubbing their eye glands on trees and rocks, or by urinating on the ground. My experiences suggest that bighorn can smell humans from as far away as 0.8 to 1.2 km (0.5 to .75 mi). Smell appears to be used to distinguish between foods, and most if not all food items are sniffed before they are eaten.

Once my wife mixed hamburger patties with her hands before preparing a bottle of milk for a hand-raised lamb. She rinsed her hands but apparently not thoroughly enough because, when she offered the bottle to the lamb, the milk was refused. After a more thorough cleansing of the bottle and hands, the lamb accepted the bottle.

Welles and Welles (1961a) describe cases of bighorn tracking other bighorn by smell, but apparently they do not depend wholly upon this method of finding others of their kind. They often rely on their eyesight. The tame ewe, Tag-along, seemed to distinguish between her lamb and a stranger by smelling around or under their tails.

I have approached bighorn to within 90 m (100 yd) without their detecting me, only to have the wind shift and take my scent to them. Then they reacted rapidly, running up onto a side hill and even moving toward me in order to be able to see what or where the smell was. When they were high enough to see me and were out in the open, they relaxed; no longer alarmed and watchful, they began to feed and some would lie down. When I got up and moved up the canyon, they seemed to ignore me, at least as long as I could see them from approximately 1.6 km (1 mi) away.

Wilson (1968) writes: "On April 21, 1966, I sighted ten mature rams [in southeastern Utah]. The wind was at my back and toward the rams . . . I was approximately 400 yards from the rams and was sure they had not seen me. After an hour the sheep became increasingly nervous, scenting the wind at shorter intervals. I lit a cigarette, and in a matter of a few seconds all ten rams were on their feet with noses in the air. In a few minutes the sheep began moving rapidly away from my location. I am sure none of the rams ever located me."

INTELLIGENCE

Learning. The hand-raised lamb, Tag-along, was easy to train and learned quickly—perhaps more quickly than most dogs. If we had known how to train her from the start, we could have taught her house manners quite easily. She learned spoken commands and obeyed them when she "knew" she had to; otherwise, she tried to get away without obeying. This lamb was raised in the house from the time she was about 3 days old until she was as big as a large dog. She was then relegated to the outdoors and only allowed in on special occasions. Even when she was more than half-grown she "expected" to be allowed to do things she had done as a lamb. She would bounce from room to room and bed to chair or couch, knocking things out of place.

Tag-along adapted quite well to a life among people, even though our house differed greatly from her natural habitat. She learned quickly to take milk from a bottle and came to my wife, although our children and other people usually fed her after my wife had fixed the bottle. Tag-along followed her foster mother everywhere during the 5 months we kept her in the house.

Almost before she learned the origin of the food, Tag-along learned where the heat in the house came from. She would lie so close to a fire in the fireplace that it would scorch the hair along her back.

Tag-along not only accepted the family as hers, but also accepted the family's Siamese cat and kittens. She would lie in the kittens' box and the four 2-month-old kittens would swarm around her. When she was too big for the kittens' box, she would lie near it, and the kittens would cuddle up to her and sleep. A male lamb raised 2 years later accepted a small dog as its companion and was "content" to be away from the family only when the dog was with him.

Tag-along learned to respond to her name and to other sounds when she was living in the house with us. She learned to get out of a room when we clapped our hands or told her to "Get out!" She also learned to come when called. But she soon became very independent and would mind only when she "wanted to."

Tag-along learned where her bedroom was and would occasionally go to bed by herself. When she was small, a room upstairs was set aside for her; later she used part of the washroom downstairs. During the period when she went to bed by herself, she would bed down and stay down even when the door was later closed. However, she usually had to be put to bed, and sometimes forcibly. She often reminded us of our own children.

This lamb learned the difference between doors and windows quite early. She would look through the windows but made no attempt to get in or out of the house that way. She ruined many screen doors with her persistent jumping and pawing; not once did she try to jump at or paw a window. Even when she could see people through the window next to a closed door, she chose to jump and paw at the door to get in instead of attempting to go through the window.

Tag-along quickly became accustomed to traveling in a car. She was taken with our children to school 25 miles away, where she was weighed periodically. She appeared unconcerned about the movement of the car. She learned that, until everyone was in it and the motor started, the car was not going to leave. She gave us the impression that at times she would rather stay home, at other times she was simply indifferent; sometimes she was eager to go. When she was indifferent, she would wait until the last minute before getting in. Sometimes it was necessary to move the car a few feet to convince her we were going.

On one occasion when she was at school, walking with her foster mother, something frightened her. With a dash, she leaped into the front seat of the car and settled down where she usually rode. Other cars were nearby, but she chose the correct one. At home she would run to the front porch for safety.

Eventually, Tag-along had to learn to live with others of her species. At first she was afraid of them and would run away as she might from a big dog.

Gradually she got used to them, but not until she had her first lamb at 3 years of age did she seem to identify with bighorn. Even after her third lamb, she seemed to "prefer" people and was rather independent of other bighorn except during the breeding season.

A male lamb taken from a penned ewe when he was about 36 hours old and raised in the house learned to follow his human foster mother in about 12 hours. Within a day or two, he seemed not to recognize his real mother, nor did she recognize him; thereafter he followed people, ignoring other bighorn.

While this male lamb was growing up, he was taken on outings into wild bighorn habitat. Once when he was about 2 months old, he was encouraged to follow me, our two children, and our dog. The five of us hiked up a heavily wooded canyon, and we occasionally encouraged the lamb to follow us. On this particular hike, my wife had stayed behind at the truck. When we were about 0.8 km (0.5 mi) away, the lamb turned back and started down the canyon. No encouragement could turn him around. We continued on our hike and returned to the truck about an hour later. There we found the lamb with my wife. He returned down the canyon without haste and walked into camp, as if it were a common occurrence to wander around new territory and find his way back to his foster mother.

Other bighorn raised in the pens at Corn Creek responded to voice and hand signals. They appeared to recognize individuals and to understand the intent of the person. Some of the penned bighorn had apparently learned their names and responded accordingly. When moved from one pen to another, they seemed to have learned that when the person doing the moving had a can of grain in his hand he was not going to upset them. So they usually cooperated and followed appropriate hand and voice signals. Occasionally they did not take the reward after they had successfully followed directions.

On one occasion, after the ewes had been separated from the rams for some time, a change of pens was needed for repairs; the rams were therefore encouraged to leave their pen for an adjacent one. This was during rut season, although the ewes had been bred about 2 months earlier. The ewes were enticed with grain into the pen which the rams had just left. The young ewe, Tag-along, went in without hesitation and began eating, but an older ewe was more wary and hesitant. She kept looking around and just smelled the grain in the feeder. After a minute or two of this uneasy action she retreated into the pen she had just left and went to the gate through which the rams had just passed. At the gate she stood and looked through the wire watching the rams running around for 2 or 3 minutes. She then turned and, without any show of wariness, went directly into the pen she had just left. Once there she unconcernedly began eating alongside the younger ewe.

My impression was that the older ewe did not want to be in the same pen with the rams, since they usually chased the ewes when they were changed from pen to pen. It appeared that she deliberately went to the other pen to make sure the rams were all there before she was content to enter the pen where the rams had been kept for several months. Since she and most of the other bighorn at Corn Creek were born and raised there, they were familiar with the pens and other bighorn.

A 6-year-old captive ram at Corn Creek Station used an almond tree about 0.6 m (2 ft) in diameter as a butting post. He would walk up to it, look it up and down, and sometimes smell it before he would butt it. Sometimes he would back away about 5 or 6 m (15 or 20 ft), then, after standing for a few seconds, would rear up on his hind feet and run at it, lowering his head at the last second before butting the tree trunk. When the tree was hit in this manner, it would shake all the way to the top. The ram would bounce off and stand for perhaps a minute or so, and then back away for another butt. Or he would walk off.

As an illustration of learning, one day in the fall while watching him butt the tree, I noticed that he was hitting it quite high and not at the usual 1-m (3-ft) level. After he hit the tree he did not stand around but immediately began picking up newly ripened almonds that had fallen to the ground. He would circle the tree, picking up almonds, and then back off and hit the tree again. He did this about five more times. When no more almonds fell, he walked away. He had learned that by butting the tree he could shake the almonds off as they ripened, because he would periodically check over the ground after butting the tree. He also learned that the tree would shake more vigorously when it was hit higher than at his usual butting level.

A similar case of a wild bighorn learning to obtain food out of reach is reported by George Welsh. "Several years ago Warren Kelly and I were running our annual bighorn surveys on Lake Mead, and we spotted an old ewe some distance above the shoreline on a low but precipitous ridge. She was attempting to feed on the top of a salt cedar *(Tamarix pentandra)* growing in a gully below her, which was just out of reach of her outstretched neck. Finally she kicked a branch with her front hoof and as it swung back toward her she bit off the tip of the branch. Finding that this method worked well, she continued kicking the branch and nibbling until she had eaten all the foliage that was available."

Memory and recognition. Lanny O. Wilson writes:

Bighorn have a precise memory. On several occasions while tracking or following desert bighorn in southern Utah across rolling ridges and deep canyons, it was impossible for me to see over many of the sharp rises in the terrain that I was traveling. When the bighorn would come upon some of these rises, they would turn off to one side or another and go around the side of the hill rather than going over. On every occasion that I walked over one of these hills and looked down the other side, there was a 50 to

100 foot drop-off. On other occasions upon coming to a sharp rise, the bighorn would travel right over it and down the other side. Whenever I went over one of these rises, there was always a gentle slope leading down the other side. There was no way that I could ever discern any difference between those hills having a drop-off and those without. There was no question that the bighorn knew where they were and remembered the terrain down to minor details.

On another occasion I tested the memory of bighorn. Nine of them had become accustomed to me, and I always made it a point to wear the same clothing when I visited these particular animals. After about a month, they became so accustomed to seeing me that they paid little attention to me. One day I located these bighorn when I had on the same clothes I had worn on previous occasions. This time I had a change of clothes along and returned to the jeep and changed. I went back to the bighorn and upon approaching them with different attire, they took immediate flight; obviously they had not recognized me.

Again I returned to the jeep, put on the clothes I had worn on previous visits, and again located the bighorn late that evening. I walked up to within 50 feet of them, and they showed no fear whatsoever. It was obvious that they had recognized me by my clothing.

Some of my own experiences lead me to believe that bighorn have good memories and recognize objects quite readily. After she had lived in the pens with other bighorn for 3 years, Tag-along remembered various people that she knew when she was a lamb. On the few occasions when she got out of the pens, she looked for human companionship and tried to get into houses to find it.

A significant behavior pattern developed when Tag-along was about to have her first lamb. About 3 or 4 weeks before parturition, she became quite belligerent toward the other ewe in the pen. Apparently this was a normal response, because ewes usually isolate themselves just before lambing. On the other hand, the ewes in the pens are not much concerned about the rams at this time. As the parturition date drew closer, my wife (who had been Tag-along's foster mother) began making visits to see how she was. Each day Tag-along became more belligerent toward my wife and, a day or two before the lamb was born, chased her out of the pen.

During this time Tag-along made only half-hearted or no attempts to butt at men who entered her pen. While the lamb was being born, she allowed me to remain 20 or 30 feet away through the entire proceeding and was not at all concerned when I approached closer. My impression was that Tag-along recognized my wife as another ewe and did not want her around while she was lambing.

Fear. Bighorn occasionally experience fear to the extent that they will run blindly in almost any direction. At times the bighorn at Corn Creek were not at ease with their handler, and they became so excited that they blindly ran into fences, trees, or other obstacles. Some individuals were more susceptible than others. Some seemed to run to get away, but others were just as likely to run toward the object of their fear as way from it.

A similar reaction has been observed in unconfined bighorn. When severely frightened, they may run blindly along cliffs and jump over rocks which they could go around by stepping only a foot or two to one side or the other. While being observed from a low-flying helicopter, ewes and lambs were seen to jump off into space, only to recover themselves by scrambling up a steep rock face which they happened to hit on their way down.

Welles and Welles (1961a) report that, at Death Valley, "... when the sheep were feeding on the shoulders [of the road] we stopped all cars, called the attention of the park visitors to their rare opportunity of observing bighorn, but curbed the inclinations of some persons to pursue them. The bighorn quickly accepted their protective status as being permanent, and appeared to take it for granted that all cars would always stop for them. ... Probably the most unexpected phase of this development ... was their refusal to be disturbed by our violent, noisy efforts to frighten them. They no longer attached a sense of danger to cars, roaring motors, honking horns, shouting people or exploding cartridges."

Wariness. Fear and wariness can only be separated as a matter of degree. Wariness has also been called fear by Welles and Welles (1961a) in association with "watering behavior," and they go on to say that, "... the degree of wariness in bighorn varies with the individual. This is evident not only in its watering behavior but in its relationship with its entire environment." In relation to age and sex, these authors say, "Nothing seems to suggest a significant variation in wariness by age, class or sex except for ewes with newborn lambs."

Lanny O. Wilson states,

> From my observations in the field, the most wary of the bighorn are the lambs between 3 and 9 months of age. I was never able to approach a young lamb in this age group when it was by itself. Always the lamb would flee immediately, and it was not to be observed again that same day.
>
> The most wary of the adult animals tend to be the rams. It is interesting that ewes when badly startled would run for 180 to 365 m (200 or 300 yd), then stand and look at me. Many times they would resume grazing, paying little or no attention to me, as long as they were in an area where they could get away into ample escape cover. Rams usually took instant flight when badly startled and were rarely sighted again the same day.
>
> There seems to be variation within herds of bighorn in response to different objects. I have seen many bighorn stand and watch large ore trucks, cars, or people walking, and without apparent alarm. At other times I have seen bands of bighorn when sighting these objects take instant flight, not to be seen again.

Senses and Intelligence

Gale Monson reports,

In July 1958 I was making one of our periodical water-hole counts at Tunnel Spring in the Kofa Mountains of Kofa Game Range in Arizona. To facilitate inspection of the spring, which is in the back of a cave up on the side of a canyon, we had anchored a wooden ladder some years earlier. Bighorn watering at the spring had become accustomed to it. One day during the 1958 count, because I did not want the ladder to show in photographs of bighorn at the spring, I removed the ladder and laid it down beside the main approach trail. Not long after, an old ewe led a group of five other ewes and yearlings into water along this trail. She immediately spied the ladder and stopped when within about 4.5 m (15 ft) of it. She gazed at it fixedly for at least 2 minutes, with the other bighorn crowding up behind her. Finally she decided it was not dangerous, and, staying as far as possible from the ladder but still on the trail, she led her band past it.

Jerome T. Light comments, "Isn't it possible that wariness is learned? At Boulder Canyon in the Santa Rosa Mountains I noticed that young, and some mature, bighorn watered in my presence while some older (experienced?) ewes and rams would not do so. In one case, an old ewe, though appearing to need a drink, would not take one, even though others of her group were watering. This ewe always had an eye on me. She was impatient and restless and would not turn her back toward me, a necessary requirement for a drink at this particular place."

Bighorn often will stand or lie and watch people go by without moving until the people are past, and sometimes not until they are out of sight. When startled while feeding along the bottom of a wash, the animals usually will run up onto the side of the canyon wall until they are above the intruder before they stop to stare back at the disturbance. If they are further alarmed, they will move on up and disappear over the top, usually with a farewell look down the slope they have just climbed. However, if they are not disturbed again, they usually will relax where they are and occasionally move back into the wash to continue feeding. While horseback riding in dense pinyon-juniper, I have known bighorn to move out ahead of me, circle, and return almost to the same spot where they were first disturbed. These actions were clear to another person on a high vantage point, although the bighorn were not seen by me as I rode through the woods.

George Welsh comments: "It is sometimes difficult to say whether or not bighorn have abandoned their caution and are not on guard, or whether they feel that they have sized up the situation and have it well in hand with an escape route already picked out."

Response to noise. Sometimes bighorn react to loud noises such as sonic booms, but on other occasions it is difficult or impossible to tell if these same animals even hear the noise. George Welsh states that he has never seen one startled by a sonic boom. Lanny O. Wilson, speaking of bighorn in southeastern Utah, says, "It is interesting to note that sonic booms sometimes startle bighorn, but on other occasions the bighorn pay no attention to them."

Curiosity. Desert bighorn in the pens at Corn Creek Field Station were very curious. They usually could not stay away from some new object placed in their pens. The less inhibited the animal, the more curious he or she appeared to be. On the other hand, certain items that were thought by us to be new to the bighorn were sometimes ignored. When investigating objects, they stood back and looked or watched until they were satisfied that they could approach closer without danger. They used their eyes, ears, noses, and even lips to investigate every detail of some object. Metal objects were usually soon ignored unless the bighorn associated them with something desirable. Sometimes new objects alarmed them, but often they returned and warily approached to investigate such things.

The following comments by George Welsh further illustrate the curiosity of these animals:

> The Willow Beach campground, 22.5 km (14 mi) below Hoover Dam, is managed by the National Park Service. Generally, each Easter, services are held out-of-doors in a dry wash in back of the campgrounds.
> On Easter Sunday in 1964 while the service was in progress a yearling ram made his way over the ridges to the south of the worshipers and stood until the service was over. He was standing about 73 m (80 yd) from where the portable organ was playing and the people were singing. Approximately 30 people were present and saw the bighorn and were amazed that he would come so close to the gathering. For many it was the first time they had ever seen a desert sheep, and their curiosity was probably as great as his.
> Many hunters have reported being "pinned down" by a bighorn for an hour or more while stalking their trophy and in fear of alerting the whole band; and on several occasions I have been "frozen" in position until my back and legs cramped. Generally when this happens the bighorn either seem to satisfy their curiosity and conclude that nothing is really wrong, after which they go on about their business, or else they become tense and run off or walk "calmly" over the ridge and, once out of sight, leave the area.

Lanny O. Wilson says: "It was apparent that bighorn are curious animals but do not let their curiosity get them into a situation where they cannot get away readily. Many times I have approached to within 15 to 23 m (50 to 75 ft) of bighorn without scaring them as long as I was in plain sight. Whenever I had to drop into a gully and come up on the other side and was out of sight for a few moments, they would take instant flight. Many times I have sighted bighorn more than 1.6 km (1 mi) away, and by waving my shirt, a red hat or coat, I attracted the animals to within 90 or 180 m (100 or 200 yd). In doing this, it sometimes would take several hours before the animals would get close to me, because they would lie down and watch me for long periods of time. It is interesting that while this particular technique would work on one band of bighorn, another band would show no apparent interest."

Welles and Welles (1961a) are more skeptical about curiosity in bighorn when they say: "Much has been said about the curiosity of the bighorn, but it is difficult to draw a line between a curiosity tinged with wariness and a suspicion tinged with anxiety. An effort to determine whether something is dangerous seems more an act of suspicion or caution than of simple curiosity. A ewe with a newborn lamb may make a great effort to keep an intruder in sight, not from curiosity but from apprehension, and yet if the observer is unaware of the lamb's presence he might be inclined to label the motivation of the ewe's action as curiosity." Further on, these authors say, "It would appear that curiosity in bighorn varies with the individual and is seldom the sole motivating factor in bighorn behavior. It is therefore difficult to discuss except in its relationship to anxiety, suspicion and fear."

9. Behavior

Norman M. Simmons

Generally, there are two methods for studying the behavior of animals: observation and experimentation (Scott, 1963). The observer systematically records basic facts which suggest explanations; these may then be tested by experiment. Few researchers or wildlife managers have systematically recorded behavior patterns of bighorn in North America, let alone patterns of desert bighorn. Much of what is said in this chapter about desert bighorn behavior is derived from studies by Ralph and Florence Welles in Death Valley National Monument, California; by Charles G. Hansen at Desert National Wildlife Range, Nevada; and from my own studies on Cabeza Prieta National Wildlife Refuge (formerly Cabeza Prieta Game Range), Arizona. This material is supplemented by information available on the behavior of Rocky Mountain bighorn, Dall sheep, and domestic sheep.

Progress in the experimental study of desert bighorn behavior is almost nil. Some field experimentation with bighorn and Dall sheep has been done in the northern Rocky Mountains, and thorough field and laboratory experiments have been conducted with domestic sheep. Some results of these experiments will be discussed briefly here.

I consider behavior as response to change. In the desert, changes in the environment of the bighorn are relatively great from season to season, and since these changes are important for the survival of the bighorn, adaptive responses to environmental changes are stressed.

SEASONAL ACTIVITIES

The rut. Desert bighorn have a fall rutting period and are relatively sexually inactive at other times. The ewes will accept the ram only during the "heat" period of her estrus cycle. If she is not bred during her first cycle, she may come into heat at regular intervals during the rut. The ewe bears the entire maternity burden, and it is from her that the lamb develops its basic learned behavior patterns. See Chapter 10 for a discussion of reproductive behavior.

The further south and the lower the elevation that bighorn are found on the North American continent, the earlier and longer is the rutting season. Welles and Welles (1961a) state that the rutting period for desert bighorn in Death Valley "begins in late June, increases in intensity rather sharply through July, maintains a fairly high level through September and October, and gradually declines through November to subsidence sometime in December." Sporadic rutting activity was observed in January, February, and March. Russo (1956) notes that sexual behavior was first evident in Arizona in late June or early July. Breeding activity peaked in August and ended in December.

The peak of the breeding season in southwestern Arizona coincides with the late summer rainy season, which probably is the most favorable for successful breeding (Simmons, 1969a). Leaf formation after late July or early August rains helps insure successful lambing later by improving the physical condition of ewes (Morrison, 1951).

Ewes and lambs. The mother-young bond appears to become established immediately after birth, when the mother eats the placenta and licks the newborn lamb (Etkin, 1964; Welles and Welles, 1961a). Even a slight interference with this relationship may result in the failure of the mother to respond completely to the needs of the young lamb. The same situation has been observed in penned bighorn at Desert Wildlife Range (Charles G. Hansen).

Young lambs may learn to associate specifically with their own mothers because they are rejected by other ewes. Thus, mountain sheep herds have as their primary group the maternal family, in which the young are associated with their mothers, and these, in turn, with their own mothers as long as the latter survive (Etkin, 1964; Hansen, 1965a). The mother remains the leader of her young (Fig. 9.1). The old female with the largest number of descendants consistently leads the flock (Hansen, 1965a; Hunter, 1964; Scott, 1963).

Norman M. Simmons

Fig. 9.1. The mother ewe remains the leader of her young, as in this family group of a matriarch ewe, her yearling ram, and a 4-month-old female lamb. Agua Dulce Mountains, Pima County, Arizona.

Usually there is a temporary breakup of the original family relationship during a period including late pregnancy and parturition. The ewe may adopt a negative attitude toward her yearling when it attempts to stay close to her, and she may drive it off by threatening gestures or butting. However, V. Geist observed Rocky Mountain ewes driving off not their own but other yearlings. He states that the separation of pregnant ewes and ewes with newborn lambs from other bighorn occurred mainly because these ewes simply did not participate in the activities of the juveniles and barren females.

Seasonal changes in group size and composition. Desert bighorn usually separate into groups of adult rams and groups of ewes and juveniles shortly before the lambing season (Deming, 1953; Russo, 1956; Welles and Welles, 1961a). According to Deming, it is at this time that the largest bands form.

Groups of adult rams on Cabeza Prieta Wildlife Refuge were generally smaller than groups of ewes and juveniles, which in turn tended to be smaller than mixed groups. However, Deming (1953) states that ewe groups tended to be smaller than ram groups in Desert Wildlife Range.

Rams mingle with ewes during the late summer and fall rut, and bighorn of both sexes and all ages mingle in temporary concentrations at water holes during hot, dry summer days. Though Deming states that bands are smallest in the summer and fall in Nevada, in other areas of the desert the bands may be smallest during the cool months of the winter when the animals are widely scattered.

The average number of bighorn per group varied between two and three each month throughout the year on Cabeza Prieta Wildlife Refuge. The sizes of the groups ranged from one to sixteen bighorn through the year, but variability was low. Of 502 observations, 65 percent were of single bighorn or pairs; 78 percent were of groups of three or less. On Kofa Game Range and Organ Pipe Cactus National Monument, both in Arizona, the majority of bighorn also were found in groups of three or fewer individuals. However, the yearly average size of the groups was significantly greater than that on Cabeza Prieta Wildlife Refuge (an average of 1.4 animals per group larger). One reason for the greater average size was that, on Kofa Game Range and on the Monument, bighorn gathered into much larger bands of rams and bands of ewes and juveniles in the spring than they did on Cabeza Prieta Wildlife Refuge. Larger average desert bighorn group sizes in areas north and east of Cabeza Prieta Wildlife Refuge reflect better habitat conditions. Group sizes are probably good indicators of population "health" and the condition of bighorn range (Fig. 9.2).

Rocky Mountain bighorn in British Columbia were found by Geist (1968b) to be separated out in groups on the basis of their position in a gradient ranked mainly on horn and body size (Fig. 9.3). Individuals at the two ends of the gradient form separate bands. Mature rams group themselves in bands that do not associate with bands composed of ewes, yearlings of both sexes, and lambs, except during the breeding season, when there is interaction between adult rams and estrous ewes. This explanation of seasonal changes in group composition is strengthened by the observations of Augsburger (1970), in his work with desert bighorn in New Mexico.

SEASONAL MOVEMENTS AND HOME RANGES

Seasonal drift. Desert bighorn in most or all of their present-day range may be nonmigratory. However, there is a gradual seasonal movement of some, but not all, members of each band between seasonal ranges. The term "seasonal drifts" has been used to describe such movements (Honess and Frost, 1942; Simmons, 1969a).

During much of the year, groups of ewes with juveniles and groups of rams exhibit different seasonal movement patterns. The range of the adult ewe is considerably reduced just before and for a short time after her lamb is born. In Nevada (Deming, 1953), Arizona (Eustis, 1962; McMichael, 1964;

Norma M. Simmons

Fig. 9.2. A group of sixteen rams in Burro Canyon, Kofa Mountains, Arizona, an unusually large group of rams for the southern Arizona desert. Such congregations are more often found in mesic habitat farther north.

Russo, 1956), and probably elsewhere, ewes concentrate in bands on rugged "lambing grounds." The rams may move to lower, gentler terrain, as on Kofa Game Range.

Eustis (1962) finds that during 5 years of observations on Kofa Game Range, 84 percent of the ewes and 93 percent of the lambs were seen in the upper third (elevation) of the Kofa Mountains during lambing seasons. Sixty-two percent of the rams were seen in the middle third of the mountains, and 9 percent were seen in the lower third.

On most of Cabeza Prieta Wildlife Refuge, ewes did not concentrate on "lambing grounds." Flights over the mountain ranges in helicopters and fixed-wing light aircraft during the peak of the lambing season showed that

Behavior 129

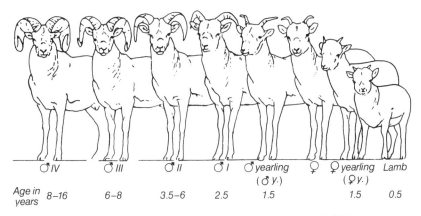

Fig. 9.3. An illustration of bighorn of both sexes and different ages arranged in a cline. From Geist (1968).

the ewes were well scattered throughout the main part of each range. Ewes on the Refuge do not need to concentrate on the escape terrain they seek for lambing areas, for each range is quite precipitous from end to end and from foot to crest, and vegetation is evenly distributed throughout the rugged slopes.

During hot and dry summer days throughout desert bighorn range, bighorn concentrate in the vicinity of perennial water. The degree of concentration or the density of the bighorn bands near water holes is related to the duration of the dry periods and the degree of heat involved, as well as the size of the population and the distance between water holes. On Cabeza Prieta Wildlife Refuge, as many as twenty-eight bighorn have lived for many days within a short distance of a water hole.

When reliance on perennial water holes is reduced or eliminated during the summer by rainfall, bighorn expand their ranges and "drift" away from the water holes. Bighorn become difficult to find during these periods and during the winter, especially in some of the sparsely populated ranges of Nevada and southwestern Arizona.

Some bighorn on Desert Wildlife Range leave the Sheep Range (Mountains) in the fall, move 16 to 24 km (10 to 15 mi) south to winter, and then return before summer. Observations of marked animals indicated that some may travel an annual circuit of 32 km (20 mi) or more. Rams that summer near Cow Camp in the Sheep Range travel to the Desert Range (Mountains) in the fall to winter. Some of these rams make a minimum annual circuit of 14.5 air-line km (9 mi) (Hansen, 1965b).

Vertical seasonal movements also have been observed in Nevada (Hansen, 1965b; Koplin, 1960). Desert bighorn in southern California move up in elevation as the summer progresses, and downward in the spring to the best watered canyons and best forage areas which seldom are above 760 m (2,500 ft) (Goodman, 1962). No such vertical seasonal movements occur on the low, rugged mountains of Cabeza Prieta Wildlife Refuge.

Dispersals. Robinette (1966) defines dispersals by mule deer as movements to new home ranges. Only one-way movement by desert bighorn across valleys between mountain ranges, which are relatively easy to identify, are considered as dispersals. The definition excludes journeys to critical resources such as water, the movements of rams between bands of ewes, and seasonal drifts involving return trips. Robinette says that dispersals apparently serve to reduce inbreeding, allow invasion of new habitats or reinvasion of formerly inhabited areas, and facilitate redistribution following localized decimations.

Desert bighorn dispersal movements have been documented in California, Nevada, and Arizona, and no doubt they occur in other states and in Mexico. Dispersals across wide valleys occur on Desert Wildlife Range, even though bighorn must cross highways in the process.

Bighorn cross between mountain ranges on Cabeza Prieta Wildlife Refuge only occasionally (Fig. 9.4). Those that make these crossings do so quickly and by the shortest routes possible across flat terrain. In the late winter of 1964, four bighorn traveled about 10 km (6 mi) from the Aguila Mountains to the Granite Mountains in a nearly straight course. They did not meander or pause to feed. A small band made the crossing between the Sierra Pinta and the Sierra Cabeza Prieta across the shortest routes between precipitous hills. Very few bighorn droppings were found along their trail. The animals were tense and sprinted in short spurts on several occasions during the 13-km (8-mi) crossing. Most of the valley crossings (67 percent) observed in southwestern Arizona were made during mild weather when environmental stresses were probably minimal (Simmons, 1969a).

Dispersal movements are uncommon, however, and the average desert bighorn probably spends its lifetime within a relatively small area. It is likely that most bighorn in Death Valley, for example, live and die within 32 km (20 mi) of their birthplaces (Welles and Welles, 1961a).

Home range. Home range is defined by Jewell (1966) as "an arena for activity with spatial qualities that vary throughout its extent, and that is subject to great irregularity in intensity of use." The area meets the energy requirements of the individual or group that occupies it. The home range therefore is not an easily delineated parcel of terrain but has ill-defined borders and shifting activity centers. Though territories (defended portions of home ranges) are established by many species of mammals, the desert bighorn, and most other North American ungulates, do not seem to be territorial.

A few generalities that describe the home range of most hoofed mammals can be applied to desert bighorn: the males range more widely than females; the areas used by adults are larger than those covered by juveniles, except when the young travel with adult females; and the range of the adult female is considerably reduced during the lambing season (Sanderson, 1966).

Norman M. Simmons

Fig. 9.4. Bighorn occasionally cross between the Sierra Pinta and the Sierra Cabeza Prieta. In this photo the observer is looking from the Sierra Pinta across the Tule Desert to the Sierra Cabeza Prieta in the distance. Cabeza Prieta National Wildlife Refuge, Arizona.

Charles G. Hansen has observed the gradual expansion of the home ranges of young captive bighorn from the small areas known to them as lambs to the relatively large areas they became familiar with as young adults. Lambs raised by the Hansens at Corn Creek Field Station on Desert Wildlife Range claimed the house as the central portion of a home range that was bounded by the picket fence surrounding the yard. The lambs would follow their foster mother (Mrs. Hansen) out of the yard but would hurriedly return when alarmed. They gradually expanded their ranges as they grew older.

According to Hansen, wild lambs on Desert Wildlife Range leave their initial range after they are about a month old and follow their mothers in large groups (sometimes as many as 100 animals) which move to higher altitudes. There they establish new home ranges. V. Geist states that 2-year-old Rocky Mountain bighorn ewes and some 2- or 3-year-old rams roam well outside their home ranges for short periods.

Wilson (1968, 1971) observed movements of individually identifiable bighorn ewes in southeastern Utah for lengthy periods in 1965 and 1966. One ewe during a 41-day period in the summer covered an 11-sq-km (4.2-sq-mi) range; in winter the same ewe ranged over 48 sq km (18.6 sq mi). Some bighorn were not observed outside a particular major drainage during Wilson's entire 2-year study; however, others left their home ranges during droughts, when all available water holes had gone dry. Wilson also observed rams that stayed within a 4.8-km (3-mi) area on Wingate Mesa during the summers of 1965 and 1966 but left this area during the rutting period.

Desert bighorn living in the Sheep Range of Desert Wildlife Range have a normal daily cruising radius of about 1.6 km (1 mi) from their overnight bed-grounds, according to Charles G. Hansen. Some bighorn travel only a few hundred meters from bed-grounds during a day, depending on the proximity of food and water.

The summer range of bighorn in Lake Mead National Recreation Area, Nevada, was a little over 0.8 km (0.5 mi) in radius in 1963 and 1964 (Denniston, 1965). During July and August in the Santa Rosa Mountains, California, adult ewes and juveniles stayed within 1.3 km (0.8 mi) of water holes. Rams traveled as far as 5 km (3 mi) from water in July and early August (Blong and Pollard, 1968).

Four ewes and three large rams traveled in an area 0.3 by 0.6 km (0.2 by 0.4 mi) during 8 days in late August 1964, in the Ajo Mountains of Organ Pipe Cactus National Monument, Arizona, when the mean daytime temperature was 27°C (81°F) with a range of 21° to 33°C (70° to 91°F). These animals were in a well vegetated area and within 1.2 km (0.8 mi) of water (Simmons, 1969a).

The activities of thirty adult ewes on Cabeza Prieta Wildlife Refuge were followed during daylight hours of the summer for periods of from 2 to 10 days. These bighorn either had been marked with dyes or could be identified by unusual marks on their bodies. The average distance between the most widely separated points in the summer ranges of twenty-one of these ewes was 0.5 km (0.3 mi) for an average period of 3 days. The observations indicated that a ewe's summer range may vary from about 0.16 to 1.6 km (0.1 to 1 mi) in diameter and that the ewes probably did not travel more than 1.6 km (1 mi) from water during the dry summer season. Little difference was observed between the ranges of different ewes. Travel usually was confined to one major drainage during periods of 10 days or more. The ewes under observation included both "dry" and "wet" ewes, yet no significant difference in the sizes of their ranges could be detected (Simmons, 1969a). It is apparent, however, that nursing ewes need water more often than dry ewes (Dukes, 1955), and this need probably is reflected in the sizes of their home ranges.

The home range limits mentioned above were both for ewes that had watered during the observation periods and for those that had not; thus, the state of hydration of the ewes is not reflected in the estimate of the size of their home ranges (Simmons, 1969a). The observations of four ewes that returned to the same water holes on separate occasions from 4 days to 1 year apart indicate that some ewes have "home waters" to which they return year after year. At least some ewes were found to be aware of watering places other than their "home water." Observations by Hansen (1965b) and others of marked bighorn on Desert Wildlife Range strengthen this hypothesis.

Observations of bighorn rams in the Sierra Pinta and Agua Dulce Mountains, Cabeza Prieta Wildlife Refuge, during June, July, and August indicate that rams are not as restricted in range as ewes and juveniles and that they move more freely throughout the main parts of each mountain range (Simmons, 1969a). Wilson (1968) and others also have found this to be true in Utah.

On occasion rams may wander widely. One about 2.75 years old was captured in the south Plomosa Mountains, Arizona, on 18 November 1977 and fitted with a collar supporting a radio transmitter. His movements were tracked for the next 12 months, during which he moved southeast through the Kofa Mountains (with one slight backtrack of a few kilometers) to the southern Tank Mountains by 9 April 1978, 70 km (43 mi) from the capture site. He remained in the Tank Mountains until 19 May, when he traveled in 5 days northwestward to the western part of the Kofa Mountains, where he was still present 26 November 1978 (James H. Witham, E. Linwood Smith). Also see Chapter 16, for an account of another well traveled bighorn, whose sex unfortunately was not recorded.

Annually or seasonally, certain populations of desert bighorn may leave a particular mountain range for another, which may be as far as 50 to 80 km (30 to 50 mi) distant, for better food and water conditions at the particular time, the range they are deserting being perhaps the better supplied at other times of the year. Some evidence for such an occurrence has been reported by Gale Monson who, when on Cerro Pico in northwestern Sonora in July 1975, found abundant evidence that bighorn were plentiful there in the preceding winter but had completely left the mountain before the spring was over. In this region summer rains are usually quite localized, and cooler average temperatures are found closer to the Gulf of California, two possible reasons for extensive movements.

DAILY ACTIVITIES

Communication. Bighorn usually are silent, especially the rams, but like most social animals they are able to maintain contact with one another at considerable distances. They have certain postures and behavioral patterns which cause other bighorn to respond, though such "communications" may only be simple expressions of emotional states. Little systematic work has been done on this subject.

Charles G. Hansen noted certain postures and other signals when he studied the behavior of penned and free-roaming bighorn on Desert Wildlife Range. When a bighorn stood with its nose extended and eyes appearing to bulge, it was signaling a belligerent emotion. A similar emotion is signaled by the stamping of a front foot (see also Welles and Welles, 1961a). Hansen says that the pawing of another animal by a bighorn may be an expression of hunger for food.

Lambs make up the most vocal segment of the bighorn band. According to Welles and Welles (1961a):

> The lamb's voice is the expected appealing complement to the general concept of the lamb's personality and is probably used more before weaning than during any other period of its life.... Once the lamb is weaned, its voice is no longer useful in securing food but is still employed in the quest for security by calling for help when lost....
>
> Returning in our jeep truck to Joe May camp in the Desert Game Range on a moonless night in March, we came directly upon a large band of ewes and lambs bedded among the Joshua trees on a knoll at the edge of the wash. We shut off lights and motor and listened for 15 minutes to a veritable din of anxious bleating, calling, and answering, which gradually subsided into quieter and plainly recognizable tones of recognition and reassurance, then finally silence when the last anxious mother touched her lamb in the dark.

The bleat of an adult ewe is deeper than that of a lamb and is "sometimes so vibrato as to be almost comical" (Welles and Welles, 1961a). Charles G. Hansen states that ewes make a mewing sound of recognition or satisfaction when they are reunited with their lambs after a period of separation.

Rams may have a deeper-toned bleat than ewes (Welles and Welles, 1961a). They make a growling sound which is low-pitched and "coarse" and, according to Hansen, may be used as a greeting or sign of recognition. Welles and Welles describe a nasal growl used by rams to communicate uneasiness. The sound is also made when two or more rams are sparring. V. Geist believes that the growling expresses dominance. An old ram held in captivity for years at Desert Wildlife Range used the call when approached by humans (Charles G. Hansen).

Wilson (1968) reports that when bighorn are nervous or frightened they emit a sound that resembles that of "two rocks being scraped together." This was usually accompanied by foot stamping. Other biologists have described this alarm sound as "coughing." Bighorn also blow sharply through their nostrils to express alarm, like deer.

I have been successful, as have others, in calling bighorn toward me over considerable distances by imitating their bleating. The greatest success occurs when a human, seeing that a ewe has been separated from its offspring, imitates a lamb, or when a human imitates a ewe when a lone lamb is in sight.

Scent undoubtedly plays an important role in communication. Charles G. Hansen says that bighorn rub their preorbital glands on rocks or trees, and other bighorn appear to seek out and smell these places. V. Geist has observed bighorn rubbing their preorbital glands on other bighorn. The preorbital gland secretes a waxy, odoriferous substance.

The communication through urination that a ewe is in estrus is discussed elsewhere. Urine "signposts" have been described by Welles and Welles (1961a). Hansen says that rams leave scent signals by spraying urine on the ground over an area about 0.6 m (2 ft) in diameter under their bodies.

Hansen notes that during the rutting season rams occasionally but purposefully spray urine on themselves. The urine is sprayed on the front and hind legs, and some is splashed on the brisket and abdomen. These wet places become caked with dirt and perhaps oil from the hair, and the animal acquires a strong odor. Similarly, dirt is deposited on the ewe during breeding, discoloring her rump. No other unusual behavior accompanies this act. Other bighorn may smell the urine-wet ground and the damp hair of the ram, and then back off and execute the lip-curl posture (see Chapter 10). Hansen says that it is not unusual for a ram to allow the urine of another ram to wet his face. The significance of these acts is unclear, but Hansen assumes that they are a form of communication.

Social dominance and leadership. Etkin (1964) says that the "great importance of dominance in social life is that it acts as an organizing principle which minimizes aggression by, in effect, securing to the dominants the fruits of victory without disrupting group life by conflict." Dominance orders also result in a conservation of energy. Although dominance orders in desert bighorn have yet to be thoroughly studied, such organization has been recognized by a few researchers.

Augsburger (1970) and Geist (1971) find that dominance orders in groups of Rocky Mountain bighorn are set up between and within the sexes and seem to be based primarily on the recognition of horn and body size. A ewe or ram that approached a dominant ram (i.e., one with larger horns and body) was often greeted with an outstretched and lowered head. This gesture may be a means of displaying horns as dominance rank symbols (Geist, 1971). It was noted in most of the contacts between rams of different horn and body sizes on Cabeza Prieta Wildlife Refuge. In all such contacts, I observed that the ram having smaller horns retreated from the dominant ram. Rams with larger horn and body sizes dominated ewes in all observations.

The dominant rams in a band seemed to be the most successful in courting ewes during the rutting season on Cabeza Prieta Wildlife Refuge. Of 39 courtship displays observed, 85 percent were performed by rams of three-quarter horn curl or larger, 8 percent were performed by half-curl rams, and 7 percent by quarter-curl rams. Geist (1967) observed that most of the rams he saw copulating were about full curl. Rams older than 7 years (the approximate time needed to produce large full-curl horns) may be the most productive members of a bighorn herd in unharvested populations, such as those found in national parks and other nonhunted areas.

Though dominant rams are usually older than their subordinates, this is not always the case. Dominant rams with abnormally massive horns for their ages have been observed on both Kofa Game Range and Cabeza Prieta Wildlife Refuge. Geist (1966) described long-term observations of young massive-horned rams dominating smaller-horned elders in the same group.

Conflicts observed on Cabeza Prieta Wildlife Refuge in which both participants were aggressively engaged usually involved bighorn of nearly equal horn and body size. If horn size is a recognized rank symbol, rams of equal horn size may have no other means of settling the question of dominance than by clashing. Clashes between both desert and Rocky Mountain rams of unequal horn size quickly ended in the smaller-horned animal's submission.

More stable dominance order probably exists among ewes and juveniles than among adult rams. Groups of ewes and juveniles may remain together for years, and group members learn their positions gradually. Banks (1964) found stable, linear dominance orders among domestic ewes, and observations on Cabeza Prieta Wildlife Refuge indicated that the same situation exists among bighorn ewes.

During pleasant weather on Cabeza Prieta Wildlife Refuge, or after rains when pothole water was plentiful, bighorn scattered throughout each of the separated mountain ranges. At such times there were few, if any, occasions when they banded together at water and set up dominance orders. During summer dry periods, however, the animals were forced to crowd around a few water holes in groups of as many as fifteen. This was one of the few occasions when family groups intermingled and dominance orders became unclear. When seeking shade near a water hole or when moving into position to enter a tank, the bighorn got in each other's way and set up hierarchies in which the "weakest" animals were compelled to make way for the dominant ones. Jostling and butting were common sights at water holes in the summer, but such activity was not often seen elsewhere on bighorn range (Charles G. Hansen; Simmons, 1969a).

Dominance should not be confused with leadership, though with some animals like red deer *(Cervus elaphus)* and bighorn, older females who are leaders are sometimes also dominant. Leadership among desert bighorn seems to be a changing phenomenon that can be recognized with certainty only when movements over a considerable distance are observed and when the leader can be identified over a long period of time. Observations in Canyonlands National Park, Utah, identify the largest-horned rams as dominant and the most commonly followed (Dean and Spillett, 1976).

As mentioned earlier, several authors suggest that groups of bighorn are formed around family units, with those ewes followed by the most offspring becoming group leaders (Hansen, 1965a; Hunter, 1964; Scott, 1963). No evidence of band leadership by desert bighorn rams has been described. Welles and Welles (1961a) vividly describe examples of leadership by ewes and state that rams of all ages appear to accept the temporary leadership of ewes whenever both sexes travel together.

Aggression. The major function of aggressive behavior is to determine or maintain rank or territory (Etkin, 1964). With desert bighorn, aggression usually concerns dominance orders, since these animals do not maintain territories. Aggressive behavior between desert bighorn, particularly the spectacular clashes between rams during the rutting season (Figs. 9.5, 9.6) so well described by Welles and Welles (1961a), has received much attention but is not common; in some areas that are sparsely populated it probably occurs only rarely (Simmons, 1969a).

Ewes sometimes fight viciously with each other, hooking and butting with their horns. Sometimes they seem to imitate rams in head-on charges, but such bouts are uncommon (Simmons, 1969a).

DAILY MOVEMENTS

It is not easy to generalize about daily movements of desert bighorn. As some frustrated observers have said, one of the few things you can predict about bighorn is their unpredictability. The timing of their daily activity depends on numerous variables: the temperature, the amount of feeding and resting the previous night, the density of palatable vegetation present, and so forth.

A general description of daily activity would run as follows (Welles and Welles, 1961a): "The bighorn are likely to rise at dawn; if young are present, they may play for a short time and on occasion be joined by adults; they may feed for a period of from 1 to 3 hours, then rest for 1 to 3 hours. Alternating of feeding and resting periods continues throughout the day, with the band usually climbing to a higher elevation for bedding at dusk."

Tommy Hailey, Charles G. Hansen, and Wilson (1968) report that the bighorn they studied in Texas, Nevada, and Utah, respectively, had fairly regular and similar daily activity patterns. During 39 days of observation in southeastern Utah, Wilson found that bighorn began rising to feed before daylight and would move downhill until between 8:30 and 10:00 A.M. Then they would bed down for 1 to 3 hours. Another 1- to 3-hour period of bedding would occur sometime between 1:00 and 3:00 P.M. In the evening they would feed and drift uphill to a bedding area, then bed down just before dark. This

Philip L. Shultz

Fig. 9.5. Two rams about to "clash" at the Red Rock big game research area, Grant County, New Mexico.

pattern was similar to the activity pattern of Rocky Mountain bighorn in Colorado (Simmons, 1961) but less irregular than the activity patterns observed in the hotter deserts of California and Arizona.

Wilson (1968) agrees with most observers when he states that bighorn observed in Utah would rest at irregular intervals throughout the day, often with no effort being made to seek a regular bed site. The bedding periods last up to 45 minutes.

Watering activity follows no discernible pattern during the cool or rainy periods of the year. A fairly predictable pattern can be observed on hot, dry days, however.

Tommy Hailey, Charles G. Hansen, and others have observed that bighorn usually paw out a small oval area for a bed before lying down. Jack Helvie observed captive day-old lambs pawing beds. At that age they usually made only two or three swipes at the ground before bedding. Hansen's captive ewe lamb was a week old before she began pawing beds. The older she

Jack R. Cooper

Fig. 9.6. Occasionally combat horn-clashing does not result in a head-on collision. Injury to one or both contestants may result from such "misses." River Mountains, Nevada.

became, the more consistent was her bed-pawing. Hailey observed bighorn using their horns to dig in the loose dirt of a prospective bed. He also noted that the beds usually were on a ridge crest or on the adjacent slopes.

Charles G. Hansen made detailed observations of bedding activities of captive and free-roaming bighorn in Nevada. He writes:

> Bighorn often use the same bed time after time, or else use other beds that have already been made. They will scratch out the center of a bed, regardless of whether the bed is new or old and well-used. Nighttime beds usually are on slopes just below a ridge top. The surrounding vegetation and terrain allow for good visibility and quick or easy escape. The ground is usually covered by coarse gravel that makes it almost impossible for predators to move about easily or silently. The beds are often made within a few yards of one another, and the young lambs often bed down against their mothers.
>
> In winter, night beds are most frequently found on the south slopes. In the summer, day beds are made in the shade of trees or rocks. As many as three or four beds may be found on the north side of a tree, indicating that one bighorn moved three or four times to stay in the shade as the sun moved across the sky.... During winter feeding activities ewes may move

frequently, and resting bighorn often make (and abandon) new beds every few minutes in an effort to keep up with the band.... Bedding may occur any time of the day even while others are feeding or moving around. It is not unusual to see one animal lie down and the rest of the band do the same. One bighorn may force another from its bed and not make a bed for itself.

Bighorn often use caves and rocky overhangs for shelter. Some caves on Desert Wildlife Range have pellets several inches thick on the floors, apparently the accumulation of many years.

Usually bighorn lie down with feet tucked up under the body and the head erect. (When completely relaxed, they rest the chin on the ground. Rams rest the chin and one horn on the ground.) When relaxed, bighorn frequently extend one front leg forward and out from under the body. When they lie down, they drop on their front knees first and then settle back on their hindquarters. The front part of the body is then lowered to the ground. The four legs drawn up under the body allows the animal to rise rapidly and be gone from the bed in the shortest possible time.

I do not recall ever seeing an adult bighorn lie on its side, with its head on the ground and feet outstretched, although lambs occasionally lie in this manner.

It is possible that bighorn may sleep standing up, and with their eyes open. It is not unusual to see them standing still, apparently staring out into space. I was able to approach an animal that was standing like this until I was about 4.5 m (15 ft) away. From that distance I spoke quietly several times. Although its eyes were open all the time, only after I had spoken to it did it turn its head and look at me. It had the appearance of a person being awakened in a strange place. It looked at me, then opened its eyes wide, and began to look for a way of escape. I walked past it slowly while it watched me. Then it moved away along the trail I had come up. It did not run or hurry, but acted like it was confused by my presence so close to it. Lambs raised by hand would appear to be asleep with their eyes open, and did not blink or react when a hand was moved in front of them."

Welles and Welles (1961a) noted that playfulness may begin when a bighorn lamb is only a few hours old. Playing among young bighorn reaches its peak in the spring when food is easiest to get, in the cool hours of a day, and when playful companions are at hand. They write:

On February 9, 1958, a band of 14 blond fat bighorn were reveling in the lush vegetation of a "good flower year" on Death Valley Buttes. All afternoon they had basked in the sun, but as the day had cooled off their activity increased. The two older rams began to challenge each other in true bighorn fashion; and the little ones even tried now and then to emulate them, standing on their hind legs, cocking their heads sideways, and advancing a few halting steps toward each other, and even bumping horns a little as their forefeet dropped back to earth.

As darkness approached, their playing increased still more. All ages and sexes began butting playfully at each other, bouncing into the air, twisting and kicking as they sailed downhill for 15 or 20 feet at a bound. Sometimes they went nowhere, and just jumped up and down in one place, twisting and turning as they did so.

Presently they began destroying the lush new green of the desert shrubs in their exuberance, trampling, pawing, and thrashing their horns in them and sometimes grabbing large mouthfuls before darting away. This activity continued until after dark and began again at daybreak, when for over an hour we were treated to the greatest playing exhibition we have ever seen. Round and round on a steep loose talus slope they raced, leaping, bounding, kicking, butting the air, each other, plants, rocks, or just butting. The 3-year-old rams backed off and "let each other have it" from about 10 feet apart with a clonk perfectly audible from where we watched three-quarters of a mile away.

They found a crumbling cliff at the edge of the slope and there the youngsters gathered, seeming to enjoy the feel of the edge giving way under their feet, leaping off into space with the greatest delight, and scampering around and back up again. There was much shoving going on too, apparently in an effort to shove each other over the edge. The two 3-year-olds joined in this and were pushed about even by the youngest of the merrymakers.

The six older ones, the ewes, disappeared into a ravine leading up to the top of the ridge, where first the six lambs and then the two rams soon followed. Within what seemed no more than a few seconds we first heard, then saw, all 14 racing pellmell across a talus slope a quarter of a mile east of the ravine into which they had disappeared. Even the old "ladies" were playing then, as round and round they raced for several minutes. Then as though at a signal they all stopped and began to graze.

V. Geist notes that such exuberant play indicates the well-being of the bighorn, reflecting an abundance of food and a lack of significant stress (Fig. 9.7).

THE ENVIRONMENT AND DAILY AND SEASONAL ACTIVITIES

Heat of summer, limited supplies of water, and sparsely distributed vegetation are the environmental factors that most seriously restrict the activities of desert bighorn, especially in the hottest deserts of southwestern North America.

Individuals of the genus *Ovis* seem to rely upon panting as the primary means of heat dissipation (Brook and Short, 1960a). In order to survive hot, dry summers, desert bighorn must combine evaporative cooling through panting and sweating with avoidance of solar radiation and reduction in metabolic heat gain. To do this, bighorn bed in the shade or in a cave (Fig. 9.8), and usually remain relatively inactive during the hottest part of each day. Sometimes they feed at night.

During hot days on Cabeza Prieta Wildlife Refuge, over 90 percent of observed bighorn bedded between noon and 5:00 P.M., the period of greatest solar radiation and highest temperatures. During daylight hours before 8:00 A.M. and after 7:00 P.M., few were seen bedded. The bighorn spent an average of 7 hours and 5 minutes in bed each hot day during daylight hours. On cool days, they bedded an average of 4 hours and 35 minutes.

Robert M. Craig

Fig. 9.7. A ram lamb exhibits playful behavior by leaping in the air before his keeper at the Arizona-Sonora Desert Museum.

Norman M. Simmons

Fig. 9.8. A yearling ram at entrance to cave used for shade and relief from the sun and dessicating winds on hot days. Near Eagle Tank, Sierra Pinta, Yuma County, Arizona. The plant is brittle bush *(Encelia farinosa)*.

When bighorn did not bed in caves, they most often bedded on light-colored rock or, if that was not at hand, on light-colored, coarse soil. The temperatures of these surfaces were much lower than those of dark-colored material and silt.

During hot, dry periods of the summer much feeding was done at night. Nighttime movement, which commonly occurred on the Wildlife Refuge during both moonlit and moonless nights, is another method by which bighorn avoid solar radiation, minimize water needs, and still satisfy food requirements. The bighorn were rarely relocated at first light where they were left bedded the previous evening. However, Gale Monson reports finding individual ewes in the early morning on at least two occasions at the same location on Cabeza Prieta Wildlife Refuge where they were last seen at dusk the previous evening. Monson (1964) and others have seen or heard bighorn moving about late on moonlit and moonless nights in both summer and winter on Kofa Game Range and Cabeza Prieta Wildlife Refuge.

Charles G. Hansen has observed nocturnal movement by bighorn in Nevada. Tommy L. Hailey and Wilson (1968) report that nighttime movement was uncommon and that the bighorn usually were found in the same area they bedded in the previous evening. Welles and Welles (1961a) report almost no nocturnal movement in the Death Valley region. Wilson believes that bighorn in Utah rarely move about on dark nights but occasionally move on moonlit nights.

BEHAVIOR AND TOPOGRAPHY

Because slopes of the narrow, rugged mountains of Cabeza Prieta Wildlife Refuge face east and west, the difference in temperatures on either slope in the morning and afternoon is quite pronounced. Most bighorn responded to the earlier warming of ambient temperatures on the east-facing slope by watering sooner in the morning than the bighorn on the west-facing slope. Bighorn on the east-facing slope began moving around earlier in the evening than those on the west-facing slope, and those that watered on the east-facing slope began watering about 2 hours earlier. The peak of the watering activity on the east-facing slope was reached in the morning; the peak on the west-facing slope was reached in the evening.

Bighorn moved from slope to slope on Cabeza Prieta Wildlife Refuge to take advantage of exposure-influenced temperature differences during the months before the summer dry season drew them down to perennial water holes. Charles G. Hansen observes that they seek breezes on a ridge top when they are hot, just as domestic livestock do.

Bighorn in southwestern Arizona tend to live in isolated groups, largely because of the physical discontinuity of the habitat. Food and water resources in suitable habitats are confined to small mountain ranges, separated by wide valley expanses. As a consequence, bighorn have developed local behavior patterns suited to the island-like habitats they occupy.

The pronounced relief of the mountain ranges of the Mexican border area and the characteristics of the desert climate make long seasonal migrations unnecessary. The abundance of precipitous terrain in this area alleviates the need for seasonal movements to lambing grounds and permits a scattering of ewes during the lambing season. Only a short distance north of the international boundary in the Kofa Mountains, however, Eustis (1962) noted a definite altitudinal differentiation between terrain occupied by ewes and lambs and by adult rams during the lambing seasons. Deming (1953) and others noted concentrations of ewes on lambing areas in the more massive mountain ranges in which they worked.

When the precipitous slopes in bighorn habitat are climbed, the easiest way up is along well-worn trails used by bighorn. (Conversely, bighorn frequently use man-made mountain trails like those to artificial water holes.) Although the bighorn are noted for their agility when alarmed, they most often use established pathways dictated by topography.

10. Reproduction

Jack C. Turner and Charles G. Hansen

The management of any animal species is facilitated by knowledge of its reproductive history and reproductive capabilities. Unfortunately, little is known about any phase of the reproductive biology of most wild animal species, the desert bighorn being no exception. Much of what is purported to be known of the breeding biology of the desert bighorn has been surmised from studies on breeds of domestic sheep, which are of greater economic import. Such surmises occur at the expense of desert bighorn investigations.

PUBERTY

Because of its relation to a population's reproductive capabilities, reproductive maturity in bighorn has been the subject of considerable conjecture. For bighorn rams, a distinction is made between physiological sexual maturity or puberty and behavioral puberty. Rams to the age of 3 years travel with ewes or ewe herds, after which time they are incorporated into distinct ram groups or herds. Rams within these bachelor bands are generally considered mature. Social pressure from the larger mature rams during the breeding season, or rut, restrains young rams from copulating, although they are perhaps physiologically capable of doing so. The exact age of physiological sexual maturity is not known. Rams as young as 6 and 17 months are known to have bred successfully.

Less is known of the behavioral and physiological sexual maturity of the desert bighorn ewe than is known for the desert ram. Studies of mortality and natality on various desert bighorn herds suggest that only ewes of at least 18 months of age, and more commonly 21 months of age, are capable of pregnancy, although copulation by mature rams with ewes younger than 18 months is often observed. Such observations suggest ewes younger than 18 months of age possess sufficient hormones to cause behavioral estrus but insufficient hormone levels to initiate ovulation or maintain a pregnancy. There does not appear to be a behavioral suppression of younger ewes from entering the breeding population by the older ewes.

REPRODUCTIVE SEASON

The reproductive period (rut for rams and estrus for ewes) is seasonal for most desert bighorn populations. Because the ewe has a well defined estrous period followed by a pronounced anestrous period, or season of nonbreeding, seasonality is more evident in ewes than rams. Although mature rams are suspected of being capable of breeding all year, reduced frequency of rut activity in spring and early summer suggests a nonbreeding season for rams as well.

The rut period is considered to be that season in which breeding activity results in 70 percent of the lamb recruitment for the following lambing season. Considerable variation is seen in the length and season of rut (Fig. 10.1). Geographical elevation and latitude of individual animals appear to be major determining factors. Generally the length of the rut period is greatest in the lower southern elevations and shortest at higher northern elevations.

The estrous cycle in desert bighorn ewes is about 28 days. The behaviorally receptive period extends for only about 48 hours. During the anestrous period, the ewe is generally ignored by the ram. However, nonestrous ewes have displayed aggression toward rutting rams. Both the rut and estrous periods are coincident in the same locality and are directly related to the period of lambing. Lambing occurs at that time of the year which offers the greatest opportunity for survival.

GESTATION

The generally accepted gestation period for desert bighorn has been 6 months. Records from ewes that were pen bred indicate the gestation interval to be 179 ± 6 days, which is consistent with the estimate determined from the elapsed time between the rut and lambing season.

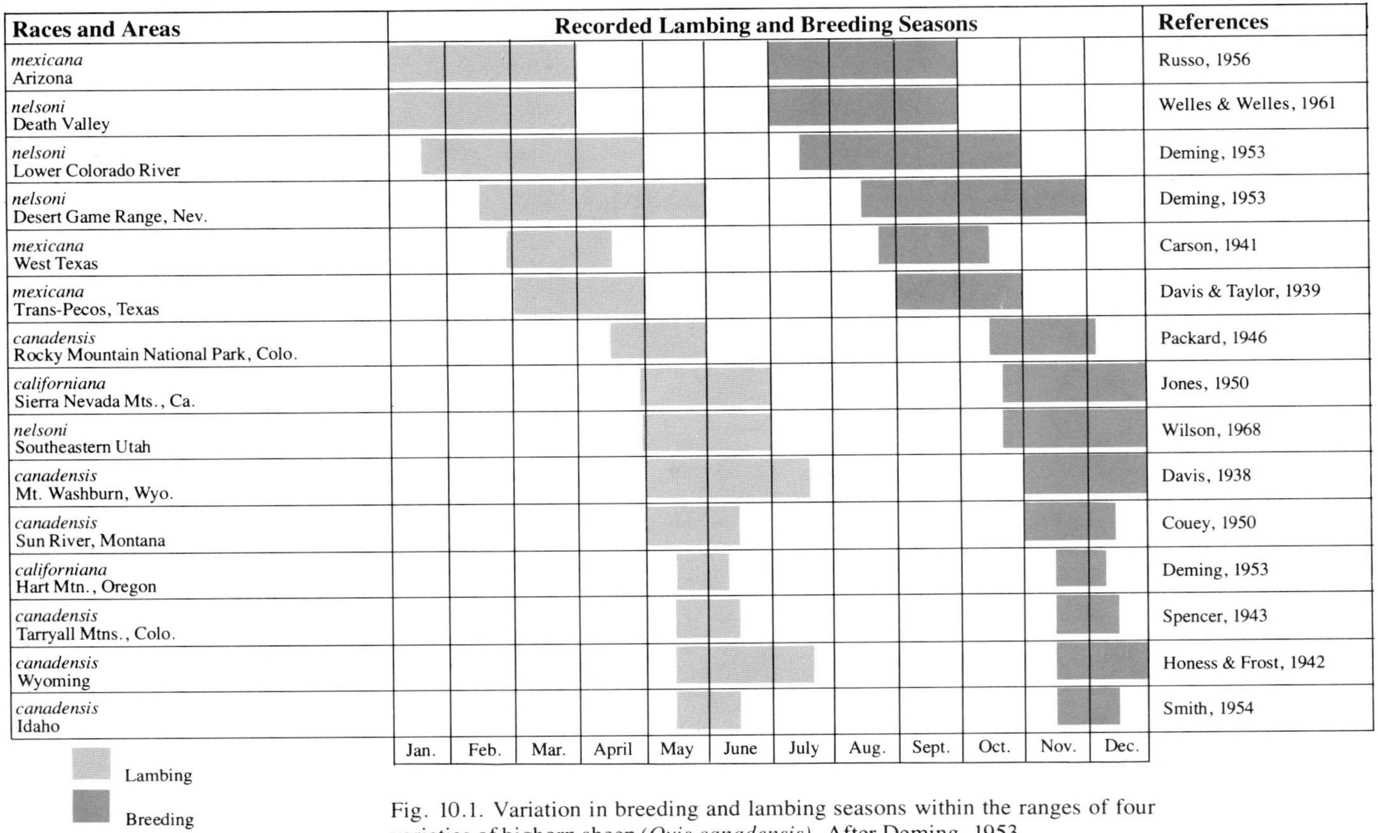

Races and Areas	Recorded Lambing and Breeding Seasons												References
	Jan.	Feb.	Mar.	April	May	June	July	Aug.	Sept.	Oct.	Nov.	Dec.	
mexicana Arizona													Russo, 1956
nelsoni Death Valley													Welles & Welles, 1961
nelsoni Lower Colorado River													Deming, 1953
nelsoni Desert Game Range, Nev.													Deming, 1953
mexicana West Texas													Carson, 1941
mexicana Trans-Pecos, Texas													Davis & Taylor, 1939
canadensis Rocky Mountain National Park, Colo.													Packard, 1946
californiana Sierra Nevada Mts., Ca.													Jones, 1950
nelsoni Southeastern Utah													Wilson, 1968
canadensis Mt. Washburn, Wyo.													Davis, 1938
canadensis Sun River, Montana													Couey, 1950
californiana Hart Mtn., Oregon													Deming, 1953
canadensis Tarryall Mtns., Colo.													Spencer, 1943
canadensis Wyoming													Honess & Frost, 1942
canadensis Idaho													Smith, 1954

Lambing
Breeding

Fig. 10.1. Variation in breeding and lambing seasons within the ranges of four varieties of bighorn sheep (*Ovis canadensis*). After Deming, 1953.

LAMBING

Within northern bighorn races environmental conditions detrimental to lamb survival maintain a narrow range for the lambing season. Environmental factors determine the time of lambing season by eliminating off-season lambs that would have become additional off-season breeders.

Within desert bighorn populations, lambing seasons show considerably more variation from year to year than the more northern races. Indeed, in some desert ranges as much as 10 to 35 percent of the lambs are born out of what is considered to be the normal lambing period. Mild climates with 200 or more frost-free days a year allow for a wide lambing interval. Both early and late lambs have favorable survival chances. Lambs born during the normal interval and those born outside of the lambing interval will reach puberty at different dates. Succeeding generations could spread the breeding and lambing seasons. The odd-season lambs allow for population plasticity relative to lamb recruitment and act as a buffer against the loss of a total age class due to drought or other environmental catastrophe during the lambing period.

PARTURITION

Ten to 14 days prior to parturition, pregnant ewes leave the bands to become solitary. Traditional lambing areas are chosen on the basis of isolation, shelter, and an unobstructed view (Simmons et al., 1963; see also Chapters 9 and 11). In older, mature ewes, the udder and teats become enlarged and darkened about 10 days before parturition. Hoggets, or first-lamb ewes, do not show comparable signs until 3 to 5 days before parturition. One to 2 days prior to parturition the ewe's vulva becomes noticeably swollen and her irritability increases (Welles and Welles, 1961a).

Parturition commences with the rupture of the amnion. However, birth may not occur for several hours. Generally the ewe remains sedentary and does not move until parturition is complete. Placental retention for periods ranging from 3 to 5 hours has been recorded. The ewe generally eats or at least chews the afterbirth and then conceals it. Such concealment might preclude attracting predators by odor to the scene where the young would be endangered (Kennedy, 1948).

REPRODUCTIVE POTENTIAL

Twinning in desert bighorn is subject to much conjecture. Numerous reports of two lambs nursing from the same ewe exist. And much of the older literature credits bighorn with twinning but offers no data to confirm this view. Eleven pregnant non-desert bighorn from British Columbia were examined and four contained *in utero* twins. At best, if twinning occurs in the wild, it is a rare phenomenon that has yet to be observed in desert bighorn.

"Reproductive senescence" has been applied to describe the retardation of reproduction at the aged end of the life cycle. No wild animal is known to have an established menopause or cessation of ovarian function. Penned rams have sired lambs at 11 years of age and have been known to have viable sperm at 16 years of age. Bighorn ewes can be expected to decrease in fertility with age, but they are capable of some reproduction in their declining years.

PERCENTAGE OF PRODUCTIVE EWES

The lamb : ewe ratio (lambs/100 ewes) is suggestive of the percentage of productive ewes within a population. This ratio varies considerably on a yearly basis. On Desert National Wildlife Range in Nevada in 1949 this ratio was as high as 81 lambs /100 ewes.

Since some lambs could have been born as much as 2.5 months prior to an observation of mortality occurring in the first few months of life, there is probably some unobserved loss. Therefore, it is most likely that the potential productivity of the wild ewe is closer to 90 : 100. This value includes young ewes not yet of breeding age. Consequently, the potential productivity may be closer to 98 : 100.

During water-hole counts in 2 different years on the Pintwater Range of Desert National Wildlife Range, the lamb : ewe ratio was 74 : 100 and 81 : 100 at a time when the lambs were expected to be about 6 months of age. The normal mortality by that age is about 45 percent. Thus, during the 2 years for which there are good records, the product of production and survival was extremely good. Additional lamb survival data are found in Chapter 11.

SEX RATIO

The normal sex ratio of adult desert bighorn has been determined to be 100 : 100 by Russo (1956), Welles and Welles (1961a), and Monson (1963). The same ratio has been observed by Sugden (1961) for the California bighorn *(Ovis canadensis californiana)* and by Murie (1944) in Dall sheep *(O. dalli)*. Sex ratio figures for Desert Wildlife Range are given in Table 10.1. The data for the years before 1948 were taken from observations made primarily in the Lambing Study Area of the Sheep Range Management Unit where ewes and lambs predominate. The figures for 1948 and 1949 show more rams than ewes. A preponderance of rams was observed in Arizona by Russo (1956) from 1950 to 1954, and a slight preponderance by Monson (1963) from 1955 to 1962.

On Desert Wildlife Range the sex ratio figures have been quite variable. Before hunting started in 1954, the sex ratio was close to 100 : 100; by 1963 it was about 23 : 100.

TABLE 10.1.
Numbers and Sex Ratios of Bighorn Observed on Desert National Wildlife Range

Year	Total	No. Rams	No. Ewes	% Rams	% Ewes
1969	286	59	227	21	79
1968	218	50	168	23	77
1967	377	103	274	28	72
1966	408	116	292	28	72
1965	291	73	218	25	75
1964	326	67	259	20	80
1963	441	79	362	18	82
1962	522	97	425	19	81
1961	509	132	377	26	74
1960	736	212	524	29	71
1959	398	116	282	29	71
1958	446	100	346	22	78
1957	795	291	504	37	63
1956*	1549	611	938	40	60
1955	398	164	234	41	59
1949	131	78	53	60	40
1948	355	188	167	53	47
1947†	608	198	410	33	67
1944 to 1946†	856	289	567	34	66

*The high total number of 1549 for 1956 is due to an increase in manpower available to make observations that year.
†These figures were chiefly from Desert National Wildlife Range Lambing Study Area where ewes and lambs predominate. Figures for 1948–64 represent data from the total bighorn habitat on the Sheep Range Management Unit and the Pintwater Management Unit.

RELATION TO NUTRITION

Successful reproduction in domestic sheep is dependent upon the quality of the food and the resultant condition of the ewes during and after pregnancy. In deer, Taber and Dasmann (1958) concluded does in poor condition do not breed as young, and do not produce as many fawns, as deer in good condition; in undernourished does a higher death loss in fawns occurred shortly after birth.

Green forage furnishes a more suitable diet for ewes and lambs than does dry feed. Of importance is the higher protein content found in green feed that is needed in increasing amounts during late pregnancy and lactation. Although the calcium content of green forage is not consistently as high in cured plants, the calcium-phosphorus ratio can be balanced from limey soils eaten by the bighorn. Vitamin D from sunshine allows for a higher amount of calcium in proportion to the phosphorus in the diet and may assist in balancing the two minerals when phosphorus intake is low. Green forage is typically low in crude fats, but this is not critical if adequate carbohydrates, proteins, and vitamins are present. An enforced dry diet due to adverse weather conditions would cause primarily a reduced protein content and reduced vitamin D content in the same diet, which might influence the calcium-phosphorus ratio.

Cold, cloudy, and wet springs influence the health standards of the animals by restricting the amount of sunshine and vitamins needed by lambs to control scours; by driving ewes and lambs to sheltered spots and promoting the multiple use of bed grounds until they become contaminated; by lowering the protein intake of the animals; and by promoting lamb diseases such as dysentery and pneumonia.

Climate may also have other effects on the mortality potential of lambs. During warm, open springs and before sufficient lambs appear to form a lamb nursery, predation on lambs is more possible as bighorn are then using the open slopes and canyon bottoms for feeding rather than being entrenched in the rough country. An open spring also decreases the disease potential in supplying adequate green feed for health and by increasing favorable sanitary conditions by the continual change of bed grounds and feeding places. A wet, cold spring has much the reverse effect.

11. Growth and Development

Charles G. Hansen and O. V. Deming

The most popular image of a bighorn is the silhouette of an impressive ram posed on a rugged cliff. How did that ram look as a lamb, and how did he develop from the embryo stage to the patriarch of the mountain? Although certain phases of the growth pattern of bighorn have received intensive study, much is still unknown. Most work on growth and development of ovines has been with domestic sheep. However, there are fundamental differences between domestic sheep and bighorn, such as the shorter gestation period of 150 days in domestic sheep compared with the 180-day period for bighorn. The observations that follow were made primarily at Desert Wildlife Range, Nevada.

FETAL DEVELOPMENT

Fetal studies of deer and elk have produced many usable data (Giles, 1969). However, an insufficient number of desert bighorn fetuses has been available for intensive study of fetal growth in that species. Russo (1956) describes a female fetus that weighed 1375 grams (3 lb) and measured as follows: total length, 345 mm (13.6 in.); hind foot, 94 mm (3.7 in.); ear, 29 mm (1.14 in.); tail, 21 mm (0.8 in.); height at shoulder, 178 mm (7 in.); and girth, 194 mm (7.7 in.). "The characteristic rump patch of the bighorn was sharply defined by the pigmentation of the skin, although no hair was present except about the muzzle. Pronounced difference in the size of the front and hind feet seen in adult bighorn was lacking in the fetus. A profile view showed a definite and pronounced undershot facial structure. Examination of the mouth revealed ridges present in the roof of the mouth and a well-formed upper dental pad. Soft, horny formations were present in the location of the incisors; however, there was no evidence of molars or premolars.

"No accurate age of the fetus was determined, but assuming the gestation period of bighorn sheep to be approximately 170 days, it was estimated that the age of the fetus was 130–140 days as based on the development of the deer fetus (Armstrong, 1950)."

LAMBS

Full-term lambs are completely covered with hair. The tips of the hoofs are white and very soft, grading up to the darker, harder parts near the hair line. The soft white hoof parts dry and harden within an hour or so after birth. The soft hoof tips are an adaptation in sharp-hoofed animals to prevent damage to the mother when the lamb moves about before it is born. The cheek teeth are occasionally exposed, but normally teeth are not through the gums at birth. Within a week some of the teeth have broken through. Lambs at birth weight about 3630 grams (8 lb) and measure 420 mm (16.5 in.) high at the shoulders, other measurements being ear, 77 mm (3 in.); hind foot, 175 mm (6.9 in.); and tail, 58 mm (2.3 in.).

The arrival of lambs in a herd of desert bighorn does not necessarily signify the beginning of spring, because in the Southwest lambs can appear any month of the year. However, the bulk are born in the late winter and early spring. They are active within a few minutes of the time they are born, and within a couple of days are able to follow their mothers almost any place the latter can go. Usually the ewe isolates herself when she gives birth to her lamb and remains away from the herd for a short time. Consequently, these wobbly little bighorn are not commonly seen. But after a week or two the mothers rejoin the herd, and frequently leave the lambs in a community nursery.

Young lambs are mouse-colored. This grayish coloration lasts for several months. On Desert National Wildlife Range the pelage of a lamb 24 hours old showed the back to be a light gray, with a light brown stripe from the head to the end of the tail. The rump was light cream color, with no clear line of demarcation between it and the back sides. The sides were brownish gray, with somewhat wavy fleece between the shoulders and rump patch. The belly was brownish white. The front legs were gray from the shoulders to the knee, and light chocolate brown from the knee to the hoofs. The leg stripe was just beginning to show. The hind legs were gray with the stripe indistinct.

The head was uniformly gray. The ears were light inside and darker behind, with long stiff hairs around the margins and the outer ends. The nose, eyes, and hoofs were black. (As the lamb grows older the leg stripes and rump patch become more defined, and by summer these markings are as conspicuous as those of the adults.)

The bleat of the lamb and the lower-pitched bleat of the ewe are very similar to those of domestic sheep. Lambs are heard often during the first month, but bleating decreases as they become older and by winter such sounds are seldom heard. On one occasion when a band of 69 ewes, lambs, and yearlings was under observation in a short, blind canyon, the sound of the feeding lambs and ewes as they kept in contact vocally resembled a mild form of bedlam. Normally, though, these sounds are obvious only in the early morning when small groups of ewes and lambs are moving away from the bed grounds, or when a ewe is returning to or looking for her lamb, after having left it behind for "safekeeping."

The following physical requirements apparently must be present in a lambing area: (1) Water needs of ewes must be met through snow, springs, natural tanks, potholes, and succulent plants. (2) There must be an adequate food supply near escape cover. Early green foods, which are higher in food value and tend to encourage an ample milk supply, are necessary. (3) Rough, high, broken country, facing south or southeast with trees, shrubs, or caves to offer protection from inclement weather and possible predatory eagles are characteristics of good lambing grounds.

Lambs have a pronounced tendency to pair off and remain together a large portion of the time during the first part of the lambing season, or where and when lamb numbers are small and scattered. Such lambs, when alarmed, often will follow one ewe. At times two lambs will be found with one ewe, while the other ewe is feeding or resting a considerable distance away—sometimes out of sight of her lamb. These two lambs will continue as almost inseparable companions and playmates, as described by Welles and Welles (1961a), as long as the two ewes remain together.

Wilson (1968) writes that,

> After lambs are about 6 to 8 weeks of age the ewes will often leave their lambs alone for long periods of time during the day. The procedure for the ewe leaving and returning to her lamb rarely varied, and is one behavior trait which I believe favors lamb survival.
>
> When a lamb would tire of following its mother during the day, it would generally move off laterally through the rocks and at a right angle from the direction its mother was traveling. The mother would stand and watch the lamb until it chose a spot to lie down. The mother would then appear to forget about the lamb and continue grazing. Usually the lamb was left ... between 9:00 and 11:00 A.M. until ... 4:00 to 6:00 P.M.
>
> Signs of the ewe showing concern about the lamb were always obvious. The ewe would stop grazing and stare back at the area where the lamb was bedded. She would then travel a distance of a few yards or a hundred yards at a fast walk or trot toward the lamb, stopping for several minutes

periodically to graze. Upon approaching the lamb, she would generally stand a few yards below it for several minutes, looking in all directions. After a short period of time the ewe would make the coarse, burping, frog-like sound. The lamb would rise from its bed and run to its mother. Once the lamb is bedded, it will not leave its bed unless it is badly frightened. All ewes I observed returned to their lambs by a completely different route than the ones they had taken to leave them.

Because lambs leave their mothers in a lateral direction, most predators would likely miss detecting the small tracks of the lamb, since they leave little or no scent. Because of the extreme caution of the ewe when returning to the lamb, and of the different routes taken by the ewe, a predator would not see the lamb until it was at its mother's side.

The nursery system. As the lambs increase in number and the band sizes grow, such expanding groups spend the nights on the high ridges and begin feeding there at daybreak. If green feed is present at lower elevations, the group will begin working down the ridges, feeding as they go. When the lower edge of the rough terrain is reached, the ewes often will establish a nursery, with one or more ewes remaining with the lambs and the other ewes continuing on down to the bottom. This system allows most of the mothers more freedom to seek out succulent spring foods. With the exception of occasional visits to the lambs for nursing and an exchange of nurses, the mothers remain in the washes or canyon bottoms until the evening shadows fall. Then they begin working upward toward the lambs, and all return to the rough ridge tops for the night.

Lambs in the nursery are surprisingly obedient to the ewes in charge. The nurse or nurses appear to keep them well grouped with very little difficulty. On one occasion when two ewes were moving the nursery to keep up with the feeding ewes below, a ewe went ahead to a vantage point of rocks, scanned the country for 15 minutes, then turned her head and uttered a blat that brought the lambs running and leaping in their eagerness to reach the new area of security. The second ewe trailed behind, following the last lamb.

The nursery often is mobile during the course of a day. When the feeding ewes are below and moving, the nursery will keep pace with them by occasionally moving parallel with them. The feeding ewes appear to have full confidence in the ewes left in charge of the lambs. Feeding or resting, often a considerable distance away from their offspring, the mothers seldom give them more than an occasional look. As the lambs become older, larger, and more able to partake of solid foods, they begin trailing along with their mothers, and the nursery system is abandoned for that year.

Weaning. Lambs usually are weaned by the time they are 6 months old. Lambs are never allowed to nurse for more than a few seconds at a time. When the lamb is quite small, these short nursing periods are quite satisfactory. As the animal gets older, it depends less and less on its mother's milk because of the solid food it is eating. By the time the lamb is 5 or 6 months old, it is getting only a fraction of its total diet from its mother.

Occasionally a young lamb is separated from its mother and becomes a "bummer" lamb, trying to "bum" a meal from any ewe that comes along. It is possible that a ewe with a newborn lamb will accept a lost lamb, since any given ewe doesn't seem to establish a specific association with her own lamb before it is a few days or a week old.

Raising lambs artificially. Desert bighorn lambs were raised in captivity at Desert National Wildlife Range several times in the last 24 years. Deming (1955) and Patricia A. Hansen (1964) published their experiences and advised on how to feed and care for such lambs. Much useful information has been gathered on behavior, growth, tooth development (Deming, 1952), and diseases from raising captive lambs.

It appears that canned cow's or goat's milk, or powdered skim milk, are suitable, depending upon the individual lamb. Karo syrup was used in varying amounts to control constipation and diarrhea. One or more ounces of syrup was added as needed to loosen the bowels. Rolled oats, cracked corn, and a calf supplement were made available, and good alfalfa hay was provided. Lambs that had a wide variety of vegetation to choose from appeared to encounter few gastric problems.

Hand-raised lambs readily accept people, dogs, and cats as a part of their family. They follow and play with people as they would with other bighorn. If permitted to do so, they will maintain this relationship with these unnatural companions until they become adults; if not "mistreated" after they are 6 or 7 months old they may retain this close attachment all their lives. Hand-raised bighorn that have been released into the wild rapidly revert to the wild state (Forrester and Hoffman, 1963). Bighorn raised in pens quickly become wild if harassed severely by people.

Lamb survival rates. Wilson (1968) presents survival rates for Utah: "In the summer of 1965 the yearling : ewe ratio was 41 : 100. By mid-July, 1965, shortly after the lambing season, the lamb : ewe ratio was 37 : 100. By mid-July, 1966, the yearling : ewe ratio was 20 : 100 with a 49 percent lamb loss noted for the 1-year period. The lamb : ewe ratio by mid-July, 1966, was 60 : 100, and 42 : 100 on November 15. This gives a mean mortality rate of 30 percent for the 5-month period for lambs in the White Canyon area of southeastern Utah."

Additional lamb survival data are found in Chapter 10.

GENERAL GROWTH AND DEVELOPMENT

Hansen (1965a) describes and illustrates the characteristic growth and development of desert bighorn from 1 day to 1 year of age. Figures 11.2 to 11.4 show the change in body conformation of a lamb from birth to 4 months. Figure 11.5 compares the changes that occur in the head and horns of a male and female lamb from 4 months to 1 year.

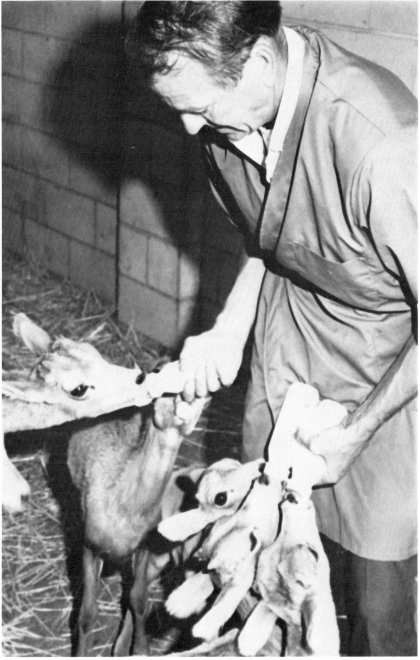

Robert E. L. Taylor

Fig. 11.1. Desert bighorn lambs being bottle fed at the University of Nevada Veterinary Medical Laboratory at Reno. The lambs are being used in a study aimed at learning more about bighorn diseases.

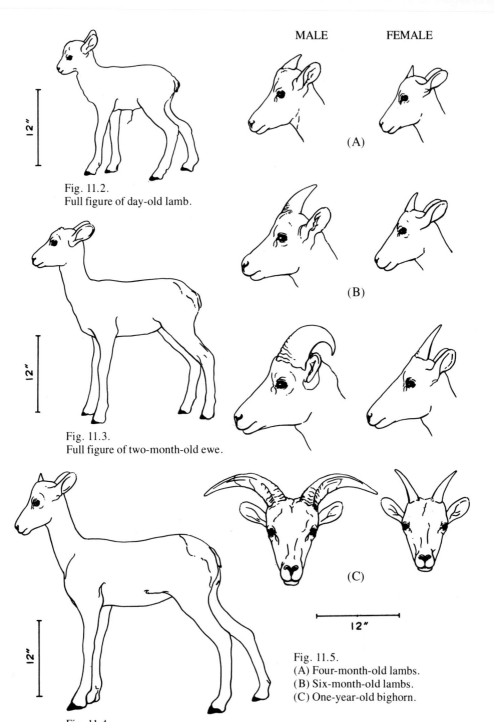

Fig. 11.2.
Full figure of day-old lamb.

Fig. 11.3.
Full figure of two-month-old ewe.

Fig. 11.4.
Full figure of four-month-old ewe.

Fig. 11.5.
(A) Four-month-old lambs.
(B) Six-month-old lambs.
(C) One-year-old bighorn.

Drawings by Patricia A. Hansen

The following descriptions are based primarily on two captive lambs, a male and a female. Italicized portions often are the best field characteristics for distinguishing the age or sex of lambs.

DAY-OLD LAMB. *Umbilical cord evident.* Hind legs large, appear to be out of proportion to the body; animal wobbly on its feet. *Face in profile is concave, and the distance from the nose to the eye is shorter than from the eye to the back of the head, thus giving the animal a short-nosed, big-headed appearance. Height at shoulder,* about 380 mm (15 in.). Lamb is able to walk under the ewe with its head lowered only slightly. *Weight,* approximately 3.6 kg (8 lb). *Horns,* none.

WEEK-OLD LAMB. *Umbilical cord* may still be evident. Hind legs large, still appearing to be out of proportion to the body; animal not wobbly on its feet. *Face in profile very concave.* The penis may show below belly hair but is not readily apparent. *Height at shoulder,* 406 mm (16 in.). Lamb must duck its head to walk under belly of ewe. *Weight,* about 3.6 to 3.9 kg (8 to 8.5 lb). *Horns,* none; slight tufts of hair about 25 mm (1 in.) long appear like horns.

4-WEEK-OLD LAMB. Trim, well proportioned miniature bighorn without horns. *Face in profile concave and eyes bulgy.* The penis is often evident below the belly hairs. *Height at shoulder,* 460 mm (18 in.). *Weight,* approximately 5.5 kg (12 lb). *Horns,* none; hair tufts look like horns about 25 mm (1 in.) long.

2-MONTH-OLD LAMB. About the same as at 4 weeks in appearance. *Face profile almost straight; not at all "dished"* as in Figure 11.2. The distance from the nose to the eye is the same, or greater, than from the eye to the back of the head. Nose no longer looks short. Eyes bulge only slightly. *Height at shoulder,* 535 mm (21 in.). *Weight,* approximately 9 kg (20 lb). *Horns,* hard knobs appearing just above the hair tufts, about 25 to 50 mm (1 to 2 in.) long from the base to tip.

3-MONTH-OLD LAMB. *Body* filling out and more in proportion to the trim legs. Neck in the female is quite slim; in the male it is thicker, giving a short-necked appearanace. Nose noticeably longer than in the 2-month-old lamb; eyes not bulgy. *Height at shoulder,* 560 mm (22 in.). *Weight,* about 14.5 to 16 kg (32 to 35 lb). *Horns,* 25 to 38 mm (1 to 1.5 in.) *above the hair tufts.* Female horns thin, pointed; male horns thick, triangular.

4-MONTH-OLD LAMB. Female lamb slim and trim in body and legs; male lamb blocky and short-necked. Nose longer than in the 3-month-old lamb; face from the front has the appearance of an even-sided triangle. Ear is nearly full grown at 100 mm (4 in.) from tip to notch and extends about 64 mm (2.5 in.) above the head. *Height at shoulder,* 610 mm (24 in.). *Weight,* 18 to 20.4 kg (40 to 45 lb). *Horns,* male, 50 to 75 mm (2 to 3 in.) above the hair, bulky; female, 25 to 40 mm (1 to 1.5 in.) above the hair, thin.

5-MONTH-OLD LAMB. No significant change in general appearance. Top of the head is about even with the back of an adult ewe. *Height at shoulder,* 635 mm (25 in.). *Weight,* approximately 23 kg (50 lb). *Horns,* male, 75 to 100 mm (3 to 4 in.) above hair, bulky; female, 50 to 75 mm (2 to 3 in.) above hair, thin.

Charles G. Hansen

Fig. 11.6. "Frisky," a male bighorn lamb born at Corn Creek, Desert National Wildlife Range, Nevada, at the age of 3.5 months.

6-MONTH-OLD LAMB. Male blocky; neck short and thick. Female trim, long-necked. Nose, both sexes, quite long but does not approach adult proportions. *Height at shoulder,* male, about 690 mm (27 in.); female, about 660 mm (26 in.). *Weight,* male, about 29.5 kg (65 lb); female, about 27 kg (60 lb). *Horns,* male, 100 to 150 mm (4 to 6 in.), thick and bulky; female, 75 to 100 mm (3 to 4 in.), long and thin.

8-MONTH-OLD LAMB. No significant change in *body* proportions. Nose continues to get longer in relation to the rest of the face. *Height at shoulder,* male, about 760 mm (30 in.); female, about 660 mm (26 in.). *Weight,* between 32 and 36 kg (70 and 80 lb). *Horns,* male, 125 to 200 mm (5 to 8 in.), thick; female, 100 to 125 mm (4 to 5 in.), slender.

1-YEAR-OLD BIGHORN. Ram is blocky and appears to be the equivalent of an adult ewe except for the face, which has a shorter nose and is shaped more like that of the lamb; it appears triangular when compared to the long-nosed adult ewe. Yearling rams are most readily identified by their *facial characteristics* when their male organs are not evident. Yearling ewes are thin and sleek-looking with long-thin necks and slim legs. They are noticeably smaller throughout than the adult ewe. *Height at shoulder,* male, about 915 mm (36 in.); female, about 690 mm (27 in.). *Weight,* male, 52 to 59 kg (115 to 130 lb); female, 41 to 50 kg (90 to 110 lb). *Horns,* male, 200 to 305 mm (8 to 12 in.), bulky. The yearling ram horn appears very much like that of the adult ewe horn but is thicker at the base than the ewe horn and is blue-gray in contrast to the light brown of the old ewe. Female horns are 125 to 178 mm (5 to 7 in.) long, thin and sharp-pointed. The yearling ewe horn is quite distinctive, in that it is much shorter than the adult ewe horn and is comparatively thin, straight, and sharp-pointed.

Growth to maturity. There is a certain amount of controversy as to when a bighorn is mature. There is evidence that a ram was sexually mature at 6 months of age (Deming, 1963). Yet the horns, a secondary sexual characteristic, do not stop growing noticeably, at least in normal rams, until the rams are over 10 years old. Geist (1968a) concludes that *Ovis dalli* and *O. canadensis* continue to mature in body size, proportions, and behavior for 5 to 7 years after sexual maturation. He believes that rams continue to grow until they reach maximum proportions some time after 9 years of age.

Clark (1964) summarizes the weights and measurements of the desert bighorn: "Desert sheep will of course vary in height and weight, but as a group they will run from 36 to 40 inches [915 to 1016 mm] at the shoulders, the latter height being about maximum. In weight a big ram would range from 180 to 200 pounds [81.6 to 91 kg], with 225 pounds [102 kg] as the maximum—all of which measurements would make them about the size of the Alaska Dall sheep."

Russo (1956) took weights and measurements of adult bighorn at various ages once a month for 1 year from the Kofa Game Range in Arizona. A summary of these data appears in Table 11.1. He also describes the average size of rams taken in the Arizona 1953 hunting season:

> ... measurements and weights were taken from several rams from widely separated areas in four mountain ranges. Average height ... 37.2 inches [970 mm]; average dressed weight 111 pounds [50 kg]; the smallest ram 85 pounds [38.6 kg], the largest 135 pounds [61 kg]; average total length ... 61.5 inches [1562 mm]; average girth 39.6 inches [1006 mm]. [He presents the following measurements for comparison.] Measurements (after Cowan, 1940): Adult male from type locality: Total length 60 inches [1524 mm], tail 8 inches [203 mm], hind foot 17 inches [432 mm], height ... 35 inches [889 mm]. Adult male from Guadalupe Mountains, Texas (measured dry skin): Total length 58 inches [1473 mm], tail 2¾ inches [70 mm], hind foot 14½ inches [355 mm], ear ... 3¾ inches [95 mm]. Young adult male from Pinacate: Total length 41¼ inches [1048 mm], height ...

TABLE 11.1.
Measurements of Individual Bighorn from Kofa Game Range, Arizona

Age Years	Month Taken	Whole Wt.		Girth		Height		Total Length	
		kg	lb	mm	in.	mm	in.	mm	in.
				Rams					
3.3	Sept.	69.4	153	991	39	895	35.25	1524	60
4	Mar.	64.9	143	997	39.25	933	36.75	1530	60.25
5	Feb.	68.9	152	1016	40	914	36	1537	60.5
5 to 6	Aug.	82.1	171	1054	41.5	876	34.5	1518	59.75
8 to 9	Jan.	78.5	173	1010	39.75	914	36	1537	60.5
not aged	July	90.1	200	1130	44.5	1003	39.5	1702	67
9 to 10	June	86.6	191	1044	41.12	889	35	1626	64
				Ewes					
3	Dec.	45.8	101	851	33.5	699	27.5	1378	54.25
4	Apr.	55.6	122.5	965	38	832	32.75	1359	53.5
6	Oct.	50.3	111	908	35.75	813	32	1403	55.25
8 to 9	May	57.2	126	876	34.5	870	34.25	1499	59
8 to 9	Nov.	52.2	115	953	37.5	813	32	1461	57.5

Source: Russo (1956).

36¼ inches [921 mm]. Adult male (type of *sheldoni*): Total length under 52 inches [1321 mm], tail 3⅞ inches [98 mm]; hind foot 14 inches [356 mm]; height ... under 30 inches [762 mm]. Hornaday (1908) recorded measurements of a large ram shot in the Pinacate area (Mexico) in 1907, which are: Weight 192½ pounds [87.3 kg]; total length 59 inches [1499 mm]; height ... 37 inches [940 mm], tail 5 inches [127 mm]; girth 42½ inches [1080 mm]....

The combined data suggest that a large adult desert bighorn ram would weigh about 90.7 kg (200 lb); an average ram about 81.6 kg (180 lb). The girth would vary from 940 to 1118 mm (37 to 44 in.); the height from 787 to 991 mm (31 to 39 in.); total length from 1321 to 1702 mm (52 to 67 in.); the tail from 64 to 114 mm (2.5 to 4.5 in.); and the ear from 89 to 108 mm (3.5 to 4.25 in.).

The average adult ewe would weigh about 48 kg (106 lb) with a range of 33.6 to 57.2 kg (74 to 126 lb). The girth would vary between 838 and 1016 mm (33 and 40 in.), the height from 711 to 864 mm (28 to 34 in.), the hind foot from 318 to 368 mm (12.5 to 14.5 in.), and the total length from 1372 to 1499 mm (54 to 59 in.).

With such a wide variation in sizes, it is evident that a large young animal would approach the dimensions of a small older animal. Therefore, it would be difficult to state categorically that a 3-year-old bighorn will be so big and a 10-year-old will be so much bigger.

Considering all the available data it appears that the ewe is mature at 2.5 years and the ram at 3.5 years. This conclusion is based on the age at which these animals normally enter the serious aspects of breeding. This does not mean that they are incapable of breeding at an earlier age. It does imply that at these respective ages the animal has reached a physical and mental stage that will vary only slightly in successive years, compared with the amount of growth and development before that stage.

Both sexes will fill out during the next year (after 2.5 years for ewes, 3.5 years for rams). However, at 2.5 years the ewe normally will give birth to her first lamb and will accept this responsibility as a mature animal. The ram will begin conditioning his skull and horns in a more serious manner at 3 years, and enter into the rut at 3.5 years, when he will try for horn "clashes" in earnest. He may not be able to stand up to larger animals at that age, but by the time he is 4.5 years old he will have had the experience and developed enough bulk to take on more vigorous adversaries.

GROWTH OF THE SKULL

There is a great variation in desert bighorn skulls, mainly due to age differences. Baker (1967) states: "A brief discussion of growth rates of the skull based on the analysis of Baker and Bradley (1966) ... illustrates the size variation in the skull attributable to age. The percent of growth at different age intervals is presented for ewes in Table 7 [11.2 in this text] and for rams in Table 8 [11.3 in this text]. The percent of growth for the six month to one year age interval in rams is underestimated since the sex of lambs was not determined and their mean values were used in calculations for both sexes.

"Rapid growth is exhibited in lambs from birth to one year of age. Most of the growth, as calculated from the standard measurements, has occurred by five years of age. Growth seems to continue, at greatly reduced rates, into the terminal age class for most of the skull measurements."

There is a lack of change in the zygomatic width of the ewes after 5 years as compared with the rams. Continued growth in the zygomatic width of the rams is probably due to the large size of the rams' horns and their continued growth throughout life.

The skull of a young bighorn ram is very thin compared with its thickness at maturity. The ewe's skull does not thicken, materially, but a ram's skull will become 25 mm (1 in.) or more thick over the top by the time he is 4 or 5 years of age.

Within a few days of birth, a ram lamb begins to butt at things such as another animal, trees, or rocks. As the lamb gets older, he butts more vigorously and more often. While his body is growing, the skull is also growing to accommodate the adult-sized brain of the future. Skull sutures must remain articulate for the skull to grow and adjust to this enlargement. While these

TABLE 11.2.

Percent of Growth Occurring During Different Age Intervals in Desert Bighorn Ewe Skulls from Desert National Wildlife Range

Measurement	N	6 months to 1 year	N	1 year to 5 years	N	6 to 9 years	N	10 to 14 years
Nasal length	2	21.8	4	54.5	8	18.2	5	3.6
Orbital width	4	33.3	6	58.3	20	5.6	16	2.8
Zygomatic width	5	37.5	6	59.4	17	0.0	10	0.0
Palatal length*	5	23.3	4	53.5	8	18.6	7	4.7
Post dental length	3	22.6	8	64.5	18	6.5	15	16.5

Source: Baker and Bradley (1966).
N = number of samples.
*Cowan (1940).

TABLE 11.3.

Percent of Growth Occurring During Different Age Intervals in Desert Bighorn Ram Skulls from Desert National Wildlife Range

Measurement	N*	6 months to 1 year	N	1 year to 5 years	N	6 to 9 years	N	10 to 14 years
Nasal width		23.5	5	44.1	19	26.5	51	5.99
Orbital width		23.1	7	53.8	20	15.4	64	7.7
Zygomatic width		27.9	7	55.8	14	14	54	2.3
Palatal length†		19.6	6	62.7	15	11.8	47	5.9
Post dental length		17.9	6	56.4	19	15.4	62	10.3

Source: Baker and Bradley (1966).
N = number of samples.
*Not known.
†Cowan (1940).

sutures are articulate, a young ram might crack his skull if he entered into a serious butting contest with a larger, older ram. Consequently, the ram cannot enter into the hard-hitting rutting activity until his skull is strong enough to take the blows.

After the skull bones have reached their full growth the sutures begin to ossify. Ossification may be due to stimulation along the sutures when the bone moves as the ram butts or shoves with his horns. The older the ram, the more the ossification, until most of the sutures on the top and sides of the skull are cemented together. The last sutures to become cemented are those between the nasal and premaxilla bones. The distal portion of these bones does not ossify, and so there usually are some short stretches of open suture in that area. Excessive blows on the nasal bones has caused an extra growth on the noses of some rams, creating the appearance of a small extra horn. It may grow 25 mm (1 in.) wide and as much high, but no more than 6.4 mm (.25 in.) thick.

GROWTH OF THE HORNS

Bighorn horns are permanent and continue to grow throughout life. The horn consists of a bony core covered with a permanent corneous sheath formed by the deposition of keratin, a horny material similar to that in hoofs. The horns grow from the base and from the inside around the bony horn core. Both sexes have horns. Each year during the rut period in the fall, horn growth subsides for several months but resumes again, usually in January. A ring-shaped depression called a horn ring is formed when the horn stops growing.

The amount of horn growth is variable. During the first 2 or 3 years, young animals may have two or three periods when the horn growth slows or stops completely; as a result, two or three rings may be found close together during these first 2 or 3 years of life. However, up to the age of 3.5 to 4 years, the annual growth usually is rapid and, with rams, when measured on the front of the horn, normally exceeds 150 mm (6 in.) for each of the first 2 years. This is the period in the ram's life when the longest growth of the horn takes place. It is not unusual to find 3- to 4-year-old rams with horns over 510 mm (20 in.) long. The new and rapid horn growth in rams under 4 years is a powdery blue-gray and relatively smooth. Deep, dark annual rings and rough wrinkles do not occur until after the ram is 3.5 years old. Since these animals can be born any month of the year, a bighorn born in early winter could be a 3-year-old during its third rut period. If this occurs, the third annual ring may be quite prominent.

At about the age of 3.5, when the ram begins butting seriously, a protective knob of spongy tissue develops at the back of the head. This has been identified as erectile tissue and becomes very distinct as the ram gets older. The knob becomes inflated during prolonged butting bouts (Welles and Welles, 1961a). The horns of young rams usually are very widely spread, and in almost all cases the widest spread from tip to tip occurs in animals by the time they are 6 years old.

Baker (1967) says "... in male lambs the measure of horn growth in the first year is somewhat less than in female lambs. This is primarily due to the higher degree of brooming of the first year's growth in later years by rams. Hansen [1965a], in a study of captured lambs, stated that the rate of horn growth is slightly higher during the first year in males than in females. The second year's growth is more rapid in rams than in ewes. After the second year, the rate of horn growth in ewes declines rapidly and by the six-year interval growth of the horn is negligible. Horns of rams grow at a reduced rate after the second year interval, but the decline in horn growth is slight when compared to ewes. The horn growth rate continues to decline slowly until the nine and ten year age intervals and then becomes insignificant.

"The diameter of the horns in rams increases rapidly for the first four years of growth. After the fourth year, there is a pronounced decline in the rate of increase and growth appears to almost cease by the sixth year. In ewes, the diameter of the horns increases rapidly for the first two years. After the second year, the rate of increase in horn diameter declines rapidly and by the end of the fourth year is insignificant in ewes.''

The percent of horn growth of the first year that is broomed in later years was determined by Baker (1967): ''It is readily evident that rams broom their horns to a much greater extent than do ewes. By the end of the second year, rams have broomed over 50 percent of the growth produced during the first year, while ewes have only broomed about 30 percent. By 10 to 12 years of age, rams have broomed over 95 percent of their horn growth from the first year. Some individuals are beginning to broom into the second year's growth at this age. Ewes of the same class have broomed less than 50 percent of their horn growth from the first year. When it is considered that the horn growth for the first year is about 254 millimeters for rams and 153 millimeters for ewes (Hansen, 1965[a]), then rams in most cases have broomed at least 240 millimeters from the horn while ewes have only broomed 64 millimeters from the horn. Brooming of the horn in older individuals commonly exceeds growth of the horns which results in a reduction of length with age.''

The term ''brooming'' is applied to bighorn horns that have become worn or broken at the tips. This occurs in the desert forms of bighorn more than in the northern ones. It is mainly the result of abrasion against rocks and dirt. Horns are sometimes broken during the rutting season head-butting bouts.

The thought has been advanced that the horn rings are the result of poor diet and vigorous rut activity during the critical winter period. Desert bighorn enter the rut season in July and August, when summer rains usually provide easy living. The peak of the rut is in September and October, when temperature and food are not critical. Desert bighorn born and raised in pens on Desert National Wildlife Range have good feed in front of them all day long. They eat an increasing amount of food from August through November and continue this high consumption through December when their food intake gradually decreases (Hansen, 1962). These artificially fed bighorn also form horn rings and go through the same sequence of growth from about February through May, followed by slowing or complete cessation of horn growth. The new horn growth is very evident on young animals because of its powdery blue-gray color and softness. From this evidence and because horns are secondary sexual characteristics, it appears that periods of horn growth are governed by sex hormones.

AGE DETERMINATION BY HORN GROWTH

The age of an animal often can be determined with reasonable assurance by using horn size (or lack of horns on lambs), horn shape, and annular rings. Previous sections of this chapter describe the growth of the horns, and Figures 11.2 through 11.5 illustrate the size and shape of horns for bighorn up to age

Growth and Development 167

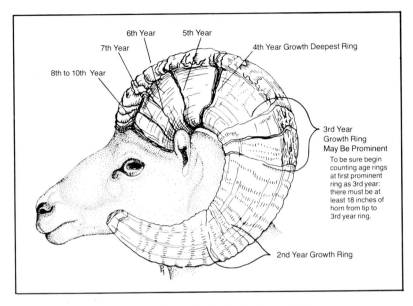

Fig. 11.7. Horn ring method of aging desert bighorn.

of 1 year. Figure 11.7 depicts how the age of a bighorn ram may be determined by the horn ring method. After about the seventh year it is not possible to determine the age of a ewe from observation of the size or shape of the horns.

Horn characteristics used to determine age.

0 to 1 week: no horns.

1 week to 1 month: tufts of hair look like short horns about 25 mm (1 in.) high.

1 to 2 months: horns just appearing above the tufts of hair, color dark blue gray.

3 months: horns 25 to 40 mm (1. to 1.5 in.) above hair.

4 months: horns 50 to 75 mm (2 to 3 in.) in male; 25 to 38 mm (1 to 1.5 in.) in female.

5 months: horns 75 to 100 mm (3 to 4 in.) in male; 50 to 75 mm (2 to 3 in.) in female.

6 months: horns 100 to 150 mm (4 to 6 in.) in male, thick and bulky; 75 to 100 mm (3 to 4 in.) in female, slender.

8 months: horns 125 to 200 mm (5 to 8 in.) in male, thick; 100 to 125 mm (4 to 5 in.) in female, slender.

1 year: horns 200 to 305 mm (8 to 12 in.) in male, appearing much like adult ewe horns but thicker at the base and still a blue-gray color in contrast to the light brown of the adult ewe; 125 to 178 mm (5 to 7 in.) in female, thin and sharp pointed.

18 months: horn growth stops or slows to the point where a depression or series of them are left on the horn. Figure 11.5 shows this as the second year growth ring.

30 months: horn growth may stop two or three times during this and the next rut period. This often leaves one prominent and one or two less prominent rings on the horn.

3.5 to 7.5 years: horn growth stops each rut period and forms a deep dark ring. These rings are separated by a portion of light brown horn. The distance between each ring is narrower each year, until only about 25 mm (0.5 in.) separates the 7.5-year ring from the 8.5-year ring.

8.5 to 16+ years: This portion of new horn is usually covered by hair in the live animal, and it is difficult to determine how many rings are present. These rings may be 3.2 mm (0.125 in.) or less apart. It is almost impossible to age ewes beyond 7.5 years because the rings are usually so close together.

It is sometimes difficult to distinguish between annular rings and other corrugations on the surface of the horns. The annual ring forms a deep continuous groove around the horn. Other corrugations consist of smoothly rounded ridges and grooves (Hemming, 1969).

If the animal is injured, sick, or feeding on an improper diet, the horns appear to be the first to suffer by becoming poorly cemented to the core. The hoofs are the second place to look for evidence of poor nutrition. On the horns the inadequate adhesion of the keratin layers will cause the horn to become extremely loose at the rings. These will separate and pieces will slough off when butting at other rams or trees and rocks. In extreme situations the horn may wear away and a portion of it can and occasionally does break off. Sometimes this leaves a large ram with as little as 150 mm (6 in.) of horn on one side. If injury or an inadequate diet occurs during the season of horn growth and is severe enough, a cessation of growth may take place, forming a ring that is neither annular or secondary. Such an abnormal ring may occur on only one horn.

Some ram horns show one or two extra rings in the vicinity of an annular ring (Fig. 11.7). If horn rings are a result of hormonal fluctuations, then it is possible that these extra rings are evidence that some rams have two or three periods of intense rutting activity between growth periods. These extra rings can be followed on the horns until the animal is 5 to 6 years old before the horn growth is so retarded that all rings for 1 year appear as one. The greatest distance between the extra rings occurs during the second, third, and fourth rut seasons, which are the years of greatest horn growth.

All of these marks on the horns make the counting of horn rings difficult to the uninitiated observer. However, most ram horns show the standard horn ring characteristics well enough to make ring counting a reliable method of aging up to 8 or 10 years. By contrast, the small size and slow growth of ewe horns reduce the reliability of this method for females.

TABLE 11.4.
Tooth Eruption and Replacement in the Lower Jaw of Bighorn

Age	Incisors*			Canine*	Premolar*			Molar*		
	1	2	3	1	2	3	4	1	2	3
Birth	D	D	D	D	D	D	(D)			
1 month	D	D	D	D	D	D	D			
6 months	D	D	D	D	D	D	D (P)			
12 months	(P)	D	D	D	D	D	D	P	(P)	
16 months	P	D	D	D	D	D	D	P	P	
24 months	P	D	D	D	D	D	P	P	P	
30 months	P	D	D	D	(P)	(P)	D	P	P	(P)
36 months	P	(P)	D	D	(P)	(P)	(P)	P	P	(P)
42 months	P	P	D	D	P	P	P	P	P	P
48 months	P	P	P	P	P	P	P	P	P	P

Source: Cowan (1940), Deming (1952), and specimens at Montana State University.

*D = milk or deciduous tooth; P = permanent tooth. Parentheses indicate that the tooth is in process of replacement at that age.

TOOTH DEVELOPMENT

Tooth development appears to be the same in all forms of North American wild sheep. At birth the entire deciduous series of teeth are present, although they are not usually exposed above the gums. Giles (1969) presents data on tooth eruption and replacement in Table 11.4 (his Table 20.42).

Both Giles (1969) and Cowan (1940) indicate that the new bighorn lamb has six teeth present and exposed. However, this is not the case according to Deming (1952) and Hansen (1964b). Hansen found that at about 4 days of age a ewe lamb had four incisors. At 1 week she had added two more incisors and a day later a pair of molars or premolars appeared. The bottom teeth developed faster, and before she was 2 weeks old she had about 25 mm (1 in.) of grinding surface, probably two full molar-type teeth. At 14 days she had four sets of incisors showing.

Deming (1952) states that "Throughout the period of tooth development and addition there is a steady growth of the tooth bearing sections of the skull. The first full set of deciduous or milk teeth are present at approximately one month of age, and consist of four incisiform teeth on each side of the lower jaw accompanied by three upper and lower premolars."

Tooth anomalies. Allred and Bradley (1965) state that "Cowan ... Dalquest and Hoffmeister ... and Deming ... all indicate that the complete set of cheek teeth are present by four years of age. They all point out the presence of a number of aberrations, damaged and lost teeth, shattered teeth and other irregularities of the tooth row. Our study fully supports these findings and indicates that deviations from the (normal) dental formula of I-0/3, C-0/1, P-3/3, M-3/3 are common."

TABLE 11.5.
Aging Bighorn According to the Replacement of Incisiform Teeth

Age	Incisiform Teeth*			
	1	2	3	4
Birth	D	D	D	D
1 year	(P)	D	D	D
2 years	P	D	D	D
3 years	P	(P)	D	D
4 years	P	P	P	P

*Parentheses indicate that the tooth is in process of replacement at that age.

TABLE 11.6.
Characteristics That Can Be Used To Roughly Distinguish Young From Old Bighorn in the Field

Young	Old
Pelage light.	Dark.
Ram horns widespread or triangular when seen from front.	Squarish from front.
Body well-rounded.	Angular.
Rump rounded.	Sloping and thin.
Ram horn, at lowest point of curl, less than 38 mm (1.5 in.); thin and pointed.	More than 38 mm (1.5 in.); thick and blunt at tip.
Ram horn, at base, less than 100 mm (4 in.) in diameter.	More than 100 mm (4 in.) in diameter.
Ram head with little or no knob on back of head.	Large knob on back of head, often calloused and hair rubbed off.
Diameter of curl of ram horn usually less than 300 mm (12 in.)	Curl usually more than 300 mm (12 in.)
Lowest portion of curl of ram horn not extending below lower jaw.	Lowest portion of curl extending below lower jaw, so that lower portion and back of jaw is obscured.
Nose relatively short; forehead broad.	Long.
Face profile between eyes concave.	Between eyes relatively straight.
Ram horns 50 to 100 mm (2 to 4 in.) above high point of back when head turned.	Horns touch back when head turned, especially in rams with large horns.
Front legs straight.	Bowed to front at knees.
Head usual held erect.	Head held low.
Attitude alert and flighty.	Slow, sedate; not flighty.

According to Bradley and Allred (1966) 29 percent of the bighorn over 4 years of age that were examined had at least one second premolar unerupted. This anomaly was far more prevalent in ewes, in which 54 percent had an incomplete molar series, as compared to 13 percent of the rams.

AGE DETERMINATION BY TOOTH REPLACEMENT

Bighorn can be aged up to 4 years with some accuracy by using the replacement of teeth as a criterion. Table 11.5 shows how incisiform teeth can be used to determine age up to 4 years, by which time the animal has a "full" mouth of permanent teeth including the cheek teeth and incisiform teeth.

Tooth wear has not been studied in relation to determining age of bighorn after the first 4 years. Allred and Bradley's 1966 study shows that osteonecrosis and osteolysis occurred in nearly 80 percent of all the teeth of Nevada bighorn that they examined. Second in frequency of tooth decay was Arizona, with about 35 percent; California followed, with less than 20 percent of the teeth affected. This would indicate that the most healthy teeth are from California and the least healthy from Nevada. It seems likely that the least healthy teeth may wear faster. Therefore, it could be postulated that animals of the same age in California would show less wear than those in Arizona or Nevada. Assuming this to be the case, tooth wear would have to be studied in each area before it could be used as a criterion for aging bighorn after 4 years of age.

AGE DETERMINATION BY SIZE AND SHAPE OF THE WHOLE ANIMAL

Table 11.6 presents several other characteristics of bighorn that can be used to determine age. Although not accurate to the year, they indicate whether the animal is young or old. If they are used in combination they may assure a more accurate determination.

FACTORS INFLUENCING GROWTH AND DEVELOPMENT

Genetics. The wide variation in the size and shape of desert bighorn is determined in part at least by heredity. This wide range is demonstrated on Desert National Wildlife Range where horn shapes and sizes vary from a wide spread with little curl to a narrow spread with a tight curl. A heavy set of horns that was inherited could cause the animal to develop its neck muscles and perhaps hold its neck more erect in order to compensate for and balance the added weight. Thus a genetic factor could influence the growth and development of related body structures.

Environment. Poor range conditions do not appear to influence the size and shape of individual bighorn, but they may cause or allow genetic factors to survive or disappear eventually in the normal process of evolution. In artificial situations, nutrition may be found to influence certain characteristics. An animal raised under conditions of poor nutrition will probably be sickly and die of the first serious disease that it contracts. Horn size and leg length do not appear to be influenced by nutrition. Bighorn raised in pens at Desert Wildlife Range were fed similar diets, and attained horn and body sizes and shapes that conformed to their parental inheritance (Hansen, 1970).

12. Natural Mortality and Debility

Rex W. Allen

Extensive pathological surveys based on necropsies have been made in the Big Hatchet Mountains in New Mexico, on Kofa Game Range, Arizona, and on Desert National Wildlife Range, Nevada. From other localities inhabited by desert bighorn, information is still sketchy, or entirely lacking, and is based principally on fecal examinations for parasitic lungworms.

PARASITES AND DISEASE

Bacterial diseases. Bighorn of various races apparently are susceptible to lung diseases. Desert bighorn are no exception: abnormal pneumonic conditions appear to be widespread among them. Lung adhesions were present in twelve bighorn collected on Kofa Game Range and in one ewe from the Gila Mountains near Yuma, Arizona (Russo, 1956). Pneumonia was found in one ewe from the Lake Powell study area of southeastern Utah (Wilson, 1966). Pathological conditions in another ewe from southeastern Utah were apparently related to streptococci, staphylococci, and other pathological forms; the lungs were pneumonic and yielded on culture *Escherichia coli*, a streptococcus (beta), and an unidentified bacillus (Wilson, 1968). Helvie and Smith (1970) noted that *Pasteurella* and *Corynebacterium* were causes of pneumonia in Nevada bighorn, and they attributed deaths to enterotoxemia *(Clostridium).* Welles and Welles (1961a) observed a cough, nasal discharge, and swollen eyes in a lamb in Death Valley, California, suggesting lung involvement. Pneumonia was considered by Johnson (1957) as a possibly important cause of death in bighorn on Desert National Wildlife Range. *Pasteurella hemolytica* was isolated by Taylor (1973) from Nevada lambs suffering from pneumonia. He considered these bacteria to be the cause since he was unable to isolate any other organisms, including viruses, chlamydiae, or mycoplasmas.

Despite the importance of bacteria as causes of pneumonia, these organisms appear to have been identified as causative agents of lung conditions in desert bighorn only in a limited number of instances. In addition to the findings of Wilson (1968), Russo and his co-workers (1956) in Arizona isolated and identified strains of the genus *Corynebacterium,* which are involved in lung and pleural cavity lesions in ovines. According to Runnells (1954), these organisms affect adults primarily. The lymph nodes and lungs are involved and have characteristic greenish-yellow, caseous, or purulent foci. The lesions may resemble those of tuberculosis, except in color and in the absence of calcification.

Organisms of the genus *Corynebacterium* have also been isolated from lung lesions in Rocky Mountain bighorn. An organism of this genus was isolated by C. W. Jenner from the horn core of a ewe found in Anza-Borrego Desert State Park, California. A "streptococcus-like" organism was also cultured from the horn core. At necropsy this animal showed embolic pneumonia, with pleuritis and pericarditis. Organisms of the genus *Pseudomonas* and a "Mimae-Moraxella type" were isolated from lung lesions and nasal exudate. Possibly viruses and fungi are involved in lung abnormalities of desert bighorn, but this has not been confirmed in the laboratory.

A report (N. Papez) of necropsies conducted by R. E. L. Taylor on two lambs from Desert Wildlife Range lists pneumonia as the cause of death in both instances. *Corynebacterium* sp. and organisms tentatively identified as *Hemophilus ovis* were isolated from the lung lesions.

Although occurrence in capitve animals may be misleading, fibrinous pericarditis, a condition usually associated with bacterial infection, was observed by Pournelle (1964) in a ram that was captured on Desert Wildlife Range and died after several years in the San Diego Zoo. This animal also showed carcinoma of the lung and an ulcer of the colon.

Organisms of the genus *Actinomyces* are important as a cause of malformation of the bones (e.g., lumpy jaw of cattle); they are suspected by Allred and Bradley (1965) as a cause of abnormalities in bighorn skulls collected on Desert Wildlife Range. Welles and Welles (1961a) conclude that there must be a low incidence of *Actinomyces* organisms in Death Valley because skulls were remarkably free of dental abnormalities. An adult ewe skull collected in the Red Canyon area of southeastern Utah by C. A. Irvine appeared to have been affected by these organisms (Juan Spillett). Apparently laboratory confirmation of *Actinomyces* organisms from desert bighorn is still lacking.

Brucellosis and leptospirosis have been suspected as occurring in desert bighorn, but there is no supporting evidence. Serological tests for the former disease were made on at least fifty-four bighorn in Arizona (Russo, 1956; Reed, 1960); seven in New Mexico (Allen, 1955); and seventeen in Nevada (Johnson, 1957; Taylor, 1973). Seventeen tests for *Leptospira* have been made in Arizona (Reed, 1960) and fourteen in Nevada (Taylor, 1973), but seemingly none elsewhere. All tests were negative.

Viral disease. The only laboratory diagnosis of a virus disease in desert bighorn appears to be that reported by Hailey (1966). A necropsy performed by R. M. Robinson on a lamb from the Black Gap area of Texas indicated that the animal died of bluetongue. This is documented by Robinson et al. (1967), who isolated the causative virus and subsequently supplied it to Vosdingh et al. (1968) for transmission studies involving bighorn and whitetail deer. These investigators produced fatal bluetongue in the one deer receiving it. They also noted a strong similarity between bluetongue and epizootic hemorrhagic disease of deer. Bluetongue in domestic sheep is characterized by localized inflammation, necrosis of the mouth and tongue, and scab formation on the lips and nostrils. Cattle also are susceptible to bluetongue. The disease is transmitted by biting midges *(Culicoides)*.

Serological evidence of bluetongue in New Mexico bighorn is reported by Trainer and Jochim (1969). However, Trainer subsequently stated that the report was in error and that no New Mexico samples from desert bighorn were involved.

Taylor (1973) reported finding antibodies to myxovirus parainfluenza-3 in Nevada desert bighorn, noting that this virus has been associated with "shipping fever" pneumonia in cattle.

Fungal disease. Apparently fungi have not been observed as a cause of disease among desert bighorn. Russo (1956) records two cases of "horny dermatitis" in Arizona bighorn, but a causative fungus could not be isolated (Fig. 12.1).

Internal and external parasites. Ransom (1911) reported the first record of parasites from desert bighorn, involving the gastrointestinal nematodes *Haemonchus contortus, Trichostrongylus extenuatus (= T. axei),* and *Trichuris ovis.* W. W. Becklund reports the parasites were collected in 1904 from a host animal in the National Zoological Park, Washington, D.C. In 1912, Bishopp (Bishopp and Trembley, 1945) described the tick *Dermacentor hunteri* from bighorn near Quartzsite, Arizona. This was followed by H. E. Kemper's discovery (see Allen and Kennedy, 1952) of the spinose ear tick, *Otobius megnini,* on a bighorn near Roswell, New Mexico. Some 25 years elapsed before additional published records became available.

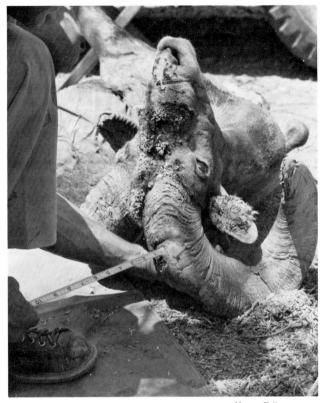

Homer Erling

Fig. 12.1. "Horny dermatitis" in a ram collected from the Kofa Game Range, Arizona.

Becklund and Senger (1967) have published a comprehensive checklist of parasites reported from *Ovis canadensis*. Their list includes all but a few of the reports concerning desert bighorn but gives no quantitative data. This information may be found in the various publications cited in the accompanying Table 12.1, which lists the parasites reported and gives the scientific and common names, location in the host, and the geographical distribution.

Surveys in Arizona, New Mexico, and Nevada show that the predominant external parasites are hard ticks, *Dermacentor albipictus* and *D. hunteri*. The former has a wide distribution and occurs on many hosts besides bighorn. The distribution of the latter, however, is much more limited (Brinton et al., 1965). Until recent years, it was known only from bighorn in the original collection area adjacent to the Colorado River in Arizona. In 1953 it was collected from bighorn on Desert Wildlife Range by O. V. Deming, who submitted specimens to E. R. Quortrup for identification. This tick has been

TABLE 12.1.

Parasites Reported from Desert Bighorn

Scientific Name	Common Name	Location in Host	Geographical Distribution
Eimeria granulosa	Coccidum	Small intestine	Utah—Lake Powell area (Wilson, 1968)
Eimeria pallida	Coccidum	Small intestine	Utah—Lake Powell area (Wilson, 1968)
Sarcocystis	Sarcocyst	Heart muscle	Nevada—Desert Wildlife Range (Deming, unpubl.)
Cysticercus tenuicollis	Bladderworm	Peritoneal cavity	Arizona—Kofa Game Range (Russo, 1956; Allen & Erling, 1964) New Mexico—San Andres Refuge (Allen & Kennedy, 1952); Big Hatchet Mts. (Allen, 1955)
Thysanosoma actinioides	Fringed tapeworm	Bile ducts and duodenum	Arizona—Kofa Game Range (Russo, 1956; Allen & Erling, 1964) California—Anza-Borrego Desert State Park (J. C. Turner in Russi & Monroe, 1976) Nevada—Desert Wildlife Range (Allen, 1962, 1964) Utah—Lake Powell area (Wilson, 1968)
Wyominia tetoni	Wyoming tapeworm	Bile ducts and duodenum	Arizona—Kofa Game Range (Russo, 1956; Allen & Erling, 1964) Nevada—Desert Wildlife Range (Allen, 1962, 1964) Utah—Lake Powell area (Wilson, 1968)
Moniezia sp.	Common tapeworm	Small intestine	Utah—Lake Powell area (Wilson, 1968)
Haemonchus contortus	Large stomach worm	Abomasum	Captive animal (Ransom, 1911)
Haemonchus placei	Large stomach worm	Abomasum	New Mexico—Big Hatchet Mts. (Allen, 1955)
Pseudostertagia bullosa	Medium stomach worm	Abomasum	New Mexico—Big Hatchet Mts. (Allen, 1955)
Trichostrongylus extenuatus (=axei)	Hairworm	Abomasum	Captive animal (Ransom, 1911)
Nematodirus spathiger	Thread-necked worm	Small intestine	New Mexico—San Andres Refuge (Allen & Kennedy, 1952)
Cooperia sp.	Cooperia	Small intestine	Utah—Lake Powell area (Wilson, 1968)
Oesophagostomum sp. (larva)	Nodular worm	Large intestine	New Mexico—Big Hatchet Mts. (Allen, 1955)
Trichuris sp.	Whipworm	Large intestine	Nevada—Desert Wildlife Range (Deming, unpubl.) New Mexico—San Andres Refuge (Allen & Kennedy, 1952)
Trichuris ovis	Whipworm	Large intestine	Captive animal (Ransom, 1911)

Trichuris discolor	Whipworm	Large intestine	Nevada—Desert Wildlife Range (Allen, 1962, 1964) New Mexico—Big Hatchet Mts. (Allen, 1955)
Skrjabinema sp.	Pinworm	Large intestine	Arizona—Kofa Game Range (Russo, 1956; Allen & Erling, 1964) California—San Gabriel Mts. (Jenner, unpubl.); Anza-Borrego Desert State Park (Russi & Monroe, (1976) New Mexico—San Andres Refuge (Allen & Kennedy, 1952) Utah—Lake Powell area (Wilson, 1968)
Skrjabinema ovis	Pinworm	Large intestine	New Mexico—Big Hatchet Mts. (Allen, 1955) Nevada—Desert Wildlife Range (Allen, 1962, 1964)
Protostrongylus stilesi	Lungworm	Lung parenchyma	California—Death Valley (Welles & Welles, 1961a); Santa Rosa & Old Woman Mts. (Buechner, 1960) Nevada—Desert Wildlife Range (Allen, 1962, 1964; Buechner, 1960; Johnson, 1957)
Muellerius capillaris	Lungworm	Lung parenchyma	California—Anza-Borrego Desert State Park (Russi & Monroe, 1976)
Dermacentor albipictus	Winter tick	Skin	Arizona—(Bishopp & Trembley, 1945); Kofa Game Range (Russo, 1956; Allen & Erling, 1964) New Mexico—San Andres Refuge (Allen & Kennedy, 1952); Big Hatchet Mts. (Allen, 1955)
Dermacentor hunteri	Hunter tick	Skin	Arizona—Kofa Game Range (Russo, 1956; Allen & Erling, 1964); Quartzsite (Bishopp & Trembley, 1945) California—San Gabriel Mts. (Jenner, unpubl.); Anza-Borrego Desert State Park (J. C. Turner in Russi & Monroe, 1976) Nevada—Desert Wildlife Range (Allen, 1964; Brinton et al., 1965; Deming, unpubl.)
Dermacentor variabilis	Dog tick	Skin	California—Anza-Borrego Desert State Park (Russi & Monroe, 1976)
Ixodes sp.	None	Skin of abdomen	California—San Gabriel Mts. (Jenner, unpubl.)
Otobius megnini	Spinose ear tick	Ear	California—San Gabriel Mts. (Jenner, unpubl.) New Mexico—San Andres Refuge (Allen & Kennedy, 1952); Roswell (Kemper, 1947); Big Hatchet Mts. (Allen, 1955)
Psoroptes sp.	Scab mite	Ear and body	New Mexico—San Andres Refuge (Lange et al., 1980)
Psoroptes cervinus (sic)	Scab mite	Ear	Nevada—Desert Wildlife Range (Cater, 1968)
Psoroptes ovus (sic)	Scab mite	Body exterior	Nevada—Desert Wildlife Range (Cater, 1968)

found in California desert bighorn along with *Ixodes* sp. and *Otobius megnini* (C. W. Jenner). *Dermacentor hunteri* is reported from mule deer on Kofa Game Range (Russo, 1956; Allen and Erling, 1964), and it is reasonable to assume that deer are instrumental in the propagation of these parasites among desert bighorn.

Dermacentor variabilis was recorded from bighorn sheep for the first time by Russi and Monroe (1976), whose observations were made in Anza-Borrego Desert State Park, California. These investigators mention that Jack C. Turner found *D. hunteri* in the same locality.

Spinose ear ticks, *Otobius megnini,* have been collected only from desert bighorn in New Mexico and the San Gabriel Mountains of California. Since cattle are the predominant hosts of these ticks, it seems probable that bighorn ranging with or near cattle might be more heavily infested. Ticks may harm their hosts not only by sucking blood but by serving as vectors of disease-producing organisms or by creating openings which would allow invasion by other pathogens.

Whether nose bots, *Oestrus ovis,* occur with any frequency in desert bighorn is not known. Despite Capelle's (1966) report indicating prevalence in Wyoming and Montana bighorn, there is little or no valid information in the Southwest, because hunters and few biologists are willing to sacrifice bighorn heads for the type of examination required to make a reliable search for nose bots.

The nose bot fly is suspected (Bunch et al., 1978) of being the cause of chronic sinusitis with subsequent bone disease and skull anomalies in some desert bighorn from at least Arizona and Nevada (in this connection, see Chapter 4, page 59). Sinusitis is believed to be a serious disease with a relatively high incidence in these states, as much as 41 percent in individuals more than one year old in a captive flock. It is thought to be terminal. The initial infection brought on by the bot fly larvae is aggravated by secondary infection caused by *Corynebacterium*. However, the exact nature of this sinusitis requires further investigation.

Screw worms were observed in the nose of a desert bighorn in the Black Gap area of Texas by Hailey (1966) but the species was not identified.

Although scabies has been reported from bighorn in various localities, its distribution in desert bighorn has not been widely documented. Cater (1968) notes several cases on Desert Wildlife Range and lists the causative mites as *Psoroptes cervinus* (sic) and *P. ovus* (sic). Scabies mites, apparently *P. cuniculi,* were found on bighorn in southwestern Utah by Irvine (1969a). Becklund and Senger (1967) refer in a general way to reports of scabies on bighorn in California, and it is quite possible that desert bighorn were the hosts involved. A report (1980) by Lange et al. describes extensive and serious scabies in desert bighorn of the San Andres National Wildlife Refuge in southern New Mexico.

Protozoan blood parasites may occur, but this problem has been only partially investigated in bighorn in the Southwest. It is complicated by difficulty in securing satisfactory blood smears (Allen, 1961). Suitable smears have been examined from five bighorn in the Big Hatchet Mountains and two from Kofa Game Range, but these were negative.

Coccidia are protozoan parasites of wide distribution among animals; they may seriously affect animal health. They usually occur in the intestines and are transmitted directly from one host to another. The only record of these parasites in desert bighorn appears to be that of Wilson (1966), who found two species in bighorn from the Lake Powell area of Utah. The two species concerned are found only in ovines, and both are of common occurrence. Three fecal samples collected from bighorn of San Andres Wildlife Refuge in November 1968 contained only one oocyst in the three samples, and it resembled *Eimeria crandallis* (D. M. Hammond).

Cysticerus tenuicollis is the intermediate stage of a tapeworm of carnivores. Presumably the principal definitive host in bighorn habitats is the coyote *(Canis latrans)*. This bladderworm is prevalent in desert bighorn of Arizona and New Mexico and probably occurs elsewhere.

Two adult tapeworms occur in the liver (bile ducts) of bighorn. One of these, *Thysanosoma actinioides*, is a parasite of practically all ruminants except cattle. The other, *Wyominia tetoni*, has been reported only from bighorn. There is evidence that the former species is transmitted by an insect vector of the order Corrodentia, but the life cycle of *W. tetoni* is completely unknown. Both species of tapeworm are prevalent in desert bighorn of Arizona and Nevada, but neither has been found in New Mexico. *Thysanosoma actinioides*, *Wyominia tetoni*, and *Moniezia* sp. have been found in desert bighorn of southeastern Utah by Wilson (1968). Tapeworms of the genus *Moniezia* occur only rarely in desert bighorn; they have not been recorded from this host in New Mexico, Arizona, or Nevada.

The most numerous and widespread gastrointestinal roundworm is the pinworm *(Skrjabinema* spp.*)*. These have been found in large numbers in bighorn of New Mexico, Arizona, and Nevada. They were also recently found in bighorn of Anza-Borrego Desert State Park (Russi and Monroe, 1976). They have a direct life history and are considered relatively harmless.

Less numerous and widespread are roundworms of the genera *Haemonchus, Trichostronglyus, Nematodirus, Oesophagostomum*, and *Trichuris*. They occur in both wild and domestic ruminants and have direct life histories.

Present knowledge indicates that the lungworm, *Protostrongylus stilesi*, may be of minor importance as a cause of pneumonia in desert bighorn. These parasites have not been found in bighorn in Arizona despite diligent searching. They occur in New Mexico, but only in animals descended from importations of Rocky Mountain bighorn; the desert bighorn of Big Hatchet and San Andres mountains apparently are free from lungworms. It is noteworthy that a lungworm new to bighorn sheep was reported by Russi and Monroe (1976) from Anza-Borrego State Park; this lungworm is *Muellerius capillaris,* according to a finding of first stage larvae in fecal samples.

The report of Wilson (1966) indicates that few, if any, bighorn in the Painted Desert area of Utah harbor lungworms. Although Welles and Welles (1961a) stated that these parasites were found in bighorn in Death Valley, Buechner (1960) reported that twenty-seven fecal samples from Death Valley were negative. The same author discovered lungworms on Desert Wildlife Range, but E. R. Quortrup, who made the laboratory determinations, described the infections as of low degree. Quortrup also described infections as light in samples collected on Desert Wildlife Range by Johnson (1957), who also observed several cases of pneumonia in which lungworms were absent. Allen (1962, 1964) found lungworms in ten of twenty-three sets of lungs from Desert Wildlife Range, but observed only very small areas of infection. Buechner (1960) and Jones et al. (1957) reported larvae in two of five fecal samples from the Santa Rosa Mountains, California, and Buechner (1960) reported larvae in one out of eight samples collected in the Old Woman Mountains, California, all identified by Quortrup. Thus, the information available at this time suggests that agents other than lungworms are more important as causes of lung abnormalities in desert bighorn.

It is difficult to estimate the effect of the various parasites on the health and well-being of bighorn, because evaluations based on experimental infections of bighorn have not been carried out. Under the circumstances, reliance is placed on field or zoo observations, and these may be misleading. Most of the information is derived from experimental infections of domestic sheep, and this information may be useful as a guide in interpreting findings on parasitism. Bloodsucking ticks and the large stomach worms cause anemia; lungworms may be a primary or a secondary cause of pneumonia; coccidia may injure the intestine, interfering with digestion and causing diarrhea; hairworms (also known as bankrupt worms) may injure the gut lining, causing an abnormal loss of fluids; several other kinds of worms injure the tissues during the course of their normal migration; and one species causes sizable tumorlike nodules in the walls of the gut.

All parasites are potentially harmful; the degree of harm depends on the numbers present. In turn, animal crowding, their nutritional status, exposure to domestic stock, and climatic conditions also affect the host-parasite relationship. Optimal conditions of warmth and moisture favor the survival and propagation of the free-living stages of the parasites, as well as the survival and propagation of any invertebrate vectors they might utilize. These factors markedly influence the transmission of parasites to new hosts.

For descriptions of most of the parasites mentioned in this report, the reader is referred to Honess and Winter (1956). Present knowledge concerning lungworms and tapeworms in desert bighorn is reviewed by Allen (1971).

OTHER MORTALITY AND DEBILITY FACTORS

Tumors. Tumors have been found in desert bighorn on more than one occasion. In addition to Pournelle's (1964) report mentioned previously, Deming (1963) referred to a diagnosis made by Quortrup on material evidently from Desert National Wildlife Range. The diagnosis was adenocarcinoma of the lungs, resulting from the metastasis of an ovarian tumor.

Mineral and dietary deficiencies. There is a dearth of information concerning mineral and dietary deficiencies in the various desert habitats of the bighorn. One might assume, therefore, that these are relatively unimportant factors. Johnson (1957) suggests that there might be phosphorus and iodine deficiencies and that bighorn advantageously might be given a phosphorus supplement. He pointed out also that the animals might be affected adversely in some areas by an excess of selenium and magnesium.

Moore (1961) notes that death losses in the Black Gap area of Texas were possibly due to mineral or vitamin deficiencies. According to Wilson (1966), much of the Lake Powell study area of Utah is deficient in phosphorus and iodine. Irvine (1969a) found plants of southeastern Utah to be deficient in phosphorus, particularly for lactating ewes.

The real effects of mineral deficiencies and toxicities on bighorn are as yet unknown. Apparently there has been no experimentation along these lines. We may infer from what is known about other animals that these factors do at times exert an influence on the health of bighorn. In support of Johnson's (1957) view, studies by Watkins and Repp (1964) of domestic stock and their ranges in New Mexico showed that the most important mineral deficiency was in phosphorus at times of the year other than the growing season. Deficiencies observed with respect to calcium, protein, and vitamin A were relatively unimportant.

Proper evaluation of mineral and dietary deficiencies in bighorn will require suitable chemical analysis of blood samples from these animals, as well as analysis of the plants which make up their food.

Poisonous plants. Comparatively few published reports throw light on the importance of poisonous plants on desert bighorn ranges. Of the few surveys of such plants that have been made, the most comprehensive list appears to be that compiled by Tom Allen and reported by Moore (1961) for the Black Gap area of west Texas. This list included *Conyza canadensis, C. coulteri,* nightshade *(Solanum eleagnifolium),* desert marigold *(Baileya multiradiata),* lechuguilla *(Agave lechuguilla),* paper-flower *(Psilostrophe tagetina),* snakeweed *(Gutierrezia microcephala),* wild tobacco *(Nicotiana trigonophylla),* cloak fern *(Notholaena sinuata* var. *cochisensis),* and centaury *(Centaurium calycosum).* Moore did not attribute death losses to any of these plants. Hailey (1962) states that death losses at Black Gap could have been caused by centaury, but in 1964 the same author said that there had been no ill effects from poisonous plants for the past 3 years.

Buechner (1960) provides quite strong evidence that beargrass *(Nolina microcarpa)* caused heavy losses of bighorn on San Andres National Wildlife Refuge, and Deming (1964) mentions Jimson weed *(Datura* sp.*)* as a possible cause of death of a lamb on Desert Wildlife Range. Deming refers to the presence of other poisonous plants in the same locality, including spineless horsebush *(Tetradymia* sp.*)*, desert almond *(Prunus fasciculata)*, loco *(Astragalus* sp.*)*, and species of the genera *Oxytropis, Delphinium, Asclepias, Nicotiana, Baccharis, Actinea, Aplopappus, Datura,* and *Linum.* However, he observes no positive evidence of plant poisoning during the period of his field observations.

Over twenty species of poisonous plants on bighorn ranges of Utah were reported by C. A. Irvine (Juan Spillett). Irvine observed bighorn eating some of these plants, but their actual effects are not known. The following poisonous plants were listed from the White Canyon area of San Juan County, Utah, by Wilson (1968): threadleaf snakeweed *(Gutierrezia microcephala)*, pingue *(Hymenoxys richardsoni)*, Washington lupine *(Lupinus polyphyllus)*, king lupine *(L. kingi)*, tailcup lupine *(L. caudatus)*, prickly acerosa *(Oxytenia acerosa)*, larkspur *(Delphinium* sp.*)*, foothill deathcamas *(Zygadenus paniculatus)*, and greasewood *(Sarcobatus vermiculatus)*. In this instance, too, no information is available concerning bighorn mortality attributable to these plants.

At present, evidence that plants are poisonous to desert bighorn is largely circumstantial, and precise evaluation of their importance is lacking. Elucidation of the problem will require more extensive plant surveys, and detailed experimental studies to determine the toxicological effects of the various plants on the desert bighorn itself under controlled conditions.

Lowell Sumner suggests that on an open range under normal conditions, bighorn, like other native animals, are unlikely to find poisonous native plants sufficiently attractive to eat them in dangerous quantities. The survival mechanisms of the bighorn probably are oriented toward avoiding such plants and situations where alternative plants are not available. This self-protective situation conceivably could break down on ranges that have been excessively denuded by domestic livestock and feral burros. Even in such cases the bighorn most probably would move on in search of other food, rather than linger in an area of nearly unpalatable, poisonous plants, because, unlike livestock, the bighorn would not be confined to such an area by fence.

Porcupine damage. Porcupine *(Erethizon dorsatum)* damage to desert bighorn is rare, since the two mammals' ranges do not ordinarily coincide, but it is interesting. Pulling (1945) reports finding in 1944 on Desert Wildlife Range the skull and a few other bones of a desert bighorn ewe with porcupine quills embedded in a small patch of skin remaining on the forehead. He theorized that the ewe, from curiosity or in a spirit of playful pugnaciousness, threatened the porcupine and was slapped across the face by the latter's tail and was possibly blinded, causing her to become a victim of predators or starvation. Jack Hall and Gale Monson, while watching bighorn at Tunnel Spring on Kofa Game Range in June 1956, saw a bighorn ewe whose forehead and top of head were literally studded with porcupine quills. Special care was taken to note that they were not cholla cactus spines. They decided the quills accounted for this ewe's antagonistic behavior toward another ewe and a lamb that accompanied her. As far as known, only two porcupines have ever been found on Kofa Game Range by refuge personnel.

Climate. Desert bighorn habitat is characterized by extreme climatic conditions which affect the availability of forage and water. Monson (1960) describes a period between August 1955 and October 1957 when there was no rainfall of consequence in southwestern Arizona. During this period an unusual number of bighorn were found dead due to causes other than predation and disease. He concludes "...that those animals of lowest vitality succumbed during this period due to food conditions and possibly water lack brought on by the drouth." Supporting this conclusion was the fact that the majority of the dead animals exceeded 8 years of age. The impact of the drought was also seen in the ewe-lamb ratio in the summer water-hole counts. It is Monson's opinion that, compared with disease and predation, abnormal climatic conditions may have been underemphasized as causes of low population densities of desert bighorn.

Adverse climatic conditions undoubtedly played a major role in a drastic reduction in numbers of desert bighorn between 1953 and 1960 in the Big Hatchet Mountains. Gordon (1957) considers that competition with deer and livestock was the principal causative factor but implies that severe drought also was a cause. Gross (1960) states that starvation was the primary cause of the die-off.

Gross (1960) refers to a 1956 drought in southern New Mexico as being the most severe in about 700 years and mentioned a related catastrophic deer die-off as evidence of the effect of the drought. He concludes that starvation for food and water was the primary cause of the reduction in the bighorn population, with a possibilty that disease and predation played secondary roles. Likewise, Larsen (1962b) considers that a combination of prolonged drought and heavy forage over-utilization was the cause of the drastic decrease in numbers of bighorn in the Big Hatchet Mountains.

However, it seems that drastic reductions of desert bighorn populations due to climate are a rare occurrence. Even in Death Valley, according to Welles and Welles (1961a), "A general shortage of food critical enough to threaten seriously the survival of the species has not occurred here in the past 10 years, if ever." These observers note that the majority of bighorn deaths occur during the first year: "The lambs' need of water and 'emergency rations' in spring areas in hot and dry weather appears to be more urgent than in adults, suggesting at least a contribution to juvenile mortality by the harsh environment.''

Despite the aforementioned indications of a relationship between drought and the stability of desert bighorn populations, it is the opinion of Buechner (1960) that until such time as more studies are made and more data gathered, it can only be hypothesized that there is such a correlation. However, he considers that records from Desert National Wildlife Range in Nevada indicate a correlation between drought and low lamb-ewe ratios.

As for the more direct climatological effects on desert bighorn, there seems to be no evidence that extreme heat and cold are detrimental to the population stability. A tolerance of high temperatures by these animals is noted by Welles and Welles (1961a), who also states there seems to be an equal tolerance for cold. They point out that a newborn lamb seems impervious to freezing weather and that there are no records of direct activity changes because of low temperatures.

Predators. Several animals are suspected of being predacious on bighorn (see Chapter 13). Circumstantial evidence is quite conclusive that some desert bighorn losses occur because of predation, but direct evidence is almost lacking. The predator situation has been well summarized by Buechner (1960) and Goldman (1961), who indicate a general agreement that predators will at times harass or kill bighorn, but no information is available as to the extent of mortality and debility caused by these predators. It seems to be the consensus that predation is not usually an important factor in limiting populations of bighorn sheep.

Accidents. Desert bighorn are injured or killed on occasion by falls, by being trapped in forks of trees and shrubs, in depressions, and in fences, by being trapped in water holes or tanks, by fighting, and by motor vehicles. Welles and Welles (1961a) note injuries probably caused by falls. This was indicated in one case by the location of a carcass and in other cases by broken limbs. Wilson (1968) observed evidence of a fatal fall over a 100-foot cliff by a 7-year-old ram. He also lists several plants capable of causing mechanical injury to bighorn.

A report by Mensch (1969) on a "death trap" water hole is discussed in Chapter 7. Such death traps have also been found in the Gila Mountains and the Sierra Pinta of Yuma County, Arizona; in the latter case an artificial ramp stopped further losses (Gale Monson). Hansen (1960) mentions that accidents account for many dead lambs. Road kills near Lake Mead are noted by Sleznik (1963). Jousts may have serious consequences for bighorn rams, according to Honess and Winter (1956). Actual losses of desert bighorn due to accidents, and their significance to population dynamics, cannot be estimated at this time because of lack of information.

Charles G. Hansen writes: "Physical accidents can cause malformed appendages or organs on desert bighorn, but it is interesting how often bighorn recover from broken bones or tissue damage [Welles, 1968]. Legs with compound fractures heal and appear to be as good as new. In one instance I saw a ram with the right femur broken and protruding from the thigh. This animal was fat and sleek. He would butt heads with another ram and kept pace with him as they ran off. The break appeared to have been several years old. An injured or malformed horn may cause a change in the shape of the skull of older rams because of the change in weight."

Group mortality. Group mortality among desert bighorn, i.e., the finding of several carcasses simultaneously, has been recorded, including at the above-mentioned death traps. Since the numbers of these animals in their natural habitats are relatively small, group mortality of any consequence could have a telling effect on populations. Unfortunately the precise causes of mortality in bighorn are difficult to determine, so that records are woefully lacking in this respect. The fact that much of the mortality occurs in remote areas not frequently visited by observers contributes to the problem. Hansen (1961) compiles records of 242 carcasses for Desert Wildlife Range over a period of several years. In 90 percent of the cases the causes of death were unknown.

Group mortalities in southwestern Arizona were described by Monson (1965), whose records accounted for some twenty-three desert bighorn. In no case could the cause be determined, but Monson suggested that disease, bad water, or bloat might be involved. He discounted predation and lack of water.

Smith (1966) reports that fourteen carcasses were found in one area on San Andres Refuge in 1950 and suspected that plant poisoning was the cause, possibly by beargrass *(Nolina microcarpa),* also called sacahuista.

INBREEDING

It is apparent that desert bighorn herds have become increasingly isolated geographically. As a result, there is less genetic exchange from other populations (gene pools). It is possible that genetic and endocrine-immune deficiencies which do not favor survival have resulted from isolation-induced inbreeding. If so, many bighorn herds face extinction through inbreeding and therefore new emphases must be introduced into desert bighorn management (DeForge et al., 1979).

13. Predator Relationships

Warren E. Kelly

Evidence indicates that bighorn arrived on the American scene after the most recent glacial period. Studies of archaeological sites in southern Arizona produced evidence that bighorn have been present since about 8000 B.C. (Haury et al., 1950). For the past 10,000 years bighorn lived in the same environment with the wolf *(Canis lupus)*, coyote *(C. latrans)*, gray fox *(Urocyon cinereoargenteus)*, bobcat *(Lynx rufus)*, mountain lion *(Felis concolor)*, jaguar *(F. onca)*, ocelot *(F. pardalis)*, and golden eagle *(Aquila chrysaetos)*.

Spirited discussions of bighorn-predator relationships have taken place at nearly every meeting of the Desert Bighorn Council since it was formed in 1957. Five papers were presented on this subject in 1961 alone. Gabrielson (1941) states, "No topic in the wildlife field is more controversial than that of predator relationships and on none perhaps is there more loose thought and 'positive' opinion based on insufficient consideration of the evidence that is available."

Predators have been accused of depleting bighorn herds throughout the Southwest, but with little justification. During the 10,000 years bighorn and predators have coexisted, populations of both predator and prey have been maximum. Some balance of nature relationship must long have been in existence between bighorn and predators. Russo (1956) comments, "It [bighorn-predator relationship] is a complex relationship that cannot be entirely understood without the basic knowledge of the animals involved and their historical pattern of life."

From the writings of the early settlers and explorers of the Southwest, including Mearns (1907), we know there were more bighorn, spread over a wider range, than at present. The advent of the settler with his livestock and guns upset the bighorn-predator relationship, and that relationship may continue to be out of balance today. The distribution of the desert bighorn has decreased since the advent of the settler. Much of this decrease in distribution has been blamed on man and his livestock, but probably some can be attributed to predators. The bighorn have withdrawn to, and now survive in, mountain ranges where they are least vulnerable to man or predators.

WOLF

Historically, the wolf ranged over much of the North American deserts. The coming of the settler and his livestock initiated the extirpation of the wolf. It is possible some wolves still exist in the bighorn ranges of Mexico, although they are extinct in the southwestern United States.

Nothing is known about the population and distribution of the wolf in Mexico, or about the relationship between it and the desert bighorn. In the United States, bighorn have outlasted wolves and do not appear to have been affected one way or another by having one less predator with which to contend. It is logical to assume that wolves of the desert must have preyed on bighorn. This predation was most likely opportunistic; as with other large predators, bighorn were not the staple prey item.

COYOTE

The coyote is the most common predator occupying bighorn ranges (Weaver, 1961), and it has the highest population density of all bighorn predators. Coyotes no doubt are capable of killing mature bighorn under some circumstances. Conversely, Groves (1957) relates this incident of a coyote believed to have been killed by a female bighorn:

> An old gentleman of Coaldale, Nevada, reported coming up to a spring and finding a ewe and a lamb and a coyote, the coyote circling the ewe and the lamb. He (the old gentleman) watched them for a considerable length of time and went on about his business (he was headed toward a mine). When he came back the coyote was dead at the spring with its hide cut full of holes, and the ewe and the lamb gone. You can take it for what it's worth, but it is an interesting observation. Sometimes predation may work the other way.

During a water-hole survey on Joshua Tree National Monument, California, Welles (1965) made the following observation:

> Five coyotes were also seen, one in deadly pursuit of an adult ewe, being about 50 feet behind her, when they appeared on the ridge above the spring, and only ten feet behind her when they disappeared in the canyon one-half mile to the west. Subsequent search of the area disclosed no carcass. We draw no conclusions from this. We saw one coyote trying to catch a ewe who knew that that was what he was trying to do.

Welles and Welles (1961a) at Nevares Spring in Death Valley National Monument, California, observed a single coyote frighten two bands of bighorn. One band of seven was at the spring. Another band of three was approaching the spring. When the coyote appeared, both bighorn bands moved into a cliff area for safety. The coyote watched the bighorn for about half an hour, then went on his way.

While I was assisting Paul Webb of the Arizona Game and Fish Department and his crew in the 1957 bighorn trapping operations on Kofa Game Range of Arizona, Paul told me of finding the fresh remains of a ram that had been killed by a group of coyotes. From the tracks, the ram had been pursued by the coyotes across an open bajada and was overtaken and killed before he could reach escape cover. It could not be determined if the ram was sick or otherwise incapacitated.

Another instance was related to me by Dick Senecker of the Nevada Department of Fish and Game. While on the Colorado River below Hoover Dam, Dick observed a coyote stalking a bighorn lamb. The lamb appeared to be just hours old and barely able to scamper about on wobbly legs. When the coyote moved in for the kill, the lamb climbed what appeared to be an inaccessible cliff and joined its mother. The coyote moved on along the river bank and made no further attempt to pursue the bighorn.

Another observation was told to me by the late John Reed. John was making a water-hole count on Kofa Game Range. A group of four bighorn was in the vicinity of the spring when a lone coyote came on the scene. The bighorn and the coyote paid little attention to each other.

The following observation, made by Huyson J. Johnson at Dripping Springs in the Gila Mountains of Arizona, was taken from the Cabeza Prieta Wildlife Refuge narrative report for May to August 1949:

> On the morning of August 12, 1949, while on routine patrol, I visited Dripping Springs in the Gila Mountains. In this area I discovered the remains of seven Arizona bighorn sheep, three lambs, two ewes, one yearling and one 4-year old ram, apparently killed by predators as there was no other activity in the area.
>
> Five of these had been killed in the past 60 days, and one, the yearling ewe, had been killed the morning of the 12th, for the blood was still moist and fresh on the rocks. The carcass of this animal had been absolutely devoured, with only the head and a few fleshless bones remaining. The

fresh blood, pliability of a few scraps of hide remaining attached to the skull, and the freshness of the bones all point to the fact that this animal had been killed and devoured in a very short time. Five coyotes were observed in the vicinity and one was shot and wounded by me as it made its escape.

Varying interpretations can be made from these seven observations. In one case, a ram was killed by a group of coyotes in an area with no escape cover. In another, a lamb survived the coyote's attack because it was in good escape cover. It is difficult to interpret Reed's observation at the water hole when the coyote passed near the group of bighorn. The observation by Welles indicates that individual coyotes do pursue adult bighorn. Welles and Welles' observation substantiates that bighorn are wary of coyotes at water holes. And, of course, Johnson's observation further indicates that desert wildlife are more vulnerable to predation at water holes than in other portions of their range. The report by Groves leads us to believe that a single coyote may have difficulty killing an adult bighorn; however, a more likely explanation would be that the ewe's mother instinct was effective in protecting her young.

An analysis of 636 coyote scats from Anza-Borrego State Park, California, reveals three scats with bighorn remains (Browning and Leach, 1959). The scats containing bighorn remains were found in September. Jerome T. Light states this is one of the periods in this area when one would expect natural mortalities due to drought extremes.

In southeastern Utah, Wilson (1968) analyzed ninety-six coyote scats and found remains of bighorn in ten.

In Arizona, Russo (1956) found eight coyote scats containing deer or bighorn remains out of 248 scats examined. On Cabeza Prieta Wildlife Refuge, also in Arizona, Simmons (1969b) reports that bighorn remains were found in five out of over 580 coyote scats. It is difficult to distinguish bighorn hair from deer hair in the scats of predators (Bruce Browning). Predator scat analysis indicates only what the predator has been feeding on; it in no way proves that the animal consumed was killed by the predator.

In summary, coyotes will attack bighorn and are probably more successful while hunting in groups. However, it is believed the coyote predation that does occur is rare and opportunistic.

GRAY FOX

The gray fox is rather small to attack an animal the size of a bighorn. Nichol (1937, 1940) found evidence of lamb predation by gray foxes in the Plomosa and Mohawk mountains in Arizona. Proof of the Plomosa Mountains incident was based on the freshness of the lamb carcass, the sign that indicated a fox had done the killing, and the stomach contents of a fox killed in the area about an hour after the lamb had been found. These two observations were made in April and May, respectively, and both lambs were very young. Although gray foxes may prey on newborn lambs, they could hardly kill an adult bighorn. The effectiveness of the gray fox as a predator on bighorn lambs needs to be explored further.

Harold T. Coss tells me that in the summer of 1963 he watched a yearling bighorn at North Pinta Tank on Cabeza Prieta Wildlife Refuge chase a gray fox around the slopes immediately above the tank for several minutes.

BOBCAT

The bobcat is found throughout the Southwest in the same range occupied by the desert bighorn. Records of bobcat-bighorn relationships are not numerous.

Groves (1957) states that in 1942 on Desert National Wildlife Range in Nevada a bobcat that weighed 13 pounds killed a yearling bighorn that weighed an estimated 90 pounds.

Russo (1957), while working on Kofa Game Range, found an adult ewe that had been killed by a bobcat. The cat had sprung on the bighorn from a rock overlooking a game trail. He also related an incident in which he found a live lamb, two dead lambs, one live turkey vulture *(Cathartes aura)*, and one live bobcat in a natural rock tank. The tank had such steep sides that some animals could not make their way out. Evidently the lambs had followed their mothers into the tank for water and were unable to get out. The bobcat, which could enter and leave freely, was feeding on one of the dead lambs. The vulture could not get out of the tank either and had been living on the remains of the lambs the bobcat had left.

James Blaisdell relates: "An observer at Lava Beds National Monument (in California) reports seeing a bobcat sneaking on a ewe and lamb. The moment the bobcat showed his head on the horizon, the ewe saw him, and quickly took her lamb to safety among the high cliffs."

Wilson (1968), during his studies of bighorn in southeastern Utah, analyzed fourteen bobcat scats and found the remains of bighorn in one.

Under certain conditions bobcats may be serious predators on bighorn, primarily during the summer months when the bighorn need water. It is believed that some bobcats wait at water holes or along trails leading to water and ambush bighorn as they come to drink. In mountain ranges with a limited number of water holes bobcats could critically deplete already small bighorn populations.

MOUNTAIN LION

The mountain lion is the second most important of the bighorn's feline predators. In many of the bighorn ranges of the Southwest, densities of bighorn or deer are too low to support a predator of this size. But there have been reports of mountain lions in bighorn country and of kills made by lions.

Cronemiller (1948) wrote:

> Evidence of the predation on desert bighorn *(Ovis canadensis nelsoni)* by a mountain lion *(Felis concolor)* was found on the San Bernardino National Forest in Southern California in March of 1947. Forest Officer Elwood M. Stone found portions of a recently killed bighorn, including the complete head. Tracks in the immediate vicinity identified the attacker as a mountain lion and showed that a struggle had taken place before the bighorn succumbed.

Blaisdell (1961) quotes from a personal letter from former Fish and Wildlife Service Refuge Manager Gale Monson:

> We have only one record of any mountain lion staying on either of our game ranges for any length of time, and that was of an old male taken in the Kofa Mountains on February 17, 1944. I quote from the Kofa Game Range Narrative Report for the period January–April, 1944: "The skin and skull of the mountain lion taken were sent to the Carnegie Museum. Hunter Casto found the partly buried carcasses of six mountain sheep in the vicinity of Squaw Peak Tank. Four of them were rams and two of them ewes and he believes that they had been killed in late August or early September as they were, most of them, cached out in small washes and were partly covered up and had they been placed there prior to the first of August, the heavy flash rains that we had in early August would have uncovered and washed them down. The altitude is over 4,000 feet near Squaw Peak Tank which would be a fairly cool location during the hot part of the summer and Hunter Casto believes that lions must have taken advantage of this condition as they evidently hunted in other sections of the country after the fall rains and there were no winter kills found."

Monson goes on to say about this report, "With respect to this quotation, I think it is fairly well established that this lion was not accompanied by others, and was strictly a solitary individual. I also think there is a good chance the bighorn were killed later than in August or September. Besides the bighorn kills, I understand a number of deer kills were also found."

In regard to mountain lion predation in west Texas, Carson (1941) reports:

> A. L. Hall stated to me that he considered the mountain lions were the biggest factor, except man, in causing the decrease of the bighorn. He has found many carcasses of bighorn on his ranch, in the heart of the bighorn range, in the twenty-eight years he has lived on the ranch.... He says he has found the carcass partly covered with sticks and brush, and has seen where bighorn had been dragged into gulleys and thickets to hide them.

Carson continues,

> J. T. Watson told me that in the winter of 1933 a mountain lion chased a band of bighorn out of the east edge of the Baylor Mountains and killed a large ewe within a mile of his ranch headquarters.... Watson found the bones of the bighorn, and the tracks of the bighorn where they had run and

dodged, and the lion tracks where the kill was made. This was about four days after the lion had made the kill. Watson found several bighorn which had been killed by lions....

Charles G. Hansen writes that eleven mountain lion kills of desert bighorn were found on Desert Wildlife Range in the timbered country, in areas normally used by mule deer. Mountain lions, according to Hansen, appear to be attracted to deer. Where deer are plentiful and bighorn are also present, mountain lions will be found and will take bighorn incidentally, especially at or on the way to water holes in "deer country."

In 1967 the Nevada Department of Fish and Game constructed an enclosure to hold a transplant of bighorn in the Mount Grant area of the U.S. Naval Ammunition Depot near Hawthorne, Nevada. The site for the enclosure consisted of a thick stand of pinyon without rock outcrops, cliffs, or boulders suitable for escape cover (Broadbent, 1969; Charles G. Hansen).

Prior to release of bighorn into the pen, the Division of Wildlife Services of the U.S. Fish and Wildlife Service conducted a 7-month effort to rid the vicinity of predators. Broadbent (1969) reported that the continuous surveillance by three field men, operating two Compound 1080 (sodium fluoroacetate) stations and multiple traps, produced six coyotes and one bobcat, and "The Division's report emphasized that no lion sign was seen at any time during this control period." However, the individuals involved later stated the bighorn enclosure was located on a travel route of lions in the Mount Grant area (Wright, 1968).

On June 17, 1968, two adults and one lamb bighorn, all pen-reared and nearly tame, and one wild bighorn were released into the enclosure. On June 24, two wild ewes, trapped at Wamp Springs on Desert National Wildlife Range, also were released into the enclosure. The bighorn appeared to be doing well until July 30 when three of them, a mature ram and two ewes, were found dead. The ram and one ewe had been killed by mountain lions. The second ewe showed no visible evidence of cause of death. The following day another ewe was found that had been killed by lions. The remaining ewe, also showing no visible evidence of cause of death, was found dead on August 12. The lamb survived the predator attack.

Between July 29 and August 2, four lions, one adult and three juveniles, were killed in the vicinity of the enclosure, and another lion escaped from a trap. After the enclosure had been protected by an electric fence, bighorn were again released in the enclosure, and a lamb was born in early 1970. By 1974, thirteen bighorn were in the enclosure and no further losses to predators occurred. The enclosure was abandoned in 1976.

The states of Texas and Arizona have also transplanted desert bighorn to enclosures. Arizona enclosed a steep, rugged canyon in its holding pasture that afforded exceptional escape cover. Texas enclosed its pasture with an electric fence, which was successful in thwarting a mountain lion from entering the enclosure.

JAGUAR AND OCELOT

In Mexico, the ranges of the jaguar, ocelot, and bighorn possibly overlap, and rarely a jaguar appears in bighorn country in southern Arizona. While doing the initial bighorn surveys in the Sauceda Mountains of Arizona, I met a ranch hand who reported seeing a "spotted" mountain lion. From his description of the animal, it seems likely it was a jaguar. Because of their few numbers in bighorn habitat, the jaguar and the ocelot would have little effect on bighorn populations.

EAGLES

The golden eagle shares the same range as the desert bighorn. Several written accounts attribute lamb mortalities and the harassment of older bighorn to eagles. Jonez (1961) relates the following:

> As they [two Nevada Fish and Game Department employees] came around a turn in the canyon they saw a bald eagle take off. Another eagle was on the ground tearing at something and attempted to lift it in flight. Suddenly, aware of the spectators, it too took off. Both men left their vehicle and checked to see what the eagle was eating. They found a bighorn sheep lamb with a hole torn in its side and the heart still palpitating. It, however, was dying. Both men suspected the eagles killed the lamb but neither saw the actual act. The lamb was given to the Nevada State Museum for mounting. When the lamb was skinned out, it was found that the right shoulder joint was fused (apparently the scapula fused to the humerus). This undoubtedly restricted the movement of this lamb and may have been the reason it was susceptible to predation.

C. A. Kennedy (1948) found evidence that a golden eagle had killed a very young lamb on San Andres National Wildlife Refuge, New Mexico. Kennedy stated the ewe looked sick prior to the birth of the lamb. If this was so, it could explain the predation by the eagle. Either the ewe couldn't protect the lamb or the lamb was also sick and couldn't follow its mother. Groves (1957) was present when two golden eagles harassed a yearling bighorn in the Pintwater Range in southern Nevada. According to Groves, the two eagles separated the yearling from a ewe and a lamb and drove it down a canyon, striking at its flanks and riding the animal with wings outspread. Subsequent investigations were made but no kill was located.

Ober (1931), while collecting bighorn for W. T. Hornaday, made this interesting observation:

At the firing of our few shots, the rest of the band of sheep promptly scattered, jumping and running over trails that a man would not care to walk carefully over. One yearling sheep, confused and a bit inquisitive, became separated from the bunch and circled back and up toward where we were preparing our kill for transportation to camp. Sitting motionless, we allowed the sheep to get within fifty yards of our position before he became aware of our presence. Upon sighting our party of three he wheeled and began a mad flight down the long sandy draw in which we had made our kill. Suddenly out of the sky came a thunderbolt of dark hue, the wind hissing through its wings; it shot straight down to deal the yearling a resounding blow on the back. The sheep, staggered by the blow, kept gamely on. The golden eagle soared up for another attack, while its mate struck the yearling. Four times each great bird zoomed and struck, always in the same place. The sheep, badly wounded by the attacks, was plainly weakening. The final act came when the largest of the birds... flew ahead of the sheep and alighting on the rocks in front of the animal beat its wings and screamed repeatedly in an effort to head the sheep off. The yearling, awed by the terrible figure of the killer, paused to seek a new exit, that pause was its doom. Straight and as fast as a bullet came the female bird, striking this time at an angle in order to get the sheep to fall. The angle was perfect, and the blow so hard that the badly weakened sheep fell sprawling down the draw. Both eagles pounced on the carcass, tearing madly at the choicer parts of the anatomy. Upon close examination we found that the killing blows were innumerable punctures by the birds' sharp talons into the back over the kidneys, the sheep bleeding internally, scarcely a drop of blood being visible on the outside. In all my experience with the wildlife of this section I have never witnessed such remarkable teamwork or such a parallel to human intelligence in making a kill.

Eagles can and do take bighorn, but since the period when bighorn are vulnerable to eagle predation is short, this predation is believed in most instances to have little effect on bighorn populations. Moreover, the golden eagle is possibly more in danger of being removed from the ecosystem than the bighorn. Excessive shooting and use of pesticides have had deleterious effects on the breeding capabilities of these birds and their numbers have become reduced.

IMPACT OF PREDATORS ON BIGHORN POPULATIONS

None of the predators discussed is solely dependent upon desert bighorn for survival, and most predation on bighorn is opportunistic. There is little evidence that the total pressure from all of the predators (excepting man) has any impact on the total bighorn population. This may be due to the low population density of these predators in bighorn ranges. Dasmann (1964) states, "When game is scarce, the effort required to catch the last few individuals may be more than the food gain will balance. A predator could not normally survive long enough to exterminate his prey." In this same light

Errington (1963) speaks of the "threshold of vulnerability," a level of numbers below which prey animals in a given habitat are relatively secure from predation and above which they are readily taken by carnivorous animals. Cain et al. (1962) state, "It is probably valid to generalize that predation is a part of the complex of interacting mortality factors that become increasingly effective as the density of the prey population increases...."

In most of the desert ranges of the Southwest, bighorn populations are low. If delineated as animals per square kilometer, they could be classified as scarce. However, these desert ranges have low carrying capacities and may be populated to capacity. The predation that does occur may be more beneficial than harmful by keeping the bighorn population in balance with a limited food supply.

During the years I spent working with desert bighorn in Nevada and Arizona, I found no evidence they were concerned with or pressured by predators. At the first meeting of the Desert Bighorn Council, during a discussion of predation, C. E. Kennedy (1957) made the following statement that substantiates my findings:

> "The majority of the sheep on the Kofa (Kofa Game Range) have never shown signs of being afraid of anything. They have shown great curiosity, no fear, and walk around as though they were perfectly at home and had never been afraid of anything in their lives."

Simmons (1969b) comments, "I have observed 12 sheep-coyote interactions involving from one to four coyotes and small groups of sheep. The observations were made during periods when sheep were congregated at water holes and most vulnerable to predation. During none of these observations did the sheep react with alarm to the presence of coyotes."

From the observations that have been made we know all of the predators mentioned in this chapter are capable of killing bighorn under certain conditions. Many wildlife biologists believe that predation removes the old, sick, weak (e.g., Gabrielson, 1941), and stupid animals from the herd and, in a sense, assists the bighorn in a program of natural eugenics. Cain et al. (1972) contend that predation "can also result from such factors as over-population, or simply chance.... As a matter of statistics, the debilitated, the incompetent, the too numerous, and the unfortunate are weeded out, and the mechanisms of natural selection and evolution continue to operate as they have since the beginning of life on earth."

Charles G. Hansen says, "On Desert National Wildlife Range I found too many sick and dying bighorn, both rams and ewes, to believe that predation was a limiting factor in bighorn survival. I've seen too many bighorn that I could run down and catch to believe that predators are numerous enough or effective enough in bighorn habitat to be a problem in bighorn survival at any time."

We are sure that during the past 25 years the bighorn herds of the Southwest have remained fairly static. An overall predator control program to reduce all predators in hopes of increasing bighorn populations would be unrealistic. Some reports (e.g., Halloran, 1949) indicate that widespread predator control programs have contributed to increased lamb survival while others (Russo, 1956) believe that controlling predators has had little effect.

Management programs other than predator control can maintain or enhance the welfare of bighorn herds. Optimum range conditions can be maintained to ensure a healthy population. This can be done by decreasing the competition with deer, feral burros, and livestock for food, water, and space. We can develop additional watering sites to spread the distribution of bighorn, thus making them less vulnerable to predation when summer temperatures force them to the water holes.

Programs to introduce captive bighorn populations to additional ranges can expose the animals to undue predation if care is not taken in selecting sites for enclosures. Escape cover must be readily available in addition to adequate amounts of food and water.

The states of Texas and Arizona pioneered the trapping and transplanting of desert sheep. Both of these states have holding pastures and are raising bighorn in these pastures. It would be well for administrators and biologists who are thinking of transplanting bighorn to study the accounts of Texas, Arizona, and Nevada transplants. The type of holding pastures constructed, the amount of escape cover enclosed, and the measures Texas used to protect its bighorn from predation, are all important items to be considered prior to a transplant.

The predator-bighorn populations will remain at their present levels unless the worst predator of all, MAN, through his thirst for "progress," destroys his environment, his own species, and the "lesser" species with him.

14. Competition

Fred L. Jones

Direct competition from other animal species for food and water is of primary concern in management of desert bighorn. Other types of interplay between bighorn and other species will also be discussed in this chapter. Predator relationships are treated in Chapter 13, and man's relationships in Chapter 19.

There are companionable relationships when bighorn associate with deer, cattle, goats, and domestic sheep; and disturbance relationships, such as bighorn reacting to a covey of quail flushing in their faces, or being frightened away from water by swarms of bees. These are not profound interspecific ties, but they do provide insight into the daily affairs of bighorn.

DEER

General. The deer species competing with bighorn is the mule deer *(Odocoileus hemionus)* (Fig. 14.1). Overlapping deer and bighorn ranges vary from the palo verde habitat of the lowlands to the timber-brushland types of the higher desert mountains. There may be overlap only on winter ranges, or deer and bighorn may occupy the same year-round range. Again, overlap occurs only at watering places.

Norman M. Simmons

Fig. 14.1. The mule deer is one of the natural competitors of the desert bighorn. Kofa Mountains, Yuma County, Arizona.

Bighorn usually occupy rougher terrain than do deer, which affords the bighorn considerable freedom from competition. For example, both deer and bighorn are in the mountain ranges bordering the Owens Valley in California. However, in summer bighorn range mainly in the alpine areas, whereas deer range below timberline. In winter, bighorn concentrate on cliffy south-facing slopes at the mouths of the more rugged canyons, while deer range on less precipitous slopes.

Where terrain is less rugged, bighorn may be "forced" to compete directly with deer. The disappearance of bighorn from some areas of relatively gentle terrain in the southern Sierra Nevada seems to have occurred as deer herds increased. After many years of intensive study of both bighorn and deer in this area, I concluded that natural factors have restricted deer numbers where they coexist with bighorn. Likewise, groups of bighorn no longer exist where conditions have allowed large deer herds to develop (Jones, 1954).

In Joshua Tree National Monument, California, both deer and bighorn use Stubbe Spring. However, the topography of their preferred habitats minimizes significant competition under normal circumstances. All the deer come into Stubbe from the northeast, whereas the bighorn come in from everywhere else but there (Welles, 1965).

In the Santa Rosa Mountains of California, I found small numbers of deer in chaparral above the arid bighorn range, which extends from the base of the mountains (above sea level) up to about 1,200 m (4,000 ft). However, there was evidence of deer descending as low as 730 m (2,400 ft) in the winter (Jones et al., 1957), at which time they did compete with bighorn. Weaver (1968) found the same situation to exist in the mountains extending south of the Santa Rosas to the Mexican border.

Welles and Welles (1961a) found deer plentiful in the pinyon-juniper zone of the Grapevine Mountains of Death Valley National Monument, California, but bighorn were scarce. There was some overlap between the species in the Cottonwood Mountains, but the few deer were confined largely to a restricted area on Hunter Mountain.

Light et al. (1966) say deer and bighorn might occur together in the Raywood Flat Area of Mount San Gorgonio, California, though deer prefer browse and oak-conifer types, and bighorn prefer cliff vegetation types. In the San Gabriel Mountains of California, the deer range generally fringes the bighorn range from above, with some overlap in a few canyons (Light et al., 1967). It is not unusual to see deer on bighorn winter and spring ranges, but large numbers of deer are not seen in precipitous terrain. Weaver et al. (1972) reiterate that, owing to both a low deer population and differing terrain preferences, competition is minimal between deer and bighorn in these mountains. They state, however, that there is some overlap in Cattle, Coldwater, and Fish Fork canyons, the South Fork Whitewater River, and a few other small isolated locations. Bighorn appear to be able to compete with deer in these locations at present population levels.

I found that both deer and bighorn range on top of the ridge above Deep Springs, California, though the latter spend most of their time on the precipitous lower slopes, which deer avoid. I also found deer to range well above timberline in the White Mountains of California on slopes used by bighorn. Weaver and Mensch (1970b) recognize this, too. Burandt (1970) reports that both deer and bighorn water at certain springs in the nearby Inyo Mountains.

Richard Weaver writes: "I cannot prove it, but it is my opinion that deer that were introduced in the New York Mountains of California were at the expense of the bighorn. Bighorn do not now occupy some areas where old-timers saw them and I have found old horns. We have found that deer numbers and bighorn numbers declined at about the same time during a recent drought; deer numbers have built up again but bighorn numbers remain down."

In Arizona, desert mule deer inhabit a considerable portion of the bighorn range, but overlap occurs mainly along the bajadas* and at watering points (Russo, 1956).

---

*Bajada: An alluvial plain formed at the base of a mountain by the coalescing of several alluvial fans.

In Canyonlands National Park, Utah, "Deer prefer a less rugged physiography than bighorn. Consequently, bighorn ... use the marginal habitat of the rugged canyon or canyons which deer are unable to use due to limited water supply or barriers which prevent access to the river bottoms. The canyons in the Maze District are good examples of deer-bighorn interaction. Here, the sheer slick rock walls allow vegetation to occur only in the washes. Thus, grazing is limited to these washes. When domestic sheep or deer use the washes, there is no suitable terrain or vegetated habitat for the bighorn to retreat to. This has resulted in the absence of bighorn as permanent residents of the Maze District" (Dean and Spillett, 1976).

Food. Direct competition for food between bighorn and deer usually occurs on winter ranges in the Sierra Nevada and southwest California mountains.

Except for certain key browse species on some ranges, deer and bighorn have basically different feeding habits. Yoakum's evaluation (1966) of the rumen analyses of bighorn and deer collected from the same portion of Desert National Wildlife Range in southern Nevada shows that each animal prefers a different forage class, bighorn consuming 65 percent grass and deer 77 percent browse.

Other studies have shown there is competition for certain key browse species in desert bighorn range. Russo (1956) reports on stomachs of twelve deer and twelve bighorn from Kofa Game Range in Arizona. One sample was taken from each species each month of the year. Both bighorn and deer took approximately 65 percent browse of six principal plant species. Halloran (1949) states that on Kofa Game Range bighorn and deer habitats overlapped in the vicinity of major water holes, that there was some competition, and that the relationship should be watched closely.

C. E. Kennedy (1963) reports that bighorn on winter ranges in the Angeles National Forest, California, compete directly with deer by browsing on the same key shrubs.

In southern New Mexico, and specifically in San Andres National Wildlife Refuge, bighorn and deer food relationships were studied by Halloran and Kennedy (1949). The animals were using the same ranges: bighorn, the crest of the narrow north-south lying semidesert mountain range, and deer, from the crest to the edge of the nearby plains. From sight observations and a few deer stomach analyses, the two observers documented direct competition for certain dominant and codominant browse plants. Competition was heaviest during winter, and for new, tender growth following rains.

Smith (1966) shows that as deer declined in numbers on San Andres Refuge, bighorn increased. Results of rumen analyses from bighorn from the Big Hatchet and San Andres mountains in New Mexico indicated 100 percent competition between deer and bighorn (Gordon, 1957). Gordon states that livestock removal and limited deer hunts caused the range to improve considerably.

Wilson (1968) says that in the White Canyon area of southeastern Utah competition with deer is greatest during the winter, from October through March. Deer start migrating from Elk Ridge to the east down into the bighorn range in the latter part of October. The deer herd consists of several thousand animals, and competition between deer and bighorn for many key browse species can become critical, especially during hard winters when more deer migrate than in normal years. Many of the key deer wintering areas are used by bighorn year long, and browse species are becoming overgrazed owing to heavy combined use.

Buechner (1960) points out that the amount and significance of competition from deer may be greater than differences in feeding habits indicate. It is certain that where the two overlap, and deer become overabundant, deterioration of vegetation is inevitable, and bighorn are likely to suffer more than deer. Since the productivity of bighorn is lower than that of deer, they cannot maintain their numbers as well when losses due to competition occur.

Water. When water supplies are marginal, competition for use of it may become critical for all species. Since deer occur in some of the more arid regions where water often is scarce, they do on occasion adversely compete with bighorn for water. Stubbe Spring in Joshua Tree National Monument is such a place. Welles (1965) estimates that eighty bighorn and fourteen deer watered there in 1964. Stubbe Spring is now dry and has been supplemented by an artificial watering device. Rainwater collected in the tanks of the device sometimes becomes critically low because of heavy use by wildlife.

Jonez (1960) gives several examples of competition between deer and bighorn for water in the Sheep Range portion of Desert National Wildlife Range. Deer contribute at times to drying up of some water sources on Kofa Game Range (Halloran and Deming, 1958). In discussing water limitations in Imperial County, California, Mensch (1970) states that competition with burros and desert mule deer appears to be the factor limiting bighorn populations. I found both deer and bighorn watering at some higher springs in the Santa Rosa Mountains of California.

Social. From time to time, bighorn have been observed feeding or traveling amicably with deer. I have heard of a number of such incidents in the Owens Valley of California but have not seen any myself. These accounts were of rams, ewes, adults, yearlings, and large lambs—usually single, but sometimes two. Winter was the most common time of observation.

Russo (1956) reports an observation of four deer and a young ewe bedded down together under a group of palo verde trees in Arizona. Two of the deer went off in one direction, while the other two followed the ewe in another. He says that in summer, when all game is forced to watering points, it is common to see deer and bighorn in the same vicinity.

I assume that the natural gregariousness of bighorn draws them to such interspecific relationships when separated from their own kind. In a competitive situation, however, conflict arises between bighorn and deer. Those who have witnessed such encounters differ as to which species is dominant.

Ralph Welles says that from observation at Stubbe Spring he felt that deer defer to bighorn in watering order. Charles G. Hansen states that, while watching a water hole used by both deer and bighorn, he saw deer of both sexes chase ewes and young bighorn from the vicinity of water. The water supply was relatively limited, and the deer very aggressive. As bighorn approached the water when a deer was watering or was also approaching, the deer would make short dashes at them. The deer would have its ears laid back and have all the appearance of a belligerent, attacking animal.

The bighorn left the area and did not return to drink for at least 2 days. Sixteen ewes, lambs, and subadult bighorn were seen during the 3-day waterhole count. Seven of them were forced away by deer; the other nine came to drink when deer were absent. Adult rams that came in at the same time were not intimidated by the deer; they walked directly to the water and drank. The deer actually stopped drinking, and moved away about 9 m (30 ft) until the rams had finished and moved off.

BURROS

General. Welles (1962) says: "For many years, statements have been made without proof regarding burros and their exact effect on sheep in Death Valley National Monument to the present time. There is no question that the burro is bad for vegetation, watershed and wildlife in general, but I cannot find anywhere a yardstick to measure specifically and exactly any reduction yet in the Monument population of bighorn because of burros. People, yes; burros, no.... This statement was based not only on our own observations at that time, but reflected the consensus of opinion of all the field men who had participated in the 1955 census."

Welles and Welles (1961b) mention observing bighorn and burros feeding in the same area in Death Valley Monument, but with significant topographical barriers of differing relative ruggedness separating their respective sections of the foraging area.

Sleznick (1963) says that in Lake Mead National Recreation Area forage competition between burros and bighorn might occur in the washes at times. At other times the bighorn utilize the less accessible canyon walls and rimrock, while the burros use the flatter land farther from the lake.

McMichael (1964) states that in the Black Mountains of Mohave County, Arizona, burros were found to occasionally use all but the most rugged mountain areas. The area of highest combined use was in the upper foothills.

He found bighorn traveling 2.4 km (1.5 mi) across rugged mountain terrain to a seep in the mountains, rather than traveling 0.8 km (0.5 mi) across an open flat to a spring.

With the exception of former control programs by the National Park Service and the U.S. Fish and Wildlife Service, there has been little effort by any agency toward burro control. Burro regulation is seriously hampered by the controversial Wild Horse and Burro Act (Public Law 92-195) enacted in 1972, which made killing a feral burro on most lands a felony; this law has inhibited efforts at control on National Park lands, as well as on lands administered by the Bureau of Land Management. The Bureau has made some effort to reduce burro numbers through capture and sale of burros, but such effort probably has had almost no effect on burro numbers.

Food. Reports on competition for food between burros and bighorn are varied and contradictory. The consensus is that where forage is ample, direct competition for food is not a serious factor. However, when it exists, competition is the most severe during the dry season and at watering sites.

There can be little doubt that the burro, a large animal capable of consuming a considerable amount and variety of food, is an important competitor of the bighorn in arid environments where forage resources are limited. McKnight (1958) points out that the burro's range of food plants is broader than the bighorn's and also that the burro can be destructive by pulling out entire plants by the roots and then eating only one or two mouthfuls.

I entered the following in my field notes of July 13, 1955, Panamint Mountains, Death Valley National Monument:

> Burros have taken all of this over and are ruining the range. The use of *Larrea* at the lower elevations and the naturally poor quality of feed below 6,500 feet means there is a naturally poor range that has been severely overbrowsed. The end result being practically hopeless for range recovery even under complete protection....*Franseria* is nearly a remnant plant—being eaten out. Some noticeable use on *Larrea,* the first I have ever seen. Some very dry *Ephedra* around. Plant growth extremely scant on these hills, just like the floor of Panamint Valley . . . at the higher elevations burros have nearly killed out grasses and such forbs as *Phlox* (which exists only in centers of large shrubs) and are using bush lupine heavily—uprooting much of it. The vegetation is being changed to practically pure sagebrush and *Haplopappus* with some gray horsebrush and bush lupine.

Richard Weaver writes: "We have found several places in the California desert where perennial grasses have been eliminated by burros—Dodd Springs area in Inyo County, for example. Yet on Dry Mountain a few miles to the north, where it is too dry for burros, a good grass stand exists with a good population of bighorn."

Using Buechner's feeding-minutes technique, McMichael (1964) records a total of 125 feeding-minutes for bighorn and 380 minutes for burros in the Black Mountains of Arizona. He shows the bighorn fed mostly on ocotillo

(Fouquieria splendens) and catclaw *(Acacia greggii),* whereas burros fed mostly on forbs. Both animals fed on brittle bush *(Encelia farinosa).* He found that both bighorn and burros like to stand in the shade of rocks and ledges, especially during the summertime. Vegetation near these areas, and around springs, showed overuse. He concluded that there was sufficient overlap of food habits and summer range of the two animals to keep bighorn from reaching their maximum population density in the Black Mountains.

Referring to the Grand Canyon of Arizona, Bendt (1957) says that "the most significant limiting factor affecting the sheep population today involves the competition with feral burros for food and water." Carothers et al. (1976) report that in the Grand Canyon, based on an ecological analysis of study areas, their "investigation demonstrate[s] that the feral ass has a negative effect on the natural ecosystem of the lower reaches of the Grand Canyon. The principal impact of the feral ass is habitat destruction through grazing and trampling."

In his report on the bighorn of Kofa Game Range in Arizona, Halloran (1949) mentions that during the height of the dry season burros were utilizing the best feed and some of the last water on the Range.

Welles and Welles (1961a) state that the burro was a threat to the bighorn's "emergency food supply" about watering sites and along trails to them in Death Valley National Monument. Bighorn, coming a considerable distance to water and sometimes through terrain where no forage exists, depend upon the vegetation around the springs and seeps for an "emergency" ration, so to speak. It has been observed frequently that burros denude the watershed in the vicinity of water sources.

Weaver (1972c), referring to a National Park Service study in Death Valley National Monument concerning plant damage and soil disturbance caused by burros, says,

> Point transects were made on two routes reaching about 5 miles [8 km] from the water. At each point, plants, litter, or bare ground was recorded and disturbance noted. Also included was the amount of use to the plants along the transect. Radiating out from the water tank the vegetation and soil show an abundance of use and disturbance. The plants are either eaten or trampled while the disturbed soil is subject to wind and water erosion. Feral burros in the Wildrose area have virtually eliminated certain species of plants for one or more miles [1.6+ km] from the water, including four species of grasses. Five miles [8 km] away from water grasses have been seriously reduced. The study concludes that burros are not only damaging the plant community but they also affect the animal community dependent on the plants.

Charles G. Hansen comments, "In Death Valley National Monument in 1973 there are only about 450 bighorn but about 1500 burros. Over a 10-year period burros and bighorn have not coexisted; bighorn are on the way out and burros in."

Mensch (1970) reports severe competition for food and living space between burros and bighorn in northeastern San Bernardino County, California. Weaver (1972a) mentions that Sheep Spring in the Providence Mountains of California (in northeastern San Bernardino County) was free of burros in 1953 but now has twenty or more burros using it. All perennial grasses have been eliminated near the springs. It is his feeling that bighorn numbers are down as a result.

Weaver and Hall (1971) report that burros in the Whipple and Chemehuevi mountains of California are spreading to new areas, have almost eliminated perennial grasses where they range, and take the available water. They feel that burros caused the disappearance of bighorn from the Whipples and that they are severely competing with the bighorn that remain in the Chemehuevis. Another report (Weaver and Mensch, 1970a) states that burro competition is probably the main cause of extremely low bighorn populations in the Argus and Slate ranges of California. Again (1970b), the same authors suggest that burro competition may be severe in the Hunter Mountain area and in the lower elevations of the Inyo, Last Chance, Saline, and Nelson ranges. These areas are, or have been, critical wintering sites. Low bighorn populations may be the result of past competition.

Sumner (1959) reports four-winged saltbush *(Atriplex canescens)* as a burro food that is lightly used by bighorn in Death Valley. Welles (1960) states that he found an area 0.4 km (0.25 mi) long in Cottonwood Canyon that was heavily browsed by both burros and bighorn. The burros apparently concentrated on burrobush *(Ambrosia dumosa)*, while the bighorn seemed to favor sweet bush *(Bebbia juncea)*. Welles believes that this preference may be another indication of the possibility of coexistence between them. Their preference of food may at least be divergent.

Burro stomachs taken in the spring from the Argus Range, California, show the burro's adaptability to available food (Bruce Browning). These stomachs were 100 percent full of green forbs, including practically every species growing on the Range.

Water. Where bighorn and burros use the same water holes, direct competition obviously exists. It may be detrimental to the bighorn when the supply is limited. Burros tend to congregate at or near water, particularly in the evening, through the night, and during the early morning. Several observers consider them to be basically nocturnal in their watering habits whereas bighorn are basically diurnal. Being large, burros use more water than bighorn and are more dependent upon it.

In hot weather, springs with small flows are likely to dry up during the hottest part of the day when evaporation exceeds flow. During the night, water accumulates and bighorn will find a supply the first thing in the morning, unless the nocturnal burros have taken it all. Tanks and other catchments that are not replenished daily are exhausted more rapidly when both burros and bighorn are using them.

The major question has devolved upon the compatibility of the two animals. Of concern is the matter of fouling of water supplies. Some writers have taken it for granted that burros foul water by defecating and by wallowing and trampling in it to the extent that bighorn cannot tolerate it. The first danger in such an assumption is the folly of anthropomorphism—attributing human reactions to bighorn. I do not know what they can tolerate and what they cannot, nor how their tolerance relates to my own sensibilities or those of other people. However, I can pass on some observations.

In the Santa Rosa Mountains of California, I once found a small pothole of water which reeked with offensive odors, which was colored brown, and which contained numerous bighorn pellets. It was too strong for me to drink, but fresh tracks indicated use by bighorn.

On another occasion two of us found a spring near Bighorn Gorge in Death Valley National Monument. We saw several burros within 0.4 km (0.25 mi) of it, numerous trails radiating in from the surrounding gentle country, and a trail coming in from adjacent cliffs. We found numerous piles of burro dung and tracks in and around the water, which flowed for about 30 m (100 ft), but no evidence of bighorn. My companion judged the spring to be polluted and not used by bighorn.

The water was flowing strongly enough over the burro dung to be clear, and a sampling at the source, which had a little well washed dung in it too, proved it to be sweet. To my companion's surprise, I quenched my thirst, then filled my canteen. A little backtracking on the trail from the cliffs proved it to be a heavily used bighorn arterial, with sign blotted out near the water by burros.

Had the two of us visited the spring separately, our colleagues would perhaps have had two disparate reports—one of burros polluting a spring and running the bighorn out; and another of compatible use, and of a spring's ability to cleanse itself.

Welles (1960) gives a similar report. He mentions a number of springs in Death Valley reported by earlier writers to be fouled by burros to the exclusion of bighorn. Welles found these springs to be clear and clean, though burros were using them. Bighorn were using some, but not others, for unknown reasons.

Nick Papez and George Tsukamoto report that about a hundred bighorn water at the Boulder Beach sewer ponds on Lake Mead National Recreation Area in Nevada and that the ponds are very smelly.

There is some evidence that trampling around a spring by burros may reduce or stop the flow of water (Weaver, 1959). On the other hand, burros at times paw out drinking holes several feet deep at sand-covered tanks or in washes, and such holes then may benefit bighorn and other wildlife (Weaver, 1959; McKnight, 1958).

Weaver (1959) reports some damage at man-made watering sites by burros. He says that they will paw at a leaky pipe, up to 5 cm (2 in.) or more in diameter, until it is broken at a coupling, and that they can break down cement, masonry, and metal pools. Detritus can accumulate in canyon bottoms from heavy burro traffic, choking diversions built to protect water developments, resulting in storm water, silt, and debris burying the development.

Mensch (1970) reports that burros rapidly deplete the water in natural tanks in Imperial County, California, thereby limiting the range available to bighorn. He concludes that competition with burros and desert mule deer appears to be the factor limiting bighorn populations there.

Social. There is a prevalent feeling that burros dominate bighorn and drive them away from water by attacking them. Weaver (1957) saw a small group of bighorn watering when a jack burro came in. There was no animosity. The bighorn just left and the burro drank. The bighorn did not return, but others came in. He also (Weaver, 1972a) states that the burro is the dominant animal and bighorn will wait for burros to leave or will leave if burros come in while they are drinking—which he observed at Sheep Spring in the Providence Mountains of California.

McMichael (1964) saw a ram and ewe approach a herd of nine feeding burros in the high foothills of the Black Mountains of Arizona. When the burros moved toward the bighorn, the latter moved off, maintaining a distance of about 9 m (10 yd). After 20 minutes, the bighorn moved rapidly past the burros and on around the mountain. While there was no outward aggression, the bighorn did seem nervous in the presence of the burros.

According to Sumner (1959) an old-time prospector in Death Valley related once seeing a ram at a spring suddenly rush forward and catapult itself against the ribs of a burro, which took off at a high lope and did not stop for a long distance.

In summary, although bighorn and burro are often compatible, especially in some places like Death Valley and in some seasons of plenty of food and water, the consensus nevertheless is that bighorn-burro competition can on occasion be severe, generally to the detriment of the bighorn.

DOMESTIC LIVESTOCK

General. Uncontrolled sheep, cattle, and horse grazing in the 1800s created widespread destruction of range, and in the last half of the 1800s the accompanying decline in bighorn numbers was equally widespread.

Since grass is a preferred food of bighorn and since livestock are grazers, livestock are readily concluded to be significant competitors with bighorn for food. Over some of the bighorn range where livestock use is controlled, competition generally is not significant. Bighorn use the higher elevations where the terrain is steep and rugged, whereas livestock prefer the lower, more level reaches. When grass is unavailable, as is the case during parts of the year or over certain parts of the bighorn range, livestock also will compete with bighorn for certain key browse species.

Where livestock and other domestic grazing animals are not controlled, which unfortunately is the case over large areas of bighorn range (including that on Indian lands), it must be considered that cattle and other livestock are a severe and dangerous competitor to bighorn. According to Gallizioli (1977):

> If desert bighorn sheep are to continue as an important, viable component of our southwestern deserts, state wildlife agencies must take a firm stand regarding the cattle problem. While I have absolutely no objection to proper use of other ranges by livestock, I believe there should be *no* livestock, and that includes sheep, burros, horses, and cattle, on ranges as fragile as these desert regions. Precipitation in many years is virtually nonexistent and growth of vegetation minimal. Existence is a struggle for wildlife under the best of circumstances, and the competition posed by livestock is intolerable. State wildlife agencies are shirking their responsibilities by not insisting, at the very least, that livestock be kept off all bighorn sheep ranges. And they should demend [*sic*] also that cattle numbers be sharply reduced in historic bighorn habitats to allow range recovery to the point where reintroduction of the bighorn can be made with some hope of success.

A study at Canyonlands National Park, Utah, indicates that livestock grazing has significantly affected bighorn distribution in the park. Prior to heavy livestock grazing pressure, bighorn occupied much of the park. Now bighorn range is restricted to canyons which were isolated from livestock grazing or canyons where the topography prohibited livestock from grazing the entire canyon (Dean and Spillett, 1976).

Some workers have urged caution in developing water that might attract livestock. If an existing water supply is small, livestock attracted by concentration of the water into a drinking basin or tank are likely to usurp all of it and leave the bighorn worse off than before. If a developed water supply is adequate but forage is scant, the livestock attracted might damage forage around the water to the detriment of the bighorn.

On the other hand, limited cattle grazing may be beneficial to bighorn because cattle tend to open up the dense vegetation which surrounds many springs. Another benefit is the maintenance work performed by cattlemen on springs which otherwise would not provide water (Weaver, 1968). Nick Papez and George Tsukamoto report several times seeing bighorn waiting at a distance while cattle drained the water from springs and tanks in the Highland Range of southern Nevada.

Domestic sheep. Domestic sheep have similar feeding habits and harbor parasites and diseases detrimental to bighorn. Of any domestic livestock, they are believed to have created the most severe competition for bighorn, especially Rocky Mountain bighorn.

In my conversation in 1948 with Sam Cuddleback, then 62, a long-time resident in the Monolith, California, region, I was told that in 1893 and 1899 a great many domestic sheep were turned loose to run in the hills owing to low prices and infection with scabies. Many bighorn died during those 2 years. Carcasses were almost devoid of hair, and it was believed that scabies was the cause. In certain spots, such as the ridge between Chuckawalla Mountain and Cross Mountain, many horns and skulls of all ages, young and old, were to be seen.

Light et al. (1967) state that in the San Gabriel Mountains, California, bighorn mortality increased during the period 1890–1900, owing to diseases, such as scabies, introduced by domestic sheep. Because of the suspected danger, the U.S. Forest Service does not permit domestic sheep in the bighorn range of the San Gabriel Mountains.

On the other hand, Yoakum (1963) writes:

> Often the question arises, can bighorns be liberated in ranges now being utilized by domestic sheep? Although early records discussed the dangers of diseases, including parasites transmitted to wild sheep by domestic sheep, one should bear in mind that the science of animal husbandry has greatly improved in recent years, and epizootic diseases are no longer common in domestic herds. The two Oregon transplants of bighorn are right in the middle of domestic sheep ranges and there are no current disease problems in that area.

Various writers blame domestic sheep for damaging the ranges, and thereby reducing bighorn numbers, in the Sierra Nevada, San Gabriel Mountains, and San Bernardino Mountains of California; the Harcuvar Mountains and elsewhere in Arizona; Engineer Mountain and elsewhere in Colorado; the San Andres Mountains in New Mexico; the Guadalupe Mountains in New Mexico and Texas; the Trans-Pecos region of Texas; and Clark County in Nevada.

Bighorn are sometimes seen with domestic sheep. There are accounts of both rams and ewes running with domestic sheep for periods of up to several months. Musgrave (1926) reports that in the early 1920s a bighorn ram bred a Rambouillet ewe belonging to a Sand Papago Indian family living at or near the southeast corner of the present Cabeza Prieta National Wildlife Refuge in Arizona. Several generations of hybrid sheep resulted from this match (Fig. 14.2).

Cattle. Mixed opinions have been expressed by various writers concerning the extent of cattle competition, past and present. Inability of cattle to negotiate the steepest and roughest portions of bighorn range is self-evident. However, there seems to have been substantial range overlap in the past and there is some at present.

During the late 1800s and early 1900s, cattle grazing was heavy and unregulated in many parts of the West. Russo (1956) cites the history of the Harquahala Mountains in Arizona, where overgrazing depleted a bighorn range until a near absence of bighorn and a low population of deer resulted. Jojoba

M. E. Musgrove

Fig. 14.2. A hybrid 3-year-old ram at Quitobaquito, Pima County, Arizona, in 1922—the result of a bighorn ram and domestic ewe mating.

(Simmondsia chinensis) was listed as the key species in the competition of the three types of animals inhabiting the range. The area today is deficient in palatable growth, the jojoba stands suffering the most severely. Russo states that many years of adequate protection will be necessary to rebuild the range to its former grazing capacity.

He points out that cattle generally are found on desert ranges in the fall and remain until spring, when the majority are removed. During this winter period, game species seldom visit the water holes and remain somewhat scattered throughout the ranges. However, cattle utilize the water holes at this time and crop the vegetation around the watering areas. During the summer, after domestic animals are removed, game species move to water, remaining relatively close. But by this time little palatable vegetation or water is left.

A 4-year research project in the Big Hatchet Mountains of New Mexico indicated that a combination of factors limited the bighorn population. Prolonged overuse of the range by game and domestic livestock seemed to be the principal one (Gordon, 1957). Parts of the range had lost up to 90 percent of the available browse. Buechner (1960) also stresses the serious effect of cattle overuse in the Big Hatchet Mountains, especially in the gentle topography at the south end and at lower elevations. He recommends that the entire mountain range be brought under state or federal control, cattle eliminated, and the vegetation permitted to recover.

Cattle were among the contributors to heavy overuse in the San Andres Mountains of New Mexico (Gordon, 1957). With the removal of livestock and with deer control, the range is recovering. Halloran (1949) states that in narrow mountain canyons containing water at their heads, cattle on the Kofa Game Range in Arizona competed for both food and water—particularly during dry periods when bighorn were forced to use lower waters and the surrounding feed. He cites the overuse of jojoba in arroyos close to water. Jonez (1960) mentions cattle and other competition in the less rugged, lower elevations of Clark County, Nevada.

Charles G. Hansen says that on Desert National Wildlife Range in Nevada range conditions are so poor the livestock only make periodic treks into or through bighorn habitat, but the bighorn are not able to adjust completely even to this type of disturbance. When and where livestock do occur, there is considerable competition for the available grasses. Grass plants are few and far between and are easily uprooted from the loose soil by the larger animals.

In this area ranchers use water to attract their livestock, with corrals around water holes to catch the animals for removal and branding. These corrals are detrimental because bighorn are very reluctant to enter corrals to drink. Even when water is abundant, a livestock corral around a water hole will discourage most bighorn from using the water.

Light et al. (1966) include heavy range use between 1840 and 1900 among seven possible causes of bighorn disappearance from the San Bernardino Mountains of California. The cattle boom of 1840 to 1870, augmented by the gold rush, resulted in excessive grazing by both cattle and domestic sheep in the Santa Ana, Mission, Whitewater, San Gorgonio, and Mill Creek watersheds. Livestock grazing is now limited to the lower washes of the Whitewater and Mission rivers, where competition with bighorn is less severe.

In regard to the San Gabriel Mountains, Light et al. (1967) give range overuse by livestock, in part cattle, from 1890 to 1900 as one of four factors directly altering bighorn habitat. They say that livestock grazing in southern California was quite spectacular between 1850 and 1930. During this period vast numbers of domestic sheep and cattle grazed the coastal and desert foothills of the San Gabriel Mountains, the headwaters of the East Fork of the San Gabriel River, the San Diego watershed, the North Fork of Lytle Creek, and the Long Pine area. Current commercial grazing privileges on San Bernardino and Angeles national forests do not extend into key bighorn range.

Wilson (1968) says:

> ...in southeastern Utah...most of the cattle do not graze in areas utilized by bighorn. Nevertheless, it is obvious that the bighorn have been pushed back in some areas due to the utilization of the range by cattle. For example, 40 cattle utilized the upper portion of Red Canyon during 1965 and 1966. Tracks of bighorn were only noted on three occasions in this range, and there was no apparent evidence that the animals had stopped to lie down or graze.
>
> In the lower portion of Red Canyon below Warm Spring cattle were never observed. Tracks of bighorn were noted on every visit to this location and bighorn were sighted on almost every visit. There is no difference in topography, climax vegetation, or available water, between lower Red Canyon and upper Red Canyon. The lack of cattle in lower Red Canyon was the only notable difference between the two areas.

Burandt (1970) says that both bighorn and deer water at Cerro Gordo Spring and Mexican Spring in the Inyo Mountains of California, in spite of up to fifty cattle trampling and fouling the water supply.

Ferrier and Bradley (1970) and Myers (1970) found both cattle and bighorn heavily using Highland and Cow Wells springs in the Highland Range, Nevada. They conclude there is direct competition for both forage and water, particularly in the summer. The bighorn use the waterless northeastern one-half of the Highland Range where there are no cattle until forced to the two springs in hot weather.

Weaver and Hall (1972) report cattle competition for water at the lower elevations of the Clark Mountains of California, and for food and space in the Kingston Mountains, where cattle occupy good bighorn habitat.

There are a number of accounts of bighorn rams and ewes running with cattle. I obtained several photos of a ram feeding with cattle east of Monolith, California, in 1943 or 1944. Over a period of 4 years of wildlife research in Owens Valley, California, I was told of three instances of bighorn, all rams, running with cattle.

Halloran and Blanchard (1950) report two observations of bighorn with cattle on Kofa Game Range. One was of a ewe with six head of cattle, September 5, 1948. When seen, the animals were traveling slowly up an arroyo. As the men approached closely the animals bolted into the brush, led by the bighorn. The other observation was of a ewe with three cows and two calves at the same location on November 10, 1949.

Horses. Competition between bighorn and horses is of little consequence at present. However, it has occurred in some places. Earlier, San Andres Wildlife Refuge of New Mexico had a substantial wild horse population that caused serious competition with bighorn on the drought-ridden range. Buechner (1960) says that in 1945, when horses were being eradicated from the Refuge, 293 were killed and 28 captured. With the advent of adequate moisture and removal of horses, the vegetation has improved.

Halloran and Deming (1958) state that a band of ten horses on Desert National Wildlife Range, Nevada, completely utilized a small spring that had supplied thirty to forty bighorn before the intrusion of the horses. Charles G. Hansen says that in that area horses pulled up grass plants by the roots, leaving much of the plant uneaten, thus reducing forage for bighorn.

Goats. Competition from domestic goats has been of importance in some places but by 1978 was very restricted. Goats were grazed on the range more commonly at one time than at present.

Because of their affinity for rough country and their propensity for heavily overgrazing the range, goats have exerted very direct and severe competition. Gordon (1957) mentions them as one of several past competitors in the San Andres Mountains of New Mexico. Gross (1960) credits them with helping to nearly eliminate bighorn in the Guadalupe Mountains of New Mexico and Texas between 1930 and 1940. Davis and Taylor (1939) mention that few bighorn occurred on the west side of the Guadalupes where goats were run.

I found about thirty goats on bighorn range in Martinez Canyon, Santa Rosa Mountains, California, in 1953. They had similar feeding habits and heavily used a limited area. Dick Weaver and I also found a band of about twelve goats in 1953 in Hall Canyon, Panamint Mountains, California, which is bighorn range. They seemed to stay mainly in the rugged canyon itself. However, Weaver (1972b) could not find any during a diligent helicopter search 17 years later.

Halloran (1949) states that at different times within the past 50 years, and perhaps before, domestic goats were allowed to wander freely over Kofa Game Range in Arizona. By 1940, less than fifty were there. As of 1949, refuge personnel had eliminated a few and other factors also had reduced them. By 1954 they were absent, according to Gale Monson. Cursory examination of the stomachs of two of these animals revealed that during dry weather they fed extensively on the small branches of jojoba in the near vicinity of water, competing directly with bighorn and deer.

I have been told of three instances of bighorn rams running with goats in Owens Valley, California, in the late 1930s.

OTHER ANIMALS

Elk. By virtue of a distinct difference in habitat preferences, elk and desert bighorn have little opportunity for interrelating. However, in the Owens Valley of California a transplanted herd of tule elk *(Cervus nannodes)* does range high enough on the mountain slopes on either side of the valley to be occasionally on bighorn winter range. But the minor incidence of such overlap creates no problem of competition. Burandt (1970) and Dunaway (1970) confirm my own observations and conclusions on this subject.

Javelina. Russo (1956) says that javelina *(Pecari tajacu)* are present in a number of bighorn ranges and that a relationship undoubtedly exists. He quotes from an unpublished report mentioning deer and javelina in the Santa Catalina Mountains of Arizona that declares "competition from these animals is quite a bit greater than that of domestic animals."

Jackrabbits and rodents. Blacktail jackrabbits *(Lepus californianus)* and rodents occur throughout desert bighorn range and consume some of the same forage plants. However, no one has considered the degree of competition to be anything but slight. In the Santa Rosa Mountains of California I found antelope ground squirrels *(Citellus leucurus)* to be common and noted that they took considerable forage in some areas. Use on juniper was noted especially, and on jojoba around one spring (Jones et al., 1957).

Birds and bees. Birds, especially white-winged doves *(Zenaida asiatica)*, and bees congregate around some desert springs and can consume substantial amounts of water. When flows become low and competition for water is severe, these animals become competitors.

Ralph Welles tells of many birds and thousands of bees at Stubbe Spring, when it was only a drip, using the water faster than it came into the tank. Bees crawled up the pipe and obtained the water inside. Bighorn were trying to drink, but only a few drops came past the bees. He also tells of bees stinging bighorn and scaring them away.

On a cool September day, Welles saw about fifty Gambel quail *(Lophortyx gambeli)* and a dozen or so house finches *(Carpodacus mexicanus)* drinking water as fast as it came into the first drinking basin, completely stopping the overflow into the second basin. He later estimated that about eighty bighorn, fourteen deer, and two thousand quail shared Stubbe Spring (Welles, 1965).

Charles G. Hansen says "The sudden movement of flocks of birds at water will alarm bighorn, and they will jump and run a few feet away. On Desert National Wildlife Range, birds generally stay away from the water when bighorn drink. Birds can consume large quantities of water, but on Desert Range they are not sufficient in number to deplete a water supply. If the water is so scarce that flocks of birds would affect the amount available to bighorn, then there isn't enough water for bighorn, even without the birds."

Weaver and Mensch (1970a) suggest that competition may occur between bighorn and the very high population of chukars *(Alectoris graeca)* in the Eagle Crag Mountains, California, where water resources are severely limited.

WITHIN HERDS

The question of competition within bighorn herds becomes basically an inquiry into the inner workings of population regulation. This is an enormously complicated field requiring conclusive information that does not exist for desert bighorn. As with any big game animal, there is competition between individuals for food and water when these are in short supply. As is always the case, the very young, the very old, the lame, and the ill may be forced to take the leavings, if any.

Charles G. Hansen provides the following:

> When surface water on Desert National Wildlife Range was confined to an area of about 2 to 3 square feet, there was considerable pushing and shoving in order to get a drink when more bighorn were present than the area would accommodate. Certain bighorn were more aggressive than others and would butt others out of the way. The lambs would usually give way to their elders when there was too much activity at the water holes. But they would take advantage of temporary openings when adults had shoved one another away from the water.
>
> The lambs would also creep in, around and under the adults, when adults were trying to crowd each other away from the water. No injuries to lambs were seen. The adults would generally begin their crowding in a tentative manner, and when lambs were first nudged, they usually gave way to the larger bighorn. The mothers of the lambs were the most active in their attempts to get to water, but made no attempt to protect their lambs from other bighorn.
>
> On one occasion, large rams began butting heads after they had shoved and pushed each other in their attempt to drink side by side. The larger, heavier ram forced the other ram away and went over to drink among the ewes that had taken advantage of the head-butting to rush in and get a drink. They seemed reluctant to allow the ram space to drink, but did yield to him. The smaller ram waited until the larger ram was through drinking before he moved in.
>
> There was a definite "peck order" among the rams and ewes when they were in contact with each other at the water holes. The same type of activity occurs with the captive bighorn at Corn Creek when a ration of grain is put out. The dominant one of each sex will take the "lion's share" before it will leave the grain. Sometimes a dominant male will not even tolerate the females feeding in the same trough. Both sexes will tolerate young lambs under most circumstances. I do not recall seeing a lamb being chased away from food, except when it was very young and tried to nurse or feed next to the wrong female.

Competition with adults for insufficient drinking water may work to the detriment of lambs; and prolonged searches for water up and down dry canyons, which I found them making, may tax the lambs severely (Jones, 1955).

Ralph Welles tells of an older ram fighting off another and bluffing two younger ones to get the little water there was at Stubbe Spring in Joshua Tree National Monument. After the older ram left, the other old one drank a couple of minutes while the young ones stood around until he was through. Then the young ones started in, but the first ram came back and emptied the basin. He tells of several bighorn using Stubbe until the water was all gone, with only a few drops coming in past the bees. Several ewes, some with lambs, then hassled over the few drops coming in. Some bighorn waited on the slope all day for water. One ewe tried for 5.5 hours to get the water she wanted.

15. Population Dynamics

Charles G. Hansen

Population dynamics are concerned with forces that cause changes in a cohesive group of organisms, such as a cohesive group of desert bighorn. This chapter deals more with changes that occur in a bighorn population than it does with the causes of these changes.

Populations continue to survive only because new organisms continually replace organisms that are lost. The ability of organisms to multiply when conditions are favorable allows a population to expand. When the environment becomes crowded or conditions are unfavorable, mortality adjusts the population to the new situation. Thus, reproduction and death are basic characteristics of any population of organisms. These characteristics are responsible for the adjustment of population densities to the prevailing supply of things needed by the individual organism (Nicholson, 1953).

The size of a population is determined primarily by the food supply. However, other factors, such as natural enemies or the availabilty of suitable space, may keep numbers far below what the food supply will permit. A growing population progressively depletes such things as food and adequate space, which usually impairs the favorability of the environment until further growth of the population is impossible (Nicholson, 1953).

POPULATION INCREASE

Population parameters for bighorn are relatively available, because these animals are comparatively easy to age and sex with reasonable reliability. Buechner (1960), following Leopold's (1933) lead, set up a model to estimate the maximum rate of increase for a theoretical North American bighorn population. He assumed (1) an equal sex ratio, (2) one lamb per ewe per year, (3) birth of the first lamb when the ewe was 3 years old, (4) a negligible number of ewes living beyond breeding age, and (5) no mortality. These assumptions were necessary to develop a standard rate of increase with which to compare living populations to determine their state of well-being.

Perhaps the main errors in these assumptions are that ewes can and will give birth at an earlier age than 3 years, they do give birth to twin lambs on occasion, and adult ewes live only to about 10 years of age. It appears that twinning and early pregnancies occur most frequently when bighorn are introduced into an ideal pristine environment; under such conditions they seem to have the capacity to increase at a high rate (Woodgerd, 1964; Deming, 1961). Buechner (1960) calculated a maximum rate of increase with $r = 0.258$, by using the growth equation $N_t = N_o e^{rt}$ (Odum, 1959), where N_t equals the population at the end of the t th year; N_O equals the number of animals present initially; e, the base for natural logarithms, is equal to 2.71828; and r equals the growth rate. Woodgerd (1964) arrived at a figure of 0.265 for the Wildhorse Island, Montana, herd which slightly exceeds the r value for the model. Buechner obtained his r value from Leopold's (1933) charts. Interestingly, an r value of 0.305 calculated from Leopold's charts would mean that more than one young was born to each female each year, or the ewes are breeding before 2 years of age.

The bighorn introduced into Hart Mountain, Oregon, in 1954 (Deming, 1961) exhibited a rapid increase. The growth rate for 1955 to 1956 was 0.365, and for 1956 to 1957 it was 0.329. This was primarily due to the uneven sex ratio (more ewes than rams), but twinning did occur. How many more years of rapid reproduction would have occurred is not known, since an accurate census became more difficult to obtain as numbers increased and the herd extended its range. However, when bits and pieces of information are brought together, they produce a picture of desert bighorn population dynamics that could not be determined otherwise.

POPULATION COMPOSITION

Age classes. Age determinations have been made for desert bighorn populations at various times and in a number of locations. Perhaps the most numerous samples are from Desert National Wildlife Range, Nevada. Information from 374 bighorn carcasses found in the field was used to determine age composition of living animals on the Wildlife Range.

A population having 450 ewes potentially could produce 450 lambs. Some ewes can probably produce twin lambs under the most favorable conditions, while other ewes may not produce; therefore, a potential ewe-lamb ratio of 100 to 100 seems reasonable. In actuality, on Desert Wildlife Range the average annual reproduction is closer to a ratio of 100 ewes to 70 lambs. This would reduce the lamb drop from 450 ewes to about 316 animals, which is further reduced by natural mortality to 158 by the age of 1 year and to 100 by age 2 years. These figures were based on survival data from spring and summer counts. Survival beyond age 2 years was deduced from studies of carcasses found on the Wildlife Range, as summarized in Table 15.1.

TABLE 15.1.

Theoretical Age Distribution of Surviving Bighorn Based on the Percentage of Natural Mortality of Ewes and Rams on Desert National Wildlife Range for the Years 1953 Through 1967

Age	1953	'54	'55	'56	'57	'58	'59	'60	'61	'62	'63	'64	'65	'66	'67
1															
2	100														
3	99	99													
4	98		98												
5	97			97											
6	94				94										
7	90					90									
8	82						82								
9	73							73							
10	63								63						
11	51									51					
12	30										30				
13	16											16			
14	6												6		
15	2													2	
16	1														1
Total	902														
Annual population 902															

(Y-axis label: Number of individuals in a given age-class for any one year)

(Diagonal label: Number of individuals left in each age-class starting with 100 in 1953)

Source: Hansen (1967a).

Differential reproduction and survival will alter the numbers in each age class each year. However, in a healthy population, when extra losses occur one year they are usually accompanied by a good gain during subsequent years.

Differential mortality occurs between sexes, and as a result the sex ratio appears to favor the rams in an undisturbed population (Buechner, 1960; Sugden, 1961; Welles and Welles, 1961a; Monson, 1963; and Hansen, 1967a). The uneven sex ratio often found in desert bighorn herds can alter the rate of increase. In Buechner's (1960) model, the maximum rate of increase was 0.258. However, if the sex ratio was altered to 25 : 100, r would change to 0.392. The calculations are as follows:

antilog$_e$ $r = 1 +$ (ewes/adults) y   (y = number of young per ewe)
antilog$_e$ $0.258 = 1 + \frac{1}{2}$ y   (ram-ewe ratio = 100:100)
y = 0.60 lamb per ewe (including yearling and 2-year-old ewes)
antilog$_e$ $r = 1 + \frac{4}{5}$ (0.60)   (ram-ewe ratio = 25:100)
$r = 0.392$

In the above calculations, when solving for r use the natural logarithm tables as indicated by the sub e in the notation antilog$_e$ r. The antilog is the number in the log tables corresponding to the given log. In the above example, antilog$_e$ $r = 1 + \frac{4}{5}$ (0.60) equals antilog$_e$ $r = 1.48$. The antilog sub e of r from the natural log tables is then 0.392.

With data from Desert Wildlife Range, r values were calculated for selected years. These data are listed in Table 15.2

TABLE 15.2.

Selected Ram : Ewe and Lamb : Ewe Ratios from Desert National Wildlife Range

Year	Ram : Ewe	Lamb : Ewe
1948	53 : 47	80 : 100
1956	40 : 60	68 : 100
1962	19 : 81	51 : 100
1963	18 : 82	38 : 100
1964	20 : 80	54 : 100

The r values were calculated as follows:

1948 antilog$_e$ $r = 1 + 1/2.80$	1962 antilog$_e$ $r = 1 + 4/5.51$	1964 antilog$_e$ $r = 1 + 4/5.54$
$r = 1.40$	$r = 1.41$	$r = 1.43$
$r = .336$	$r = .343$	$r = .358$
1956 antilog$_e$ $r = 1 + 3/5.68$	1963 antilog$_e$ $r = 1 + 4/5.38$	
$r = 1.41$	$r = 1.30$	
$r = .343$	$r = .262$	

These r values indicate that there was a rather high rate of increase during these years, even though the lamb : ewe ratios were relatively low in the later years. However, mortality appears to have exceeded the maximum rate of increase. The result was a reduction of adults, which contributed to an overall reduction of the entire population. This will be discussed under the section on population growth.

GROUP COMPOSITION

Bighorn groups are formed at various seasons, in a multitude of combinations. However, they generally are groups of ewes with lambs and offspring from the previous year or two. For a brief time during the rut, some adult rams are with the ewe groups. Ewes about to lamb isolate themselves for only about a week; at this time the young rams often are forcibly driven away. Deming (1953) recorded group sizes for Desert Wildlife Range (Table 15.3), and Simmons (1969b) gathered corresponding data from Cabeza Prieta National Wildlife Refuge (Tables 15.4 and 15.5).

These data show that bighorn are encountered most frequently alone, or in groups of two or three. Groups of less than nine are most common. The large number of animals reported by Deming (1953) in March and April reflects the work done on Desert Wildlife Range during the lambing season. The summer water-hole count period is also reflected in his figures.

Simmons (1969b) went a step further and determined group composition. He states that

> Groups of adult rams on the Cabeza Prieta were generally smaller than ewe groups (adult ewes and juveniles), which in turn tended to be smaller than mixed groups [Fig. 15.1]. Deming (1953), however, noted that ewe groups tended to be smaller than ram groups on the Desert Game Range, Nevada.
>
> The average sizes of ewe and mixed groups appeared to increase ... from the Cabeza Prieta to the Kofa-Ajo Mountains type ... [Fig. 15.2]. This increase probably parallels an increase in the quality of sheep habitat and a consequent increase in population density. There was no significant difference in the average size of ram groups. Nevertheless, the large bands of rams that have been seen on the Kofa Game Range have not been seen on the Cabeza Prieta. Perhaps larger samples from the Kofa-Ajo Mountains habitats might show significant differences in the sizes of ram groups.
>
> Figure [15.3] reflects the seasonal changes in bighorn group composition described by Russo (1956), Welles and Welles (1961a), Simmons (1961), and others: separation into ram and ewe groups shortly before and during the lambing season, and mingling rams and ewes in the late summer and fall for rutting activity. The demonstration is somewhat obscured by the mingling rams and ewes at water holes during dry periods throughout the year.
>
> The reasons for the changes in group composition have been discussed by Geist (1968b). He stated that mountain sheep can be ranked by outward appearance from the adult ram to the lamb, and that sheep at the opposite ends of this smooth cline segregate into separate bands: ram bands and "nursery" bands. The sexually mature rams associate with all sheep, though they prefer to interact with rams of equal horn size and with adult ewes. Females and juveniles, however, interact almost exclusively with sheep of equal or smaller horn and body size. Estrous ewes change their regular behavior patterns and associate with adult rams, while anestrous ewes withdraw from interactions with adult rams (see also Banks, 1964, for descriptions of similar behavior in domestic ewes).

TABLE 15.3.

Sizes of Ewe, Lamb, and Yearling Bands on Desert National Wildlife Range Between 1947 and 1952

Band Sizes	Number of Observations												Total No. of Observations
	Jan.	Feb.	Mar.	April	May	June	July	Aug.	Sept.	Oct.	Nov.	Dec.	
1	11	21	22	16	14	11	40	7	8	6		9	165
2	11	18	20	12	16	8	29	7	6	7	9		143
3	13	13	24	11	17	8	15	13	6		5		125
4	4	9	17	10		8	19	6	6		11	6	96
5	1		14	6	5	8	4	9			3	2	52
6	1	9	14	12		1	11	1	9	3		2	63
7		6	12	7	12	7	5	7	7	4	2	1	70
8	1	9	14		2	9	5	2	2				44
9		8	10	7		6	4	2	1	1	1		40
10	1		10	8			2	1		1		1	24
11		6			3			1	1		1		12
12		5	8	7	3	1			1				25
13		1	6	1				1				1	10
14			9	10	1	1							21
15			7				1						8
16				5		1							6
17		1	3	6									10
18				1									1
19			4	1								1	6
20	1		3	1									5
21		1	6	3								1	11
22			2					1					3
23				1									1
24				1									1
25				2									2
26			1										1
28			1										1
34			2										2
35			1										1
39			1										1
50			1										1
69				1									1
Total No. of Bighorn Observed	139	549	1725	1112	226	393	493	274	213	85	126	125	5438

LIFE TABLES

A life table is a concise summary of certain vital statistics of a population (Deevey, 1948). Life tables are presented in several different ways. The following tables were developed using mortality data available from Desert

TABLE 15.4.

Sizes of Bighorn Groups in Arizona

Size Class	Percentage of Total Observations	
	Cabeza Prieta*	Kofa-Ajo Mountains†
Singles	40	37
2–9	58	52
10–19	2	10
20–29	0	1
Over 30	0	0

Source: Simmons (1969b).
*502 groups observed.
†174 groups observed.

TABLE 15.5.

Mean Bighorn Group Sizes in Cabeza Prieta and Kofa-Ajo Mountains

	Mean Group Size		Standard Error $s_{\bar{y}}$		Range of Group Sizes		Total Groups Observed	
	CP*	KA†	CP*	KA†	CP*	KA†	CP*	KA†
January	3	2	0.54	0.91	1–8	1–4	17	3
February	3	3	0.39	1.30	1–8	1–18	24	13
March	3	5	0.48	0.66	1–11	1–23	37	63
April	2	2	0.24	1.00	1–8	1–4	49	3
May	3	3	0.35	0.71	1–10	1–4	42	4
June	2	2	0.20	1.30	1–13	1–9	146	6
July	3	3	0.33	0.47	1–15	1–10	60	28
August	3	4	0.30	0.65	1–10	1–11	43	18
September	2	5	0.22	1.01	1–5	1–12	26	9
October	2	2	0.32	0.48	1–7	1–4	26	6
November	3	3	0.45	1.62	1–5	1–11	12	6
December	2	2	0.39	0.46	1–8	1–7	20	15
Yearlong	3–	3			1–15	1–23	502	174

Source: Simmons (1969b); data for Cabeza Prieta before 1961 are from U.S. Fish and Wildlife Service narrative reports (1940–61).
*CP = Cabeza Prieta, 1940–65.
†KA = Kofa-Ajo Mountains, 1961–65.

Wildlife Range (Hansen, 1967a). Bradley and Baker (1967) also use survival data for the first two age classes, 0 to 0.5 year and 0.5 to 1 year, from the same source, in order to include the most accurate data for the younger age classes.

Tables 15.6 and 15.7 show that a ram which reaches age 2 can be expected to live for 7 more years. After the age of 10, both ewes and rams can be expected to die in less than 2 years. This does not preclude the possibility that some will live to the ripe old age of 15 or 16 years.

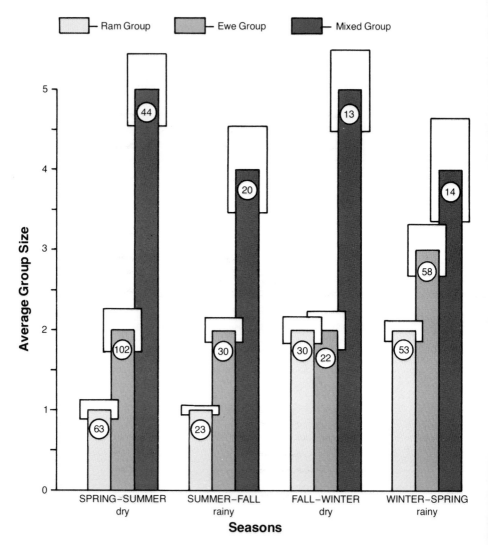

Fig. 15.1. Mean bighorn group sizes by group classes and seasons, Cabeza Prieta Wildlife Refuge (from Simmons, 1969b). Sample sizes are shown on each bar. White bars indicate twice the standard error. (Data for 1940–65; data before 1961 are from U.S. Fish and Wildlife Service narrative reports.)

Bradley and Baker (1967) include survivorship curves for both ewes and rams (Fig. 15.4). These illustrate the high lamb mortality, then the near immunity to death until the animals reach age 9 or 10 years.

In working with data from Desert Wildlife Range, Hansen (1967a) considers the actual number of bighorn instead of the theoretical number which is

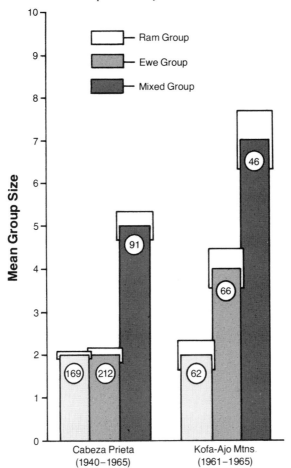

Fig. 15.2. Mean bighorn group sizes by group classes in two different bighorn refuges (from Simmons, 1969b). Sample sizes are shown on the bars. White bars represent twice the standard error. (Cabeza Prieta data for 1940–61 from U.S. Fish and Wildlife Service narrative reports.)

usually based on 1000. Table 15.1 illustrates the population structure of adult bighorn and follows their demise through the years. This table treats the animals more as individuals than as statistics.

A bighorn manager can use this method to account for bighorn each year. As an exaggerated example, if one hundred yearlings show up one year but only fifty lambs were counted the previous year, he can immediately tell that the previous count was not valid. This can be illustrated by using actual published figures. An example comes from Woodgerd (1964), as shown in Table 15.8 (his Table 1).

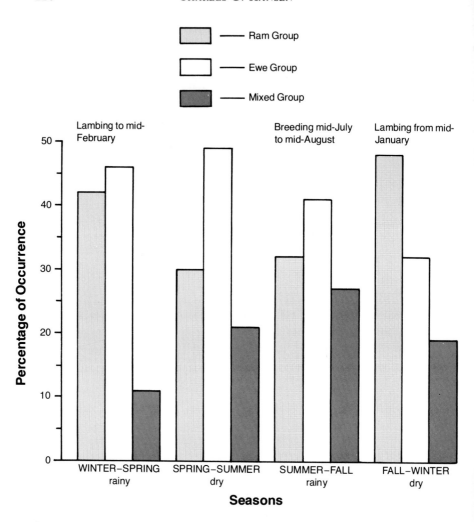

Fig. 15.3. Change in group composition by season in Cabeza Prieta Wildlife Refuge (from Simmons, 1969b). (Data for 1940–65; data prior to 1962 from U.S. Fish and Wildlife Service narrative reports.)

Table 15.9 shows that in order to get thirty-five mature bighorn (eighteen adult ewes and seventeen adult rams from Table 15.8) in 1952 there had to be twelve 2-year-old bighorn in 1951 ready to enter the breeding population in 1952. Also, to get forty-three mature bighorn in 1953, there had to be eight 2-year-old bighorn in 1952; Table 15.8 shows only three. Therefore, assuming that the sex and age classification was accurate, but some bighorn were overlooked, the minimum total population would have to be adjusted to fifty-four in 1951 and seventy-one in 1952.

TABLE 15.6.

Life Table for Desert Bighorn Rams
Based on the known age at death of 302 rams (2 to 16 years of age) dying between
1946 and 1966 on Desert National Wildlife Range

x, Age in Years	Frequency in Sample	dx, Number Dying	lx, Number Surviving	$1000\,qx$ Mortality Rate	ex, Expectation of Life in Years
0–0.5*	705	500	1000	500	2.95
0.5–1*	404	286	500	572.0	4.41
1–2	4	3	214	14.0	8.63
2–3	5	4	211	18.9	7.74
3–4	4	3	207	14.4	6.88
4–5	6	4	204	19.6	5.98
5–6	6	4	200	20.0	5.09
6–7	19	13	196	66.3	4.18
7–8	28	20	183	109.2	3.44
8–9	30	21	163	128.8	2.80
9–10	59	42	142	295.7	2.14
10–11	40	28	100	280.0	1.83
11–12	48	34	72	472.2	1.35
12–13	29	21	38	552.6	1.11
13–14	19	13	17	764.7	0.85
14–15	3	2	4	500.0	1.00
15–16	2	2	2	1000.0	0.50

*The first year age intervals are computed from lamb and yearling : ewe ratios of 70 : 100, 35 : 100, and 15 : 100 for the 0- to 1-month, 6-month, and yearling age intervals, respectively (Hansen, 1967a). Mean length of life, 2.4 years.

Survival of 2-year-old bighorn in 1953 would determine the number of adults present in 1954. If ten survived, then the total mature animals for 1954 should be fifty-three instead of forty-seven. Also, if fifteen yearlings survived from 1953 to make a total of fifteen 2-year-olds in 1954, then the total would be ninety-nine instead of eighty-seven. Woodgerd (1964) lists the estimated population as one hundred animals for 1954, which corresponds with the adjusted total (99) in Table 15.9.

Projecting these figures as they are in Table 15.1 can allow a prediction of the expected population for future reference. This method can be refined to predict the number of animals in each age class in each succeeding year. The two critical ages to consider are the yearling and 2-year-old categories. These are relatively easy to determine in the field and are subject to most variability. Therefore, if the number of animals in these two age classes is determined each year through an actual census, a population expectancy table can be developed. The reliability of such a table will depend on the accuracy of the data entered in it each year. As additional data are acquired each year, the estimated figures can be corrected. For example, in the exaggerated example above, if one hundred yearlings are counted one year then there must have

TABLE 15.7.

Life Table for Desert Bighorn Ewes
Based on the known age at death of 144 ewes (2 to 15 years of age) dying between 1946 and 1966 on Desert National Wildlife Range

x, Age in Years	Frequency in Sample	dx, Number Dying	lx, Number Surviving	$1000\,qx$, Mortality Rate	ex, Expectation of Life in Years
0–0.5*	336	500	1000	500.0	2.56
0.5–1*	192	286	500	572.0	3.62
1–2	6	11	214	51.4	6.79
2–3	3	5	203	24.6	6.13
3–4	7	11	198	55.5	5.27
4–5	11	17	187	90.9	4.55
5–6	13	11	170	64.7	3.95
6–7	14	21	159	132.0	3.19
7–8	21	32	138	231.8	2.60
8–9	14	21	106	198.1	2.24
9–10	28	42	85	494.1	1.66
10–11	9	14	43	325.5	1.80
11–12	7	11	29	379.3	1.43
12–13	7	11	18	611.1	1.00
13–14	3	5	7	714.2	0.79
14–15	1	2	2	1000.0	0.50

*The first year age intervals are computed from lamb and yearling : ewe ratios of 70 : 100, 35 : 100, and 15 : 100 for the 0- to 1-month, 6-month, and yearling age intervals, respectively (Hansen, 1967a). Mean length of life, 2.00 years.

TABLE 15.8.

Sex and Age Composition of Bighorn from 1951 to 1954 on Wildhorse Island, Montana

Year	Rams			Ewes			Lamb	Total
	Mature	2 Yr.	1 Yr.	Mature	2 Yr.	1 Yr.		
1951	12		4	11			11	38
1952	17		6	18	3	4	18	66*
1953	21	6	7	22	4	8	15	83
1954	20	4	6	27	5	7	18	87

Source: Woodgerd (1964) and Ogren (1954).
*The figures for 1951 and 1952 give us an r value for the increase from 38 to 66 animals that would require the birth of about three lambs per ewe. However, if the figures are tabulated as in Table 15.9, the reason for the big increase may become apparent.

been at least one hundred lambs born the previous year. Thus, the previous year's ewe : lamb ratio may need to be changed, and the number of lambs adjusted accordingly.

Differential mortality is shown in Figure 15.5. The great discrepancy between the years 10 and 11 could be due to aging techniques more than actual mortality rates. Figure 15.5 shows that the highest adult mortality occurs after

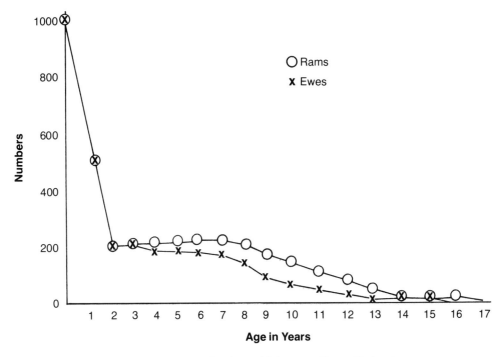

Fig. 15.4. Survivorship curves for desert bighorn on Desert National Wildlife Range (from Bradley and Baker, 1967).

TABLE 15.9.

Adjusted Age Composition of Bighorn on Wildhorse Island, Montana, from 1951 to 1954

	Total	Lambs	1-year*	2-year*	Mature	Adjusted Total*
1951	38	11	4 (8)	(12)	23	(54)
1952	66	18	10	3 (8)	35	(71)
1953	83	15	15	10	43	83
1954	87	18	13	9 (15)	47 (53)	(99)

*Figures in parentheses are adjusted to match subsequent data.

9 years of age. In the year classes 10, 11, and 12, 47 percent of the adult population dies. The percentage of deaths after 12 years is greatly reduced; however, this reduction merely reflects the number available to die rather than a reduced likelihood of animals' dying. When the number dying is plotted against the number available to die (Table 15.10), it is obvious that animals over 10 years are dying faster than those 10 or younger. The numbers avail-

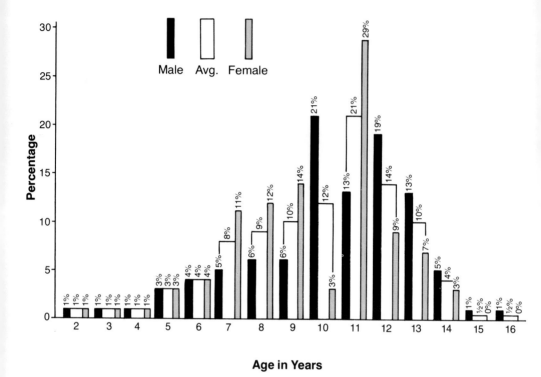

Fig. 15.5. Natural mortality of desert bighorn by percentage for each yearly age class from 2 years to end of life span. Also shown is the average mortality of both sexes. Percentages based on 374 carcasses found between 1948 and 1964 on Desert National Wildlife Range, Nevada (Hansen, 1967a).

able to die in the 15- and 16-year age classes are so small that the percentages (25 and 50 percent) in Table 15.10 do not truly represent old age mortality in desert bighorn populations.

(*Editors' note:* McQuivey [1978], using data gathered from all desert bighorn ranges in Nevada and based on 1,939 aerial observations between 1969 and 1976, demonstrates that in the interval between 1.5 and 12.5 years of age, the average age structure of rams declines gradually through each successive year of life rather than, as demonstrated by Hansen's data, the highest adult mortality occurring after 9 years of age.)

POPULATION FLUCTUATION

In a growing population, the herd composition counts will show a predominance of young animals. From Table 15.1, it can be seen that out of 902 animals 2 to 16 years old, 660 are 8 years or younger, while there are only 242

TABLE 15.10.

Proportion of Bighorn Dying in Relation to Those Available to Die

Age	No. Available to Die	No. Dying Out of Total	% Dying of the No. Available Each Year
2	100	1	1.0
3	99	1	1.0
4	98	1	1.0
5	97	3	3.1
6	94	4	4.3
7	90	8	8.9
8	82	9	11.0
9	73	10	13.7
10	63	12	19.0
11	51	21	41.0
12	30	14	46.7
13	16	10	62.5
14	6	4	66.7
15	2	½	25.0
16	1	½	50.0

Source: Records of Desert National Wildlife Range.

that are 9 years and older. This represents a stable population, whereas a growing population would have an even greater number of young animals. A declining population would have a predominance of old animals, with relatively few young animals. It is not known at what low point a declining population cannot recover. However, bands of twenty or less bighorn have been introduced into former bighorn range and have multiplied rapidly, especially where range conditions were pristine-like.

A population of predominantly old animals not only indicates a declining population but a declining habitat as well.

POPULATION GROWTH

Actual figures of population growth are not readily available and often do not cover a long period of time. Table 15.11 presents data from bighorn released on Hart Mountain, Oregon. Figure 15.6 shows the estimated growth and the potential growth of that herd. If animals were not removed for transplant elsewhere, presumably there would have been 95 in the release enclosure by 1961, after release in 1954. The number for 1961 is given as 70+, so it is possible the actual number of bighorn was closer to 95 than is suggested by the number 70. The r value for the years 1956 and 1957 would be 0.365 and 0.329, respectively. As mentioned under "Population Composition," these values indicate that there were more ewes than rams in the initial population and that more than one lamb may have been born per ewe over age 2.

TABLE 15.11.

Population Growth of Bighorn Released on Hart Mountain, Oregon, in 1954

Year	Number Present
1954	20
1955	25
1956	36
1957	48
1958	54*
1959	58*
1960	64*
1961	70*

Source: Deming (1961)
*Some animals escaped and some transplanted.

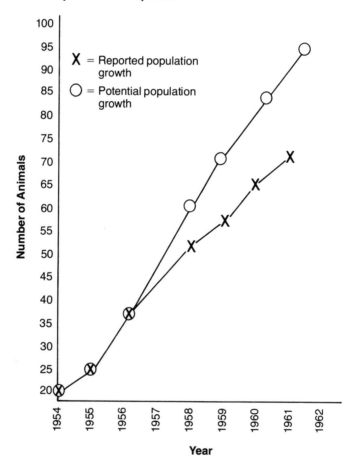

Fig. 15.6. Population growth of bighorn released on Hart Mountain, Oregon (based on Deming, 1961).

The Desert Wildlife Range population was estimated at 300 bighorn in 1936 and proceeded to grow under the protection and habitat improvements provided by the U.S. Fish and Wildlife Service. In Table 15.12, the survival rates for this population are based on census figures, as well as on mortality data gathered over 33 years of study on the Wildlife Range. Since these data were checked regularly in the field, their reliability appears to be relatively high. Starting with the 300 animals in 1936, and using the rate of mortality presented in Table 15.10 and the survival data collected during spring and summer censuses, a table of living bighorn was developed as Table 15.1. The annual fluctuation of lambs and yearlings was determined from lamb : ewe and yearling : ewe ratios. The total annual increment of 2-year-old bighorn was calculated from these ratios and entered in the table.

This number was then reduced annually according to the mortality rate determined from dead animals found on the Range. Total population estimates, made occasionally, were used to determine the reliability of the artificial population structure that was being developed. Therefore, it is felt that the figures presented in Table 15.12 are indicative of the true population.

TABLE 15.12.

Population Growth and Decline of Bighorn on Desert National Wildlife Range Between 1936 and 1974

Year	Estimated No. Present*	Year	Estimated No. Present*
1936	300	1956	2170
1937	350	1957	2140
1938	440	1958	2050
1939	480	1959	1935
1940	525	1960	1850
1941	615	1961	1750
1942	635	1962	1615
1943	675	1963	1460
1944	785	1964	1355
1945	910	1965	1240
1946	950	1966	1180
1947	1155	1967	1195
1948	1420	1968	1025
1949	1690	1969	1000
1950	1825	1970	800
1951	1895	1971	800
1952	2035	1972	500
1953	2060	1973	600
1954	2270	1974	600
1955	2280		

*Figures are estimates based on records from spring and summer censuses and general observations. Area south of U.S. Hwy. 95 (Mt. Charleston) was deleted from the Range in 1966 and not included after 1966. Summer water-hole counts based on time-lapse photography 1970–74.

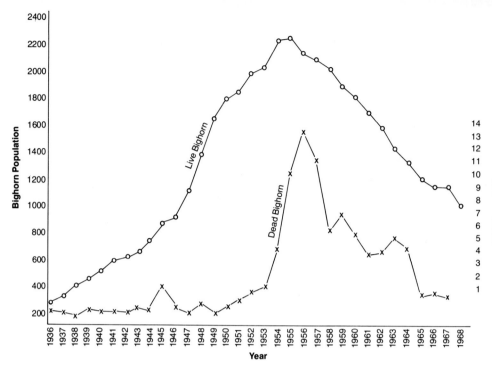

Fig. 15.7. Population growth and decline of bighorn on Desert National Wildlife Range between 1936 and 1968. The data for live bighorn are from Table 15.12 and for dead bighorn from Table 15.13.

TABLE 15.13.

Year of Death for 610 Bighorn Found on Desert National Wildlife Range Between 1936 and 1967

Year Died	Percent	No.	Year Died	Percent	No.
1936	.33	2	1952	1.80	11
1937	.33	2	1953	2.13	13
1938	.00	0	1954	5.08	31
1939	.49	3	1955	10.66	65
1940	.33	2	1956	13.77	84
1941	.33	2	1957	11.64	71
1942	.33	2	1958	6.56	40
1943	.66	4	1959	7.71	47
1944	.49	3	1960	6.07	37
1945	2.13	13	1961	4.59	28
1946	.66	4	1962	4.75	29
1947	.33	2	1963	5.90	36
1948	.98	6	1964	4.92	30
1949	.33	2	1965	1.48	9
1950	.82	5	1966	1.64	10
1951	1.31	8	1967	1.48	9
			Total	100%	610

Figure 15.7 illustrates the rise and fall of the bighorn population on Desert Wildlife Range. Included in Figure 15.7 is a curve for the animals found dead on the Wildlife Range. These latter data are presented in Table 15.13. Figure 15.7 shows that a relatively small number of bighorn died each year until 1954, when large numbers began dying. Several things are suggested by these two curves. One of these is that between 1936 and 1949 the population was made up mainly of young animals. Then between 1949 and 1953 a normal mortality rate for a normal population began to develop. However, the population appears to have exceeded the carrying capacity of the habitat and a major die-off began in 1953. Mortality did not exceed the annual increment until 1956 when the population began to decline.

POPULATION TURNOVER

The population turnover rate is equivalent to the mean annual replacement, as well as the mean annual mortality rate for a given population. Using Bradley and Baker's (1967) data (Tables 15.6 and 15.7) but excluding lamb mortality, the mean mortality rate for rams is 48.90 percent and for ewes is 40.98 percent on Desert Wildlife Range. Table 15.14 presents the figures used to arrive at these rates. The combined mean mortality rate for both sexes is 46.38 percent.

TABLE 15.14.

Mean Mortality Rate of Rams and Ewes on Desert National Wildlife Range

Age, Years	Rams				Ewes		
	Frequency of Sample	dx	lx		Frequency of Sample	dx	lx
1	404	572	1000		192	571	1000
2	4	6	428		6	18	429
3	5	7	422		3	9	411
4	4	6	415		7	21	402
5	6	9	409		11	33	381
6	6	9	400		13	39	348
7	19	27	391		14	42	309
8	28	40	364		21	62	267
9	30	43	324		14	42	205
10	59	84	281		28	82	163
11	40	56	197		9	27	81
12	48	67	141		7	21	54
13	29	40	74		7	21	33
14	19	27	34		3	9	12
15	3	4	7		1	3	3
16	2	3	3				
	706	1000	48.90*		336	1000	40.98†

Source: Bradley and Baker (1967). dx = number dying in age interval out of 1,000 born.
*Mean mortality rate for rams. lx = number surviving at beginning of age interval out
†Mean mortality rate for ewes.   of 1,000 born.

16. Sign Reading and Field Identification

Fred L. Jones

Much of our knowledge of desert bighorn comes from observing them under field conditions. Trained by experience of his own and using information imparted by others, a bighorn student can learn much from reading various signs left by the animals—tracks, pellets and urine, beds, trails, food plant clippings, signposts, remains, and so forth. The same person can learn to identify individual bighorn in the field by their natural markings (variations in horns, body marks, family characteristics, individual actions, etc.), or he may resort to the use of artificial markings (such as dye, tags, brands) to distinguish one bighorn from another.

SIGN READING

Sign reading enjoys the eminence of romanticism—the mountain men tracking for days by bent twigs, dew-free grass blades, overturned leaves, and other minuscule evidences of a creature's passing. It also suffers from charlatans who, upon close scrutiny of sign, determine just how long ago, for instance, a ram of a certain age mated a ewe of a certain color.

At its most incisive, sign reading is an art, involving intricate inductive reasoning that interprets physical evidence through exhaustive knowledge of behavior. Unfortunately, few of us acquire the scope of field experience necessary to achieve truly ingenious and insightful expertise. However, any competent field man can readily learn to recognize the more obvious sign elements and to develop sufficient interpretive ability to use the research and management tools that sign reading affords.

This chapter is drawn primarily from my own field research experience with bighorn, deer, and other big game; to that end my field notes have been thoroughly culled. It also draws upon the field experience of others, through personal discussions, notes, letters, and the literature. It does not tell the whole story, for no one has unraveled it. But it tells what I have learned from the bighorn and what others have passed on to me.

Fig. 16.1. Bighorn pellets.

Pellets

Description. Pellets may be deposited singly or in clusters of 2.5 to 4 cm (1 to 1.5 in.) in diameter (Fig. 16.1). They may be strung out if the animal was walking, or lie in a group if the animal was standing or lying down. They vary in shape, from the tiny apple seed to the large chocolate drop with an attenuated tip and an indentation on the bottom, to the flattened discoidal shapes of those in clusters. Exterior color may be black, brown, or sometimes green, and interior color may be green or brown. Some may have a pattern of alternating dark brown and light green. Pellets are moist externally when first dropped but soon dry to a varnished finish. They are dry internally, however, even when fresh.

Color, consistency, shape, size, and amount of varnish vary with differences in food and condition of the animal. For instance, in Death Valley, California, fresh *Physalis crassifolia* and *Euphorbia* sp. produced a nearly black pellet, and *Atrichoseris platyphylla* eaten exclusively produced a jet-black, soft, flat, petal-shaped pellet. Abundant green forage created a thick black varnish that lasted longer than thin brown varnish associated with dry forage (Welles and Welles, 1961a). Green forage generally results in green interiors and dry forage in brown interiors.

Similarity to other animals. Bighorn pellets are, for me at least, indistinguishable from those of domestic sheep and goats.

Likewise, I know of no method by which bighorn and deer droppings can be consistently distinguished. The average bighorn pellet of the tailed chocolate drop form can be distinguished from the average deer pellet, which is longer, more tubular, and tends to be rounded on both ends. However, there is considerable variation in the pellets of both, and appearance alone is unreliable for identification.

Lanny O. Wilson feels that he could be 90 percent correct in determining bighorn use areas from those of deer, or of deer and bighorn jointly, in southern Utah by two factors: (1) cracking that occurred throughout bighorn pellets, but not in deer pellets, and (2) the distinctive nipple on all bighorn pellets, which rarely occurred on deer pellets.

Weathering. Determining the age of desert bighorn pellets is uncertain because of the variation in color, consistency, and degree of external varnish at the time of defecation and because of variation in weathering rates. Very fresh pellets can be told by moisture on the outside, or between pellets in clusters, and by warmth. However, the exterior moisture usually dries within minutes—except for pellets in clusters, which take longer to dry and cool.

The pellet then remains firm, with a dark varnished exterior, for months. The rate of dissolution of the varnish, whitening, and deterioration of consistency depends upon exposure to sun, air, and rain, or moist soil. Green interiors gradually become brown. The bottom side remains dark longer than exposed surfaces. The fibers loosen. Ultimately the pellet becomes white on the outside, lifeless brown on the inside, and crumbles at a touch. Gerald I. Day states that in Arizona insects (usually termites) cause extensive deterioration.

Though there is no precise scale of deterioration criteria to aid a researcher, much useful information can be obtained from pellet condition. An interesting exception to the norm is reported by Welles and Welles (1961a).

Rains brought salts to the surface at Nevares Spring in Death Valley. The pellets became encased in salt crystals, which, as they expanded, disintegrated the pellets into particles. After 3 months, the pellets had completely disappeared.

Relationship to population. Pellets, because of their relative durability, lend themselves to tallying on belt and line transects. Deer pellet counting techniques have been in common use for decades, but similar techniques for bighorn have not developed comparably. However, even with crude techniques, pellets can be used to determine relative density of populations.

For instance, I tried carrying two hand tally counters while hiking through bighorn range, recording on one the number of paces and on the other the number of pellet groups within 0.9 m (3 ft) on either side of the line of travel. All groups that were green inside, or unbleached on the outside, were counted and the number of pellet groups per acre was calculated. The figures for several areas in California in the spring and summer of 1955 are shown in Table 16.1.

TABLE 16.1.
Sample Bighorn Pellet Group Counts, California

Location	Acres Sampled	Pellet Groups	Pellet Groups/Acre
Magnesia Spring Canyon, Santa Rosa Mts., Riverside Co.	3.9	142	36.4
Old Woman Mts., San Bernardino Co.	8.4	185	22.0
Intrigue Spring Canyon, Santa Rosa Mts., San Diego Co.	1.4	29	20.7
Deep Springs, White Mts., Inyo Co.	4.7	85	18.1
Cottonwood Mts., Death Valley National Monument, Inyo Co.	8.5	94	11.0
Telescope Pk., Panamint Mts., Death Valley National Monument, Inyo Co.	3.4	2	0.6

On the average, whatever factors affect weathering of pellet groups can be expected to function equally on all desert bighorn ranges of a similar type. The figure of average pellet groups per acre, then, provides a population-related index to average animal density.

On seasonal ranges occupied during only a part of the year, such as winter or summer, it is not difficult to distinguish the dark, firm pellets of the current season from the bleached ones of previous years. Use of average defecations per day allows calculation of animal days per acre from pellet groups per acre (a figure of 13.7 per day has been used for deer, but a figure has not been determined for bighorn). An estimate of the number of days most of the animals spent on the range, based on field observations, allows calculation of animals per acre. Such calculations are often used in deer management but are seldom attempted for bighorn.

Relationship to sex and age. The size of the pellets generally, but not always, is related to the size of the animal. Lambs generally deposit small pellets, and the proportion of these to larger ones around a group of beds can be useful in judging the proportion of lambs in the band. Young lambs 4 to 5 weeks old dependably leave the tiny apple seeds, according to Welles and Welles (1961a), who also observed adult rams dropping surprisingly small pellets—like apple seeds but twice as long. These observers also saw a ewe dropping big "ram-sized" pellets, and a lamb (presumably over 5 weeks of age) dropping full-sized pellets.

At beds. Welles and Welles (1961a) report that pellet groups at night beds seem to have more pellets in them than those dropped at day beds or while feeding. They report an average of 300 pellets in groups counted at night beds (range from 200 to 450), 150 at day beds (80 to 250), and 80 at groups dropped while traveling (1 to 130).

Other workers report that night beds usually have a neat group of pellets at each end of the bed and may have up to seven groups per bed. Bighorn deposit pellets while lying down and will deposit once or twice in the same pile, or will get up and turn around, thus leaving pellets at both ends of the bed. The number of fresh groups at a bed would seem to indicate the amount of time spent in it.

Relationship to food habits. Fresh pellets provide a means of field identification of some food items. Flower buds, bracts, leaves, and seeds of some plants are identifiable. R. M. Hansen (1971) provides information on relating pellets to food habits.

Ingested rocks and minerals show up also. I have found a large proportion of red rock in pellets near a small area of such rock—indicating a deliberate seeking of it.

Miscellaneous. On one occasion I noted an unusually large number of pellet groups at a spot on a ridge where a band jumped up and bunched together for a few moments before going down a steep chute one by one. The number of fresh defecations appeared to be a fear reaction.

Welles and Welles (1961a) note that excretory density increases with proximity to water and tends to reach a maximum where the bighorn stand to drink. Studies on Desert National Wildlife Range in Nevada showed that bighorn defecated within 1 to 2 minutes after drinking (Charles G. Hansen).

Urine

Urine as a basis of sign reading has received little notice. It is useful in several ways.

It generally crusts the soil soon after wetting it. Since it is so thick and viscid when first deposited, any moist urine clearly is recent. The odor remains longer than the moisture (I have judged 2 to 3 days in some instances) and so continues to serve as a reflection of recency. Once I noted urine as a yellow stain on a granite rock which would not wipe off with a handkerchief but which was still fresh enough to smell strongly.

Urine spots, in conjunction with pellets and/or tracks of front and hind feet, can identify sex. The ewes squat on their haunches and spread their legs a bit when urinating. This deposits the urine behind the hind feet and practically in the droppings. Rams, however, throw the urine forward in front of the hind feet, considerably ahead of the droppings. The relationship of the urine spot to the hind feet is sufficient to determine sex without the droppings.

Tracks

Tracks of the front feet are larger than those of the hind feet. The largest tracks, of the forefeet of rams, reach 10 cm (4 in.) in length. An average large ram track is about 9 cm (3.5 in.) long and 6.4 cm (2.5 in.) wide. The smallest lamb tracks are about 3 cm (1.25 in.) long. Tracks of small rams and large ewes overlap in size.

On a side slope, the toes may be separated 2.5 cm (1 in.) or more, reflecting the flexibility of the hoof.

I have observed tracks of bighorn that have jumped off large rocks to gravel 1.8 m (6 ft) or so below. When landing they placed the forefeet side by side so that a single large impression was made. The hind feet were placed likewise and either hit right next to the forefeet, so that a single impression about 7.5 cm (3 in.) by 13 cm (5 in.) was made, or the hind feet struck up to 10 cm (4 in.) behind the forefeet, creating two oblong impressions.

I have seen bighorn in frisky moments jumping off rocks and twisting clear around as they hit gravelly sand below, churning the surface. I have also found the earth to be distinctively churned where rams had been fighting.

Distinguished from deer. In general, bighorn tracks are larger in width and length than deer tracks. The toes are separated anteriorly, and the track appears blunt and squared off rather than pointed as in most deer tracks. The posterior of the track is squared off with less indentation between the ends of the toes than at the anterior. A marked cone in the center of the track corresponds to the depression on the sole of each toe. On receptive soil the rubbery pad on the posterior portion of each toe leaves a definite round depression not found in deer tracks.

Large deer on sandy areas leave indistinct split tracks difficult to distinguish from those of bighorn. Such deer tracks usually may be identified by considering differences between normal strides of deer and bighorn. Bighorn have shorter legs, shorter strides, and larger hoofs in comparison to leg length than deer.

Measurements I have made between tracks of twenty-seven adult California (Sierra Nevada) bighorn that were walking averaged 30 to 48 cm (12 to 19 in.), with a median of 41 cm (16 in.). Similar measurements between tracks of five adult deer ranged from 46 to 61 cm (18 to 24 in.), with a median of 53 cm (21 in.). A track interval : track length ratio resulting from measurements of tracks of four adult deer and four adult bighorn was 9.5 : 1 for deer and 6 : 1 for bighorn (Jones, 1950).

Other observations on tracks of desert bighorn recorded in my field notes give a 7.6 cm (3 in.) ram track at 46 cm (18 in.) intervals (6 : 1), and another set of bighorn tracks at 47 cm (18.4 in.) intervals.

Two tracks are visible when bighorn step in their front tracks with their hind feet because of the disparity in size. This is not true with deer.

Deer tend to move in a zigzag pattern down steep slopes, while bighorn go straight down. Deer consistently avoid the rocks by little detours, while bighorn seldom do.

Weathering. Tracks provide valuable insight into bighorn behavior when they are interpreted correctly. In order to avoid misreading, it is necessary to understand the effect of the elements and the matrix on track appearance as time passes.

On firm soils with fine particles, fresh tracks can be identified with ease. The sharp edges are finely drawn and the packed particles appear bright and shiny. Such a track appears "hot." Even hard ground may show fresh tracks by bright scratches made by the hard sides of the hoofs. Lanny O. Wilson reports that on the vast areas of sandstone rock called "slick rock" in southern Utah, bighorn leave a whitish scuff mark on the desert varnish covering the rock. These can be seen distinctly when the light is at an oblique angle.

On dustier soils any breath of wind will shift the particles and loosen them so that the bright, sharp definition is lost—the track will look fresh but not "hot." As the weathering process continues with wind and rain, the track becomes less and less distinct and becomes "old." A storm may completely obliterate fresh tracks, but in some matrices, such as firm mud, tracks may appear fresh for months.

I find in my field notes mention of tracks in gravelly soil being legible 2 weeks after I saw them made, but adjacent tracks in fine soil had been so windblown as to make them almost illegible.

Welles and Welles (1961a) describe disappearance of tracks in salty soil within 3 months after a rain. The rain carried salts to the surface, which made the ground white and fluffy. The outlines of tracks softened, blurred, and disappeared.

Fresh tracks in snow have clear, sharp definition. Grooves may occur just behind the track, from dragging the toes. Very shortly the settling and melting of the snow blurs the definition. The blurring progresses until the tracks appear only as a line of melted-out holes. The spacing continues to provide a clue to identification, however. A line of tracks in the snow can be seen a long way off and can serve to bring an observer into ready contact with the animals. Tracks in fresh snow, or right after a rain, or on ground previously brushed smooth, leave no doubt as to when made and can be studied for evidence of weathering.

Use. The following excerpts from my field notes of July 8, 1953, in Guadalupe Canyon, Santa Rosa Mountains, Riverside County, and of November 15, 1953, at Bagdad Summit, Bullion Mountains, San Bernardino County, respectively, both in California, demonstrate the importance of tracks in unraveling the intricacies of bighorn life.

[July 8, 1953] Fresh tracks of a ram, ewe, and lamb made about yesterday. They had pawed for water at a damp spot, but hadn't taken any of the stagnant water in the three holes as their tracks stopped in the mud well back from the water. They had eaten a few cattail plants growing at the water holes.

Fresh sheep tracks were continuous up the creek and there were pawed holes and other sign of their looking for fresh water. At a small hole of fresh water, sign indicated that bighorn had lain around for a day or two, drinking.

[November 15, 1953]...found recent sheep sign. Tracks were partly weathered in, but pellets were soft.... The sheep sign is probably right fresh—as rains last night would have smoothed the tracks. Some pellets were found that were moist on the underside, though the rain could have done that.... Tracks went to plants of *Encelia actoni* which was green and blooming—flowers immature yet, so growth very recent.... About ½ mile [0.8 km] down the wash the plants stopped being green—apparently local rain higher up only.... Were about 6 sheep represented where first sign was seen—one lamb, one ram at least, and 4 others."

Wilson (1968) tells of finding together the running tracks of a mountain lion and the running tracks of four ewes and two lambs in upper Blue Notch Canyon, Utah. These tracks were followed for about 1.6 km (1 mi), and a spot was found where the lion had given up the chase. Two days later he found the tracks of a running lion on the rim of one of the tributary arroyos which drain into Blue Notch Canyon and the tracks of a running mature bighorn in the bottom of the arroyo. As the arroyo widened, the distance between the lion and the bighorn increased until the lion apparently gave up the chase.

Trails

Throughout the desert mountain ranges, ancient trails beaten into the concreted sand and rock show sign of bighorn use. From an airplane an abundance of such trails can be seen on the ridges. In places the trails lead off the ridges and out onto the desert, indicating considerable movement of bighorn between mountain ranges over the centuries (Jones and Deming, 1953).

Certain routes used extensively by bighorn, such as those leading to watering spots and to salt licks, become heavily trailed, the trails being well defined and visible for long distances. Trails develop readily in sand but are also found across talus slopes. On such slopes, the larger rocks are tumbled to the side with time, so that foot-wide pathways of small stones remain. These are shifted and rolled so that a fairly flat surface results.

Bighorn trails run through rocky areas more commonly than do those of deer, but where the two species occur together they use the same trails. Man-made trails and roads are also readily utilized.

Beds

Description. Bighorn beds are oval depressions 0.6 to 0.9 m (2 to 3 ft) long by 0.3 to 0.6 m (1 to 2 ft) wide, created by the animals for resting or sleeping. Loose surface rock is cleared from the spot and a hollow is made by four or five swipes of a forefoot. As the animal settles into the loosened soil, it shapes the bed by shoveling actions of the knees, hocks, and hoofs. In rocky areas, large rocks are removed and the remaining smaller ones line the bed. On soft ground bighorn may just search for a good spot and plunk down without bothering to make a bed, but usually they make an attempt to paw. They may also lie down on hard, rocky fans or huge boulders with no effort at bed preparation.

Beds vary in depth depending upon the nature of the surface material and the number of times the bed has been used. Old beds usually are pawed deeper before being used again, sometimes attaining a depth of 0.5 m (18 in.) or more over the years. When lying down after a brief rise, a bighorn may make a new bed a short distance away, or it may resettle in the same one.

Bighorn may lie down in a hole created by pawing up a food plant. One bighorn may also force another from its bed and lie down in it without pawing.

Distinguished from deer beds. Where deer and bighorn occur together and other sign is lacking, the identity of a bed can often be told by its location. Deer generally bed under trees or in brushy areas that afford seclusion, whereas bighorn usually bed on open slopes.

Favored locations. Beds may be found almost anywhere in bighorn habitat—but factors of security and comfort in varying degrees, seasons, and other circumstances cause them to be found more in certain types of terrain than others. Day beds may be found on hillsides, along old trails and old roads, or anywhere the impulse to lie down may occur. But more than anywhere else beds seem to be found on promontories, especially above watering places; at the foot of shaded, north-facing cutbanks; on the brow and at the foot of cliffs in summer browsing areas; and in shallow caves along trails to water (Welles and Welles, 1961a). In bedding, the animals seek sun in winter and shade in summer.

Lanny O. Wilson found that bighorn in southern Utah quite commonly seek caves, mine shafts, or rock overhangs to bed in during the heat of the day. He never sighted bighorn bedding for the night in any of these situations. Gale Monson frightened a ram in daytime from the interior of Sinita Tank on

Cabeza Prieta Wildlife Refuge, Arizona; the amount of droppings indicated it had bedded there regularly for at least 2 weeks. The tank was without water, but the sand at its bottom was slightly damp.

Welch (1969) found day beds on San Andres National Wildlife Refuge, New Mexico, to be most abundant in non-bluff and shallow bluff areas, and night beds to be most abundant in deep bluff and broken bluff areas. The day beds were in the gentler terrain where most of the feeding was done, and in the non-bluff areas were always associated with isolated pinyon or juniper trees. The night beds were in the more rugged terrain.

In the winter, night beds are most frequently found on south slopes. In summer a fresh bed on the southern exposure of a promontory is very likely a night bed. On a northern exposure it may be either a day bed or a night bed, as it would be on a southern exposure with sufficient elevation to reduce the temperature to a point of tolerance for daytime bedding.

Night beds tend to be clustered closer together than day beds. Bighorn may or may not spend 2 nights in the same place, or in the same beds. Wilson (1968) reports that sometimes ewes take 10 minutes to make a bed for the night and that rams tend to return to the same night bedding locations more than ewes. He saw deep beds with numerous droppings on several occasions. In almost every case he saw a large ram either leaving or returning to one of these beds.

Weathering. Beds and tracks weather in the same way. The rate of weathering depends on the nature of the soil, the slope, and the exposure to wind, sun, and rain. Fresh beds may contain hairs that may blow away in a short time.

Dust in a bed usually is a sign of freshness. It will soon crust over or disappear, being blown away, settled back into the soil, or covered up. The presence and condition of other signs at the bed—tracks, pellets, or urine—help to determine the recency of use.

Relationship to number of bighorn. A great weakness in sign reading or interpreting behavior lies in assuming that animals "always" or "never" do a particular thing. Likewise, it is very hampering to take an isolated instance of unusual behavior to denigrate the value of the usual behavior in interpretation.

As an example, one writer never saw a bighorn band bed in the same place on consecutive nights—and might be persuaded that they never do. Another writer reports that on two occasions the same beds were used on two consecutive nights. Obviously, they sometimes are but usually aren't—and knowledge of this usual behavior may be valuable in interpreting beds.

From my own observations of a number of bighorn bands for periods of time ranging from a few hours to 2 days, and from similar but more numerous and extensive observations reported by others, a usual pattern of bed making during the day develops.

When a band feeds, a few individuals drift off by themselves, while most remain in one group. At intervals during the day, most bed down wherever they may be. The beds found on these feeding areas will be grouped together for the most part, often within a radius of 18 m (60 ft), the location of the groups of beds depending upon the movement of the band between rest periods.

While bedded, some bighorn rise to feed for a few moments or to defecate or stretch. They may lie down again in the same beds, may paw out new ones, or may lie down without a bed.

The formation of two beds at one resting area by some bighorn seems to me to compensate roughly for those bedded apart from the main group. Accordingly, the average number of fresh beds in several of these groups can be taken as a rough indication of the total number of animals in the band. Of course, the soil must be such that beds are readily identified.

Other people feel that because the number of beds is not entirely an accurate indication of the number of bighorn involved, they can be used only to tell that bighorn were there. I feel that the best information obtainable should be continually sought out but that, when rough estimates are all that can be obtained, they should not be ignored simply because of the lack of precisely accurate information.

Charles G. Hansen mentions that as many as three or four beds may be found on the north side of a tree, indicating that one bighorn moved three or four times to stay in the shade as the sun moved across the sky. This situation may lead an observer to think that not one but three or four bighorn were bedded down, so that care has to be taken to note all the surrounding conditions before estimating the number of bighorn. Hansen also states that during winter feeding activities ewes may move frequently, and resting bighorn often make new beds every few minutes in an effort to keep up with the band. An observer may be confused by this type of activity and think that not one, but many, bighorn had been bedding in the vicinity.

Welles and Welles (1961a) made extensive observations of a band of six that usually left only two or three beds. They also report one occasion of a band of seven leaving twenty-two fresh beds at a site.

Reading. Welles and Welles (1961a) illustrate a step-by-step reading of three groups of seventeen, thirty, and twenty-four beds less than 90 m (300 ft) apart, with other smaller groups more widely scattered.

> (a) The presence of the beds indicated the past presence of sheep. (b) The freshness of the beds (yellow-dusty) indicated use of the beds within the last few days. (c) More than 70 fresh beds suggested a relatively high number of sheep, certainly more than 2 or 3, probably at least 12 adults. (d) The relative size of some of the beds (as well as of the pellets) indicated several lambs, a probable minimum of six. (e) The placement of a high rim in full sun close to a water source suggested a lambing ground, which checked with (f) the season during which the beds had been and were being used. And finally, (g) the weathering of all the beds, both old and new, indicated annual use of the area, or a resident herd.

Feeding

Evidence of feeding is rarely the only sign present and rarely is distinctive enough to be diagnostic of bighorn presence by itself. However, sign of feeding can be useful.

In loose soil some forbs may be pawed up by bighorn, and the roots or bulbs eaten. Evidence of such action can be distinctive. The resulting holes pawed in the soil may give the landscape the appearance of having been shelled. These holes may be 25 to 35 cm (10 to 14 in.) long, 10 to 13 cm (4 to 5 in.) wide, and 7 to 10 cm (3 to 4 in.) deep. Fragments of the harder parts of the plants may be in and around the holes.

Bighorn may take only the tips of twigs and stems. These light nips can be identified by close scrutiny. The ragged edges of bighorn nips serve to distinguish them from rodent, rabbit, and reptile nips, which are cleanly cut on an angle, as if with a sharp knife. However, nips of deer and other ungulates are indistinguishable from those of bighorn.

I have observed that bighorn feed on barrel cactus by taking a gouge out of the side, scattering the spines. Sometimes they clean out all the inside pulp, leaving only the loose spines and the dry skin. Rabbits and rodents may do likewise. Bighorn will take bites out of the soft, growing stalks of *Agave*. My field notes of 18 July 1954, Santa Rosa Mountains, California, state: "Many *Agave* flowering shoots had been eaten. Some were short stubs, others had been nipped midway as by a beaver and the tops had folded down."

Russo (1956) describes, with photographs, the way in which bighorn feed on saguaro. The trunk is butted to remove the spiny layer and the pulp is eaten out clear around the trunk. He mentions having seen wild horses similarly eat the pulp and suspects that burros may do it also.

Signposts

In addition to the rams leaving a signpost by spraying urine on the ground, bighorn may leave other signposts. Charles G. Hansen reports that they rub their eye glands on rocks or trees, and other bighorn appear to seek out these places and smell them. The ewes urinate, and their urine is smelled by rams. Trees and bushes butted by rams may also indicate a signpost situation. Individuals have favorite trees or posts that they butt, and these may represent a certain portion of a range or territory.

Remains

On Desert Wildlife Range, Hansen (1961) judges that if the remains of an adult bighorn were from 1 to 4 years old when found, observers could be quite accurate in estimating the time that had elapsed between death and discovery. Beyond 4 years, recorded estimates became increasingly more questionable. Determination of the causes of death was very difficult, and about 90 percent of the remains were recorded as dying from unknown causes. Locations of

242 carcasses were plotted on a map by sex and age. This map showed a definite concentration of rams in some areas and ewes and lambs in others. There was an overlap in some places, which was to be expected. Determination of the ages of carcasses allowed creation of a theoretical life table (see Tables 15.6 and 15.7).

Under "Beds" is mentioned the value of hair in determining the recency of occupation. Clumps of hair shed by bighorn may be found, but microscopic comparison with known specimens is necessary to assure identification. Charles G. Hansen remarks, "shed hair in spring rubbed on rocks, etc. should be looked for. Old (1 year) hair is bleached almost white, while fresh hair (of the year) is a more natural brownish color."

Odor

Sites or beds receiving heavy bighorn use have a "sheepy" odor, which may exist even though visual sign has been obliterated. I have noticed this around springs—the most pungent being in a palm grove where there was little air circulation. Charles G. Hansen says, "I have smelled both wild and penned rams from some distance when the wind was right and the animals were 'raunchy'."

Sign Interpretation

The foregoing has dealt with specific types of sign, but generally an observer is confronted with more than one. What tracks do not tell, pellets might. The more things there are to go on, the more thorough can be the interpretation. A few examples demonstrate the process.

Reconnaissance of sparsely populated desert range may fail to disclose animals themselves, and sign may be the only source of knowledge. In November 1953 Oscar Deming and I reconnoitered the Sheephole and Bullion mountains, on the Twentynine Palms Marine Corps Artillery Training Center, San Bernardino County, California (Jones and Deming, 1953).

Fresh sign was sparse. On 12 November we found fresh tracks and droppings of about six bighorn in the Sheepholes. That same day we found an old track, some old beds, and several old trails in the Bullions. On 13 November, at another location in the Bullions, we found fresh tracks and droppings of one bighorn traveling from southeast to northwest and recent sign of two others. We also found several old trails and one old pile of droppings. On 14 November, we found fresh tracks and droppings of one bighorn traveling from southeast to northwest in an adjacent part of the Bullions, and there were several old trails, old beds and droppings, and three recent beds with abundant droppings.

On 15 November, we found fresh tracks and droppings of about six bighorn in yet another part of the Bullions where they had been feeding in a limited area of green browse resulting from localized showers. We found some old trails and beds and a few old piles of droppings. In places the trails led off the ridges and out onto the desert.

The evidence at our disposal, all from sign, led us to the conclusion that the Bullion Mountains constituted only seasonal bighorn range from about August to April. During this season of precipitation and mild temperatures, plants are growing and free water may be available periodically following rains.

A similar account is given for Utah by Wilson (1968). On 19 June 1966, fresh tracks of six bighorn were found in the mouth of Mahon Canyon. The bighorn followed the second tributary canyon leading into Red Canyon, and they followed this into Red Canyon proper where they turned west. They continued west to Warm Spring and then turned south to the talus slopes below Mancos Mesa. They traveled 11 km (7 mi) before reaching a small pool of water adjacent to Warm Spring. The tracks indicated four adults and two lambs. They were never sighted.

On forty-seven occasions, ewes and lambs looking for water were tracked in these canyons for distances from 5 to 14.5 km (3 to 9 mi). They traveled steadily, not stopping to eat. None of these animals was sighted.

At times, the age of sign can be judged sufficiently well to reconstruct the daily routine of animals observed. My field notes of 2 June 1955, in the Old Woman Mountains, read: "Above Ford Well found fresh sign. Followed it and saw three sheep after one-half mile [0.8 km]. The first sign seen of these three was about one day old. They had been feeding slowly out the ridge since then."

When attempting to estimate bighorn numbers, all signs have to be interrelated, as my field notes of 29 August 1948 in the Sierra Nevada show: "Hard to estimate numbers as tracks were all over. Possible 4–5 and one lamb.... There were two old beds (had old droppings in them) and 10 recent ones, 9 together and one, 100 feet off. Of these there were 7 that had droppings or both droppings and urine in them. There were 6 other piles of droppings among the beds, one a lamb. So I guess about 6, and a lamb to make 7."

FIELD IDENTIFICATION

The ability to identify individuals provides the opportunity to divine any number of details about bighorn life: sex and age composition, movements, longevity, band structure, turnover, etc. More is said elsewhere in this book about specific techniques that are based on field identification. Some examples that might not be discussed elsewhere are given later in this chapter.

An important aspect of field identification is the determination of sex and age. This subject is not dealt with here because it is thoroughly covered elsewhere (Chapter 11).

Much time and expense can go into artificial marking. In many instances natural markings are as good or better. Some are readily obvious, others require careful scrutiny by a trained observer. By going into detail here about the specific markings that field workers have successfully used, we may help others.

Norman M. Simmons

Fig. 16.2. This old ewe was positively identified by her broken and dangling left horn and was followed intermittently in a behavior study for a year. Sierra Pinta, Yuma County, Arizona.

Natural Markings

Horns. There is a great deal of variation in horns: amount and tightness of curl, spread, brooming, chips, nicks, breaks, ridging, and so on. Welles and Welles (1961a) refer to "hornprints" as being individualistic, and go so far as to say no two sets develop alike. Whether or not that is the case, horns often can be used to distinguish individuals (Fig. 16.2).

In my field notes of 13 July 1953, written in Guadalupe Canyon, Santa Rosa Mountains, I mention seeing a ¾-curl ram with the right horn broken off. On 3 June 1955, in the Old Woman Mountains, California, I noted a ewe with her right horn broken off about halfway and an unusual ½-curl left horn. On the same day I saw a ½-curl ram with the left horn broken in half.

Blong and Pollard (1968) saw a ewe in the Santa Rosa Mountains distinctively marked by a conspicuous hole in one horn. Wilson (1968) mentions a ewe, easily recognized by a badly broomed right horn, which was seen on almost every visit to Rainbow Canyon during two summers. He also saw a ram with badly broomed horns, making for easy recognition, and a unique old ewe having a flared right horn. In a personal communication he mentioned that he was able to identify sixty-five bighorn by distinguishing physical characteristics.

Welles and Welles (1961a) list fifty-one bighorn individually identified in Death Valley Monument. The descriptions of those for which horn characteristics constituted the main point of identification are quoted partially below:

	Rams
Flathorn	Big dark ram, aloof and suspicious, with a flat area on the top curve of his left horn.
Full Curl	The oldest ram we know. Possibly 14 years. Very heavy and badly broomed, but still full-curled horns.
The Hook	Prime ... left horn broken off leaving jagged, sharp "hook." Right horn split off in jagged point, making identification easy.
Kinky	Young ... with widespread horns with a "kink" near tips of both horns.
Nevares	Last 10 inches of curve of horns extremely thin and flat, sharper curve or "hook" toward end. On 2 September, left horn splintered off 2 to 3 inches.
Nevares II	... horns deeply corrugated ... No "hook" on left horn, but left tip turns toward his body, right tip away from body.
Tight Curl	Mature ... with peculiar tight curl of right horn, nick near tip of left horn; bases of both horns scarred by heavy fighting.
	Ewes
Brahma	Broken malformed horns, drooping ears, and the light, blue-gray color of a Brahma cow.
Brokeoff	She was tall, gaunt, and gray, with one very long horn and one—what else? Broken off!
Droopy	The Badwater contender for leadership, with the unique, down-curved horns that led many to think that she was a ram.
Longhorn	Prime ... with the longest horns we ever saw.
Old Eighty	Right horn tip missing; pronounced annular horn rings on both horns, 1 set of annual rings deeply grooved.
Old Mama	Her right horn was chipped on the inside (rare) near the base, and distinctively broomed at the tip.

Welles and Welles (1966) list twelve bighorn identified at Stubbe Spring, Joshua Tree National Monument. Those with horn characteristics predominating were:

> A 10-year-old ewe with one horn gone and both front legs appearing to have been broken and healed—one above the knee, the other below.
>
> An 8-year-old ewe with both horns broken down to about 10 cm (4 in.).
>
> A mature ram with most of his right horn missing.
>
> A mature ram with most of his left horn missing.

My field notes of 17 July 1953, in Magnesia Spring Canyon, Santa Rosa Mountains, show the value of degree of horn curl in keeping rams separate when many bighorn are being observed in rapid succession.

> At 0530 (PST) saw 5 sheep—1 ¼-curl ram, 2 ewes, 2 lambs—on top of the ridge north of the mouth of Magnesia Canyon at 1000′ [305 m].... At 0610 saw 8—2 ¾-curl rams, 4 ewes, 2 lambs—bedded on top of a hill south of the spring at 1000′ [305 m].... At 0700 saw a ½-curl ram on the south slope high up at 1200′ [365 m] walking along. At 0730 saw 2 ¾-curl rams which had been bedded on the south slope about 100 yards [91 m] above the second spring.... 0915... spotted 5 ¾-curl rams in a sandy wash.... One was bedded and four were taking turns hitting each other.... At 0930 saw a ½-curl ram watching the show from a saddle 150′ [45 m] away.... At 1155 saw the two ¾-curl rams seen at 0730 on top of the ridge on the south side of the creek at 1400′ [425 m]—so they were not with the 5 fighting.... Jumped 2 ¾-curl rams and 1 ewe at 1500′ [460 m] on a ridge about ¾ mile [1.2 km] NW of the second water. The rams could very well have been of the 5 fighters some of which went this way.... At 1600 jumped 2 ½-curl rams from the canyon bottom about ⅛ mile [200 m] above the first water.

Trophy ram determination. While this subject is treated elsewhere, it warrants discussion here also because of the interesting field identification aspects. The legality of desert bighorn rams for hunting purposes has generally been based on ¾-curl horn, or larger. Since experience showed that younger rams than desired were being taken, largely by inexperienced hunters, a new approach was devised. Jonez (1966) explains it well for the State of Nevada:

"A new definition was decided upon because we wanted the hunter to forget all about the three-quarter curl law. The new definition finally agreed upon read:

> Male Trophy Ram, Mature, Desert Bighorn Sheep, at Least 7 Years Old, or With a Boone and Crockett Measurement of at Least 144 Points, Using the Horn with the Most Points Doubled.

"The new definition would take some explaining for the hunter to use it, but this was the idea—we wanted the hunter to have an inquiring mind and be concerned about using the new definition. In order for the hunter to use the new definition, he had to be able to evaluate the head; consequently, a 15-power spotting scope was required to be in the possession of the hunter while in the field hunting.

"Successful applicants in the drawing were then sent a letter explaining that there would be a voluntary two-hour indoctrination period, and the hunters were asked to choose a date and time when they could attend.... the hunter was shown how a Boone and Crockett measurement was taken, and how to age a ram by the Horn Ring Method.

"During the second hour the hunter went through a test of looking at 10 sheep heads, one at a time, through a spotting scope. Each head was placed approximately 100 yards away.

"The hunter was asked to mark his test by using the following definitions:

1. *Definitely a trophy ram*
   (Hunter was then asked to estimate the Boone and Crockett score on any head he considered in this class.)
2. *Doubtful* if ram is the type we want harvested—Get a better look.
3. *Definitely illegal*—Don't shoot.

"The test was corrected on the spot and the hunter was given the opportunity to view at close range the heads he was tested on....

"The new system was evaluated after the hunt....

"Only one (1) ram out of 17 harvested was under seven (7) years of age, yet, this animal had a Boone and Crockett score of 158 points—15 out of the 17 sheep had green scores large enough to go in the Boone and Crockett book."

By adequate indoctrination, it proved possible to train inexperienced people to successfully make a rather esoteric distinction in field identification of trophy rams.

Body Markings

Pelage. Pelage varies with altitude, climate, foraging conditions, physical condition, seasonal shedding, and pelage renewal. Individual variations often allow identification.

Some bighorn are dark overall and others light. Some have prominent dorsal stripes which may cut across the white rump patch, joining with the tail. On others without the dorsal stripe through the rump patch, the black tip of the tail looks like a bull's-eye. There are differences in definition of leg and breast markings: the inner surfaces tend to be white and the outer surfaces dark, with varying contrast in individuals. Some are quite striking in contrast.

Some are a solid tone all over except for the white rump patch; others are lighter on the bellies and inside the legs. Some have light streaks in the natural creases behind the shoulders and in front of the hind legs. Some are predominantly light tan, others dark brown, nearly black. Many, especially the lighter ones, have dark legs below the hock. Darker ones may have a white line down the back of each leg. Some may have color patches, and there may be differences in the shedding pattern.

Careful scrutiny of a band often will disclose pelage patterns that will distinguish each individual—at least for that day. My field notes of 25 July 1948 state: "This was the same bunch I saw yesterday, lead ewe same color—dark band on shoulders and lighter body." Wilson (1968) observed a young ram with a white patch of hair between his horns.

I once observed an albinistic bighorn in the middle of winter, with the entire body (except the black tail) the color of the rump patch, which could not be distinguished. I have also seen numbers of bighorn that were nearly white just prior to shedding. Hansen (1965c) reports a pinto ewe seen at Tim Springs on Desert Wildlife Range in Nevada in 4 successive years, and 11 km (7 mi) north of there on another occasion. Welles and Welles (1961a) list a number of individuals in Death Valley Monument identified mainly by pelage characteristics, as follows (in part):

Rams	
Black and Tan	Prime. Tan with blackish mane, ears, and legs.
Mahogany	Big, mahogany red, archetype of the desert bighorn ram.
Paleface	Dark gray young ram with whitish face.
Roughneck	Not yet prime. Rough pelage on neck and shoulders fitted his aggressive nature.
Tan Rump	Prime. Brown, with tan instead of white rump patch.
Toby	Tall, bony, high shouldered, scraggly "wig" on back of head.

Ewes	
Big and Little Sandies	The two unmarked sand-colored ewes, perhaps sisters.... Big Sandy had three faint "warble" scars on the right side.
Little Ewe	Pale gray. Dainty.
Long Brownie	Named for reddish-brown color and descriptive conformation.... Only red-brown ewe seen in area. Gives impression of white-socked horse.
Pearl	Pearl was a big ewe who got her name because of the peculiar quality of the gray of her coat.
Whitehorns	She had a white patch of hair at the base of each horn, which seemed to extend her horns down the side of her head.

	Lambs
Light Neck	Named for whitish patches on both sides of neck.... Dark gray with blackish tints, light neck.
Little Whitey	A ewe lamb, light colored and with white rump, white face (relatively rare).
Marco	Buckskin color (rare) with a blackish mane.

Welles (1968) also identified an 8-year-old ewe in Joshua Tree Monument whose pelage was almost gone. What was left looked like a patchwork quilt.

Injuries. Bighorn suffer broken bones, cuts, and other damage to a degree sufficient to provide useful identifying marks. "Battle bulges" (the swellings behind the horns brought on by fighting) may be of temporary value, until they subside.

Wilson (1968) sighted, on three different occasions in 1965 and once in 1966, a ewe with five distinct scars on the left side of her neck. I found a track in 1948 with the toes splayed out at right angles to each other, and thereby judged the band it was with to be one I had not encountered on previous days.

Hansen (1965c) tells of a ram with a split ear seen three times during successive summers, and a ram with a crippled foot seen two times, on Desert Wildlife Range.

Welles and Welles (1961a) found four injured sheep in Death Valley (quoted in part):

	Rams
Broken Nose	Heavy, mature ram with crooked, humped nose. Dark chipped away patch on left horn.
Rambunctious	Scar on his back and right side.
	Ewes
Gimpy	Lame in her right hind leg but a great traveler. Observed at Furnace Creek, Big Wash, Paleomesa.
Scarface	An otherwise sleek and beautiful 2-year-old who had apparently fallen from a cliff when she was very young. That she survived the severe facial lacerations and possible skull injuries which left her face the way it was is remarkable.

Welles (1968) found several animals recognizable by their injuries at Stubbe Spring in Joshua Tree Monument:

A ewe with a stiffened right hind leg and deformed foot and ankle.
A 10-year-old ram with a stiffened and atrophied left hind leg.
A 4-year-old ram with an enlarged heel on his left hind foot.

A 12-year-old ewe with a bulging protuberance about 4 inches out from her head where her right eye should have been.

A yearling ram whose right eye was concealed by a growth.

A 12- to 14-year-old ram with a broken nose, shattered horns, and right eye blind white.

A 10-year-old ram with right hind leg broken in two places below the knee, healed with leg curved with foot in the air, and a large callus on the bottom of the curve.

Another mature ram with a gallon-sized cyst, or rupture, on the right side was reported by another observer.

Miscellaneous. Obvious pregnancies (drooping bellies and swollen udders) may have temporary utility in identifying individuals.

Wilson (1968) reports a ram with unique yellow irises.

Welles and Welles (1961a) saw several individuals in Death Valley with unique characteristics (quoted in part):

Rams	
Knocker	Young ram with outsized testicles to match his overactive ego.
Low Brow	Young ram, with heavy projecting forehead.
Slim	Gangly, high horned. About 3 years old.
Tabby	Extra "tab" of skin on scrotum. Scar on right flank. No eye rings or other facial markings except black spot on nose. No white on front legs or inside back legs. Eight to nine years old.

Ewes	
Dark Eyes	This was the only ewe we ever knew who seemed to have black eyes. We never could get close enough to her to analyze the reason.
New Mama	Slender and elegant compared with Old Mama, but a nervous leader of the new band that came in March.
Old Leader or the Patriarch	The dignified unhurried old leader of the Badwater band.

Family characteristics. Lambs and yearlings often show markings identical to their mothers, which is helpful in band identification. An albinistic ewe I saw in 1948 had a lamb that was much paler than the other lambs, both being very distinctive.

Welles and Welles (1961a) report several cases of regional or family characteristics in Death Valley. The Badwater band of six all had a rich chocolate brown color, even in a 5-month-old lamb. They were relatively stocky with heavy bodies and were short-necked compared with the rangy "ewe-necked" high-shouldered animals at Nevares Springs. Those at Navel Spring ran somewhere in between—the first band being stocky, the second rangy and slim but not ewe-necked.

In Echo Canyon a band of six was longer bodied, longer in leg and neck, and uniformly gray. The Natural Bridge band was rangy and light tan; the faces and feet were lighter—almost white. The Willow Creek herd was dark with blackish tints and distinct white markings. A band of seven at Indian Pass ranged from dark mahogany to pale cream. A band of eighteen on Paleomesa was uniformly brown. Twenty-eight at Quartz Spring were all gray.

Actions

Bighorn can be located by the sounds they make. I have located rams by the "clonking" of horns. Other observers report being able to hear a band of bighorn for as far as 0.8 km (0.5 mi). Charles G. Hansen says that ewes and lambs baa back and forth; ewes mew when satisfied; and rams make a low-pitched, coarse growl.

Behavioral Characteristics

One worker has reported identifying three bighorn by their atypical behavior. One ram, "Whitey," became attached to the observer and sought him out when he was in the vicinity. He also performed a number of unusual antics which led the observer to suspect which bighorn it was while too far distant for identification by his characteristic white spot (Wilson, 1970).

Another, a small ewe of palomino color, was called "Spooky." Each time she was encountered, she continually danced around, showing extreme nervousness. The third, called "Schizophrenia," was an extremely dark ewe who had been harassed so much by the other bighorn that she became extremely nervous and developed the unique habit of cocking her head at a 45-degree angle whenever approached by another bighorn or the observer (Lanny O. Wilson).

Artificial Markings

Chapter 18 covers the methods and equipment used in marking bighorn. This chapter will cover the characteristics of the different markings only in regard to field identification.

Dye. Dye usually is used when the animals are not to be captured and is applied by a number of self-tripping or manual spray devices. When done manually, bighorn that have outstanding physical characteristics need not be sprayed. A different color may be used for each location. The patterns of the dye stains may allow identification of individuals marked with the same color. Dye usually is intended for marking animals for short periods of time. Colors may fade considerably in 2 weeks or so, particularly if rained on, and are lost when the hair is shed. They are longer lasting on the horns.

Streamers, brands, and tags. When the animals are to be caught, various devices may be attached to them for field identification. Combinations of streamers, brands, and tags have worked well.

Aldous and Craighead (1958) report that on Desert Wildlife Range a marking technique was sought to mark each animal as a distinct individual, last the life of the animal, and be readily and easily seen in the field. A different brand number for each bighorn solved the first two requirements, using a horn brand similar to that used on cattle. Number brands were burned on the outside of each horn. Ewes were branded with a ⅞-inch iron, and rams with a 2-inch iron. Outlining the brand with paint might make it easier to read. The brands proved to be easily read some 460 m (500 yd) away with a 30-power spotting scope.

The next step was to devise something that would catch the eye, making a branded animal stand out in a group. For this they tried brightly colored plastic ear streamers, inserted in a small slit in the left ear and secured with a jess knot. If put on properly, they will last for a year or more. A stock-type ear tag was placed in the right ear for further identification.

The colored plastic streamers show up as far away as the animal can be seen. In themselves, they identify the trap in which the animal was caught—each having a separate color. Animals not marked with the plastic streamer are more difficult to locate. If the animals can be stalked, each can be identified by the brand.

Woodgerd and Forrester (1962) marked lambs with colored plasticized polyvinyl chloride tape inserted in a slit in the lower edge of the ear and attached with quick-set rivets to form a loop with a tail. Seven different colors were used: white, light blue, yellow, green, red, light brown, and blackish-violet. There was no indication of color change of the markers during the observation period of 18 months. Three colors—blue, yellow, and red—were each used on two different lambs. Individuality of each lamb was retained by marking one in the left ear and the other in the right. Over a year and a half, 147 observations on the ten lambs were recorded. No lamb mortality or loss of markers occurred. The lighter colors, white and blue, were easily recognizable. Yellow, green, and red were fair; brown and violet were poor.

The markers were mostly seen against the animals themselves, or against green or dry vegetation. Because good contrast is essential for quick and accurate recognition, the expected background should be considered when colors are chosen. Owing to the limited number of colors available, it may be necessary to use combinations of those that are most visible to mark large populations.

Other markers. A wide variety of markers has been used on other species, and new materials, devices, and techniques continue to be developed. It is not within the purview of this book to cover such practices throughout the wildlife research field. A good source is Giles (Ed.), 1969.

Hansen (1965c) reports using a bell on a collar around a ewe's neck. She was seen or heard on four occasions between 1947 and 1962, when she was found dead.

Observations and use. Some examples of observations of marked animals and use of the information derived demonstrate the utility of marking to a research or management program.

Monson (1964) tells of a 2-year-old bighorn that was sprayed with dye when he watered at North Pinta Tank on Cabeza Prieta Wildlife Refuge at 9:35 A.M., 18 June 1963. Two days later he watered at 8:40 A.M. at Eagle Tank, 16 air-line km (10 mi) away.

In early August 1963 veterinarian John B. Allen of Ajo, Arizona, saw a bighorn with a red rump about 0.4 km (.25 mi) northeast of Coffee Pot Mountain in the Sauceda Mountains. This most likely was one marked with dye at Dripping Springs on Organ Pipe Cactus National Monument the previous 9 July—50 air-line km (31 mi) away, or 70 km (42 mi) through mountainous terrain (Monson, 1964).

Hansen (1965c) states that observations of distinctive bighorn on Desert Wildlife Range are limited to a few natural markings of individuals and to many of the fifty-six animals trapped and artificially painted or branded. Two periods of trapping occurred: 1946 through 1948, and 1955 through 1959. Sixteen of the seventeen bighorn caught in the first period had their horns painted various colors for easy identification at a distance. Forty-nine were trapped during the second period. Their horns were branded or painted. Metal strap tags were fixed to one ear, and most of them had a plastic streamer attached to the other ear.

Of the seventeen trapped in the first period, only two were subsequently recorded. One was retrapped in 1956 and branded. Of the forty-nine branded and released animals, twenty-four were ewes and twenty-five were rams. Eleven of the ewes were found dead after 1 or more years, and eight of the rams were either shot by hunters or found dead a few months or several years after being released. Five ewes and eight rams were never seen after being released. Considering the small number branded, it is remarkable that so many observations were made and that so many were found after they had died.

The distances traveled are not impressive. The longest air-line distance traveled by any one animal was about 18 km (11 mi). If it had gone directly from one point to another it would have traveled about 27 km (17 mi), an easy 2-day trip for a bighorn. Some bighorn appear to have annual circuits of 32 to 48 or more air-line km (20 to 30 mi); others, particularly ewes, appear to stay closer to home or water sources.

Most of the repeat observations were made where the bighorn were originally trapped. Migration patterns around the different springs were indicated by subsequent observations. Observations indicate that the water holes in this area are the centers of use from which the bighorn move away in the cooler and moister parts of the year.

17. Population Survey Methods

Norman M. Simmons and Charles G. Hansen

Wildlife managers have numerous techniques for estimating the numbers of animals in populations; these techniques are continuously being reviewed and new ones developed. Unfortunately, however, few of these methods are completely satisfactory for application to estimating desert bighorn numbers. The habitat they occupy is just too rugged, their numbers too small, and their distribution too clumped. Efforts continue, however, to refine bighorn survey methods. We are at least able to make a rough identification of population trends in most desert bighorn ranges.

AERIAL SURVEYS

Desert bighorn are one of the most difficult big game animals to observe from the air, mainly because of hazards involved in flying close to the ground in the steep-walled canyons where they live. Nevertheless, when the proper aircraft for the job is used, aerial surveys may be the most effective means of determining total numbers of desert bighorn in some mountain ranges.

The best aircraft used in surveys of desert bighorn must be able to fly at low speeds (less than 120 km [75 mi] per hour) in rugged terrain while maintaining full aileron and rudder control. The aircraft must be able to turn and climb suddenly and sharply in narrow canyons.

The observer should have with him a large-scale map (1 : 250,000 or larger) on which to follow his course and plot his observations. A tape recorder or dictaphone is useful for recording observations of bighorn and habitat conditions without the observer taking his eyes from the terrain. A simple observation form should be used in conjunction with the recorder. Photographs should be taken when possible so that large groups of animals may be more accurately counted and identified (Watson, 1969). Usually it is difficult to identify the sex of young bighorn from the air; unless an unusually good view is obtained, they should be listed as unclassified.

The pilot of the aircraft should be skilled in mountain flying. He should be able to anticipate downdrafts and turbulence so characteristic of desert ranges. Surveys are significantly more effective if the pilot is interested in the task as well as confident of his aircraft and skill; however, he should concentrate on flying and leave the survey work to the observer. One reason for the failure of a desert bighorn survey in the Cabeza Prieta Mountains, Arizona, was the anxiety of the helicopter pilot to get the job over with and get home.

The skill of aerial observers is important. Gilbert and Grieb (1957) conclude after their study of aerial deer counts in Colorado that counts made by inexperienced observers should not be averaged with other totals. Counts by inexperienced observers average 61 percent fewer in animals counted than those of experienced observers in five trials.

"Sustained concentration is not easily reconcilable with the considerable number of potential distractions which result directly from being in a flying machine. Any aerial observer must be sufficiently familiar with the normal range of movement and manoeuvers of an aircraft as to be free of any sense of apprehension or tendency to nausea. The effects of these will be minimal upon the pilot himself, who is able to preorientate himself for his own manoeuvers, but the demands of flying on the one hand and of observing on the other are usually incompatible with the degree of concentration necessary for each as, for example, when flying accurate transect lines at a fixed height. Perhaps the best compromise is a pilot/observer team which has a bond of mutual understanding and confidence born of experience'' (Mence, 1969).

Neither the pilot nor the observer should allow himself to become fatigued or otherwise uncomfortable (Mence, 1969). Even under ideal weather conditions and in a quiet, comfortable aircraft, both men should take frequent breaks and fly for no more than 4 hours a day to maintain maximum effectiveness. Mountain flying is tiring, especially when the air is turbulent, the steep turns frequent, and the engine noisy. Noise suppressors worn over the ears markedly reduce the strain on the observer and prevent permanent ear damage.

Even with an ideal combination of pilot, plane, and observer, the weather often forces cancellation of a survey. For best results, surveys of bighorn habitat should only be flown under ideal weather conditions: in smooth air and bright sunlight (Graham and Bell, 1969). The lack of turbulence is essential mainly for the sake of safety. In the desert, where thermal air currents are common, turbulence is a serious obstacle to successful aerial surveys. Often it may be avoided by conducting the census early in the morning. High temperatures that develop later in the day may cause aircraft engines to overheat at low speeds, further reducing the effectiveness of the aircraft.

In any case, the wildlife manager should take the precaution of flying light—taking the least practical gear and, if possible, no other passengers. The lighter the load, the safer the task.

The best time of year for aerial bighorn surveys in the desert may be the cool season. Although the animals are well scattered and perhaps hard to find, the air is calmer, the plane performs better, and the bighorn are not as seriously affected by the presence of an aircraft as they are during the hot season when they need to remain calm and quiet to minimize heat gain (Simmons, 1969a).

Fixed-Wing Aircraft

Fixed-wing aircraft are not usually used to count desert bighorn, although they have been used to some extent to count other bighorn races farther north. Hence, the characteristics of various fixed-wing aircraft need only brief mention.

The military *L-19* or *DeHavilland Beaver* is the largest and most expensive of the fixed-wing planes described here. It can land and take off in a short distance but is noisy and will not perform as well in rugged terrain as some smaller aircraft.

The *Cessna 180* and the more powerful *Cessna 185* have been used on some aerial surveys because, in the hands of a skilled pilot, they respond like a short takeoff and landing (STOL) aircraft. They are far quieter than the L-19 and are usually available from private charter companies.

The *Piper Super Cub (PA 18)* is one of the best fixed-wing aircraft for bighorn surveys. Of the planes mentioned above, it is by far the most maneuverable. It turns sharply, climbs quickly, and is light and relatively safe to operate in rugged terrain. Visibility from the tandem seats is good. The Super Cub is available from many charter firms, especially in areas where aerial crop dusting is done.

A less readily available but superior fixed-wing aircraft is the *Helio Courier*. It has the most ideal combination of safety and maneuverability in rugged mountains. Though it won't climb quite as quickly, it will turn about as sharply as a Super Cub. It won't stall and spin like other aircraft. Lateral control is maintained at any speed, though the rate of vertical descent may become disconcerting at about 50 km (30 mi) per hour indicated airspeed. The

260- to 300-hp engine is fairly quiet, especially in the new, better insulated models. The Courier will carry a heavier load under STOL conditions than the Cub or Cessna. The main disadvantages of the plane are its higher rental cost and the fact that few charter companies have this expensive and highly specialized plane.

Survey methods with fixed-wing aircraft. A fixed-wing aerial survey of a sparse population of bighorn that is scattered widely in low, extremely rugged desert mountain ranges is accomplished by flying selected areas to count all animals in a sample unit. The ideal sample unit is surrounded by a barrier which a significant number of bighorn will not cross during the survey. Such ideal units exist in southwestern Arizona where small, narrow mountain ranges are surrounded by wide alluvial valleys. This is not the case in most desert bighorn ranges.

The wildlife manager should be able to estimate the error involved in aerial surveys so that a correction factor may be applied to the totals (Jolly, 1969a, b). As far as we know, insufficient data have been gathered for this to be done with desert bighorn surveys. Usually intuition is used to evaluate the precision of a survey. Sumner (1948), Edwards (1954), and Banfield et al. (1955) tested the accuracy of aerial surveys of other mammals by ground checks and found them to be about 20 percent low.

Most observers have found bighorn surveys by fixed-wing planes discouraging. Refuge Manager Cecil Kennedy flew San Andres National Wildlife Refuge, New Mexico, in an L-19 and concludes that "... a fixed-wing aircraft is of no value in censusing desert bighorn sheep...." in that area. Simmons found the L-19 to be of little value for bighorn surveys on Cabeza Prieta National Wildlife Refuge, Arizona; he was unable to try other fixed-wing aircraft there. Russo (1956) writes that "well-powered, slow speed airplanes were used experimentally in an effort to facilitate inventory (in Arizona). Because of the natural blending of the sheep's pelage color with its environment and the animal's use of caves and overhanging ledges for shelter, aerial censuses proved impractical."

However, when vast wilderness areas are involved, fixed-wing planes must definitely be considered in planning aerial surveys. They are widely recognized as the best and most economical tool for surveys of Rocky Mountain bighorn and Dall sheep in northern states and Canadian provinces and territories.

Rotary-Wing Aircraft

Disregarding expense, the helicopter is by far the best aircraft for aerial surveys of desert bighorn. With proper technique and under ideal weather conditions and with suitable terrain, aerial surveys by helicopter may be the most effective of all bighorn survey methods. With the best helicopters, an observer may look over most of the terrain in his sample area, and a total count of desert bighorn becomes feasible.

The main disadvantage of a helicopter is its adverse affect on the bighorn themselves. Observers usually yield to the temptation to fly close to the animals to classify them and take photographs, causing them to panic. Charles G. Hansen noticed that bighorn on Desert National Wildlife Range, Nevada, raced wildly about in all directions when a helicopter came close, and some even jumped off cliffs. Lambs became separated from their mothers. Because of the possibility of causing injury or death to desert bighorn lambs, surveys by helicopter on Desert Wildlife Range were discontinued during the lambing seasons.

However, Simmons, on Cabeza Prieta Wildlife Refuge, found that observers could count bighorn and classify them according to group types—adult ram groups, ewe groups (including 1- to 3-year-old rams and lambs), and mixed groups—by flying quickly past the bighorn. The animals did not appear to be affected adversely as they would be by a hovering helicopter. The observers would still profit considerably from their ability to search the terrain thoroughly, even to the point of peering into caves. Observers on Cabeza Prieta Refuge rarely paused long enough to photograph bighorn, and then only to take long-distance pictures of large groups for later classification, or pictures of adults in relatively gentle terrain.

Light and Winter (1967) of the U.S. Forest Service state that use of a helicopter was probably the most practical approach to a survey to estimate bighorn numbers. Jonez and Monson (1957) found helicopters useful in locating bighorn in the rugged mountains of Nevada and Arizona. Kiger (1970) also used helicopters to successfully count and classify bighorn on San Andres Refuge. Even advocates of helicopter surveys would agree, however, that they should not be conducted during hot, dry periods because of the serious effects they may have on bighorn remaining quietly in the shade.

Turbojet helicopters offer the best combination of safety and maneuverability for surveys in desert bighorn habitat. The extra expense of operation is considered to be offset by the safety and maneuverability factors as well as in reduced ferrying charges because of their greater speed (as much as 200 km or 125 mi per hour).

Bell, Hiller, Kaman, and especially the Hughes 500 turbojet helicopters have been used successfully for bighorn surveys (as well as in capture operations). The Kaman helicopter was used on several surveys on Cabeza Prieta Refuge (Simmons, 1969b). The twin-rotored craft was larger and more powerful than necessary, but it performed excellently during aerial surveys in rugged terrain. No canyon was too steep or narrow for the observers to search thoroughly and safely with this aircraft. On Cabeza Prieta Refuge, fuel was trucked to the survey area so that the helicopter did not have to return to its base to refuel. The Hughes 500 has been a most satisfactory helicopter in working with desert bighorn by both the Arizona and New Mexico game departments.

Bell piston engine helicopters with bubble cockpits have been most frequently used on bighorn surveys. The supercharged Bell 47G3B-1 is adequate for work in desert mountain ranges. Visibility from the cockpit is better than from any other helicopter we are familiar with. For a safe load, only the observer and pilot should fly in the aircraft, even though it will handle two passengers, and no more than 2 hours of fuel should be carried, though it can carry fuel for over 3 hours of flying.

Survey methods with rotary-wing aircraft. On Cabeza Prieta Refuge, Simmons (1969b) thoroughly searched entire sample areas by helicopter. These were canyon-by-canyon searches. Often it was best to start low in a canyon, work up to the main ridge, and then cross to the opposite slope to search the adjacent canyon. The speed with which these canyons were searched likely precluded duplicate counts. When bighorn were seen, the helicopter pilot would move away from them, but keep them in sight until they were counted and roughly classified. The helicopter would land after 1 to 2 hours of flying so that the pilot and observer could rest.

WATER-HOLE SURVEYS

For centuries, desert Indians hunted for bighorn at water holes during the dry periods when the animals needed water. Crouched behind makeshift blinds only a few yards from the water, or near a well-worn trail to the water hole, the Indians would wait for hours until one or more bighorn came within range of their arrows or spears. Though in many ways the dry periods were a time of hardship for Indians and wild animals alike, that season meant relatively easy big game hunting because of the overwhelming attraction that the thinly scattered water holes had for the wildlife of the desert.

When the first steps toward management of desert bighorn were taken, a rough idea of the number of bighorn occupying an area was gained when observers waited near water holes during the hottest, driest days of the year and counted the animals as they came to drink. These surveys were the basis for early population estimates and management practices. However, in most areas the water-hole surveys were unsystematic and imprecise. Little was known about the optimum duration of a water-hole survey under various weather conditions. Little was known about the statistical errors involved. Efforts to reduce duplicate counts were haphazard. Some game managers in the southwestern states assumed that 3- to 5-day surveys were enough. Often little effort was made to determine the number of bighorn they missed because all water holes were not covered, or because some bighorn watered at night, or did not come to water at all.

Perhaps the first concerted efforts to improve water-hole survey techniques were made by game managers and biologists on Desert Wildlife Range in the late 1950s. They began using standard forms to record identifying characteristics of each bighorn visiting water holes, such as scars and chipped horns, so that these animals would not be counted again. Bighorn on the refuge also were marked by dyes during water-hole surveys or were trapped and marked more permanently at other times. Marked animals would only be tabulated once during a survey (Hansen, 1965a).

The lead taken by Desert Wildlife Range personnel has been followed by wildlife managers on Cabeza Prieta Refuge since 1961. Surveys on the latter refuge were conducted simultaneously at all water holes in a mountain range that was isolated by wide alluvial valleys. Bighorn were dyed and their individual characteristics noted on forms like those used on Desert Wildlife Range. Surveys of 3, 5, and 10 days' duration under varying weather conditions were tried. Thus the refuge managers were able to determine the best method for a water-hole survey under a variety of conditions.

Conducting a Water-hole Count

Discussed below are the basic, minimal requirements for a water-hole survey designed to produce an estimate of the number of bighorn occupying a certain area of a mountain range, or an entire range.

A man at each water hole within a survey area. This requirement is essential, so that the entire area can be surveyed at once. Since rams and even "dry" ewes may move 8 km (5 mi) or more to visit a water hole, migrations in and out of a survey unit can cause a serious error in a count. No intermittent water in the sample unit should be left unwatched.

A definable survey unit. An example of an ideal survey unit is the long, narrow Sierra Pinta range of Cabeza Prieta Refuge, isolated by surrounding alluvial valleys. The absence of bighorn tracks in the valleys during summers indicates no movement to or from the range during that season. Watering places in the range are few.

It is more difficult to conduct a survey in such massive mountain ranges as the Kofa and Ajo mountains in Arizona, the Santa Rosa Mountains in California, or the Sheep Range in Nevada, which contain many water holes and allow the bighorn to drink at numerous places.

The right weather and no moon. A long-lasting, hot, dry period is best for a water-hole survey. One may ask, "What is dry?" Simply no rain from about 2 weeks to a month before the survey, and no rain during the survey. Simmons (1969b) found that desert bighorn on Cabeza Prieta Refuge began evading heat gain by bedding in shade when wet bulb temperatures reached 18°C (65°F). Most watered when wet bulb temperatures were above 19°C (67°F). He recommended that the daily wet bulb temperature after 0700 reach

18°C (65°F) before water-hole surveys are started. (Wet bulb temperatures were used because they reflected both the air temperature and humidity, which work in combination to affect the comfort of the bighorn.)

Simmons noticed that bighorn on Cabeza Prieta Refuge came to water on nights when the moon was full. Gale Monson reports that on Kofa Game Range bighorn rarely came to water at night, even during a full moon. It is advisable to conduct surveys during periods of minimum moonlight, or the water holes should be observed or closed to access on brightly moonlit nights, unless previous surveys convincingly show that watering does not occur on moonlit nights.

Length of time. A general rule covering the proper length of a survey cannot be made. Simmons (1969b) says that the surveys should be continued at least until the "dry ewes" (not nursing) that have watered at the beginning of the survey return to water again. This may be a minimum of 10 days.

A ewe usually will return repeatedly to the same water hole. Rams may not. Therefore the return of dry ewes, the ewes that need water the least of all ewes, to water a second time is a good index to the end points of a survey.

A way to eliminate duplication of counts. There are several ways that bighorn can be identified as individuals so that they are not counted twice (see Chapter 16). Since no one way works perfectly, a combination of identification procedures should be used.

If individual bighorn are photographed as they come to water (Helvie, 1972), the photo prints could be compared after the survey to reveal duplicate counts. Difficulties are encountered in photographing individual bighorn during a survey. Film must be kept cool during the entire survey. On Cabeza Prieta Refuge, film was kept in ice boxes, but after 10 days very little ice was left. Individual bighorn are hard to photograph adequately when they jostle around the water hole in groups of five or more.

The marking of bighorn with long-lasting ear tags, horn brands, or paint is highly recommended. One of the best desert bighorn marking programs was executed on Desert Wildlife Range and resulted in the compilation of many valuable movement data. Bighorn have been temporarily marked with dyes in the Santa Rosa Mountains, California (Blong and Pollard, 1968), on Desert Wildlife Range (Hansen, 1964a), and on Cabeza Prieta Refuge (Simmons and Phillips, 1966). Marked animals are easily identified as individuals and therefore counted only once.

Forms have been used on Desert Wildlife Range and Cabeza Prieta Refuge to standardize records of the identifying characteristics of individual bighorn; chips in horns, scars, and other deformities are recorded. One ewe with a withered ear and a blind eye, and another ewe with an oddly formed horn were seen on Cabeza Prieta Refuge on several occasions during a 3-year period. Eighteen other bighorn on Cabeza Prieta Refuge had physical

Norman M. Simmons

Fig. 17.1. A large, relatively comfortable blind made of saguaro "ribs." The natural pothole has been deepened by a cement lip, and a dye spray nozzle stands over the entrance ramp to the water. Heart Tank, Sierra Pinta, Yuma County, Arizona.

peculiarities that enabled the observers to identify them positively with the aid of the forms. Naturally white-spotted bighorn (see Chapter 4) on Desert Wildlife Range were identified easily as individuals year after year.

Comfortable blinds and enough men. One of the most important requirements of a successful water-hole survey is a comfortable blind. Such a blind should provide adequate shelter from sun and wind, and room enough to sit and sleep comfortably. Sitting under a piece of canvas on the hard, hot ground can quickly become distracting and can adversely affect the observer's efficiency. The best blinds on Cabeza Prieta Refuge were built against live trees that shaded and partly obscured the structure. They were made of saguaro cactus skeleton "ribs" (vascular bundles) bound to portable pipe frames, and they blended well with the surrounding terrain. Such blinds do not need to be camouflaged (Fig. 17.1).

Blinds permit observers to be close to bighorn and still move around without startling them. The animals quickly become used to the blinds and often bed down within them in the absence of the observer.

Even with comfortable blinds, survey hours are long (preferably from a half hour before dawn to after dark), hot, and tiring. Ideally, two men should alternate between resting and watching at each water hole to minimize fatigue and thereby maintain alertness. These men should be relieved during lengthy surveys; perhaps every 5 days. Although this is a considerable expenditure of manpower, it will pay dividends in precise counts.

Precision checks of water-hole surveys needed. Most surveyors probably will never learn the actual population of their sample areas. However, they may use the results of their surveys to establish trends in population fluctuations. To do this, they must be able to compare the results of one survey statistically with the results of other surveys conducted in the same area in the same and other years. They must be able to measure the precision of each survey.

Experiments should be conducted to develop a technique for estimating error. Perhaps a carefully controlled water-hole survey of long duration in an isolated sample unit could be compared with a simultaneous search by helicopter. The check by helicopter should be carefully conducted in the early morning hours when heat stress on the bighorn is minimal and air turbulence is not severe. No effort should be made to fly close enough to classify the animals, since total numbers are all that is needed at this time.

SURVEYS BY BOAT

Desert bighorn range borders Lakes Mead, Mohave, and Havasu in northwestern Arizona, adjacent Nevada, and extreme eastern California. During hot, dry summer months little water is available in that area, except that found in the lakes and along the Colorado River. Game managers have taken advantage of this situation to count and classify desert bighorn that come down to the lakes to drink (Welsh, 1964). This survey technique is in its infancy, but it may prove to be the most practical method of surveying this large and rugged area of Arizona, as well as those parts of Utah flooded by the water of Lake Powell behind Glen Canyon Dam.

Surveys conducted from Lakes Mead, Mohave, and Havasu are similar in nature and purpose to water-hole surveys, except that in this case the observers move about in the "water hole" and use boats as blinds. The same basic requirements should apply to both types of surveys if a precise count is the goal. The entire sample area should be surveyed at the same time, and if there are water holes (other than the lakes and river) in the area, they should be manned. The possibility that bighorn may drink from lakes and the river even on moonless nights should be examined.

The survey should be of adequate duration so that repeat observations can be made of dry ewes. Duplications of counts should be avoided by careful identification of individual bighorn. Care should be taken that the observers do not become tired and uncomfortable before they are relieved. The boats should be relatively comfortable and shaded, and enough men should be available to permit a rotation system. Finally, a method of checking on the precision of the survey should be devised.

This is a large order for a boat survey and challenges the ingenuity of the game managers. Although total counts from boats are not practical, accurate determinations of the sexes and age classes of bighorn can easily be made from boats along the Colorado River and in the lakes. This method of determining age class and sex ratios is probably the most practical for this area of bighorn range. The ratios can be meaningful indices to the health and productivity of the bighorn bands in that area.

SURVEYS ON FOOT AND HORSEBACK

Sometimes the exhausting, time-consuming survey by foot can be the most rewarding. No matter what the goal of the survey, the well-trained hiker can record valuable information about the condition and distribution of vegetation and the presence of species other than bighorn, and he can deduce useful information from the tracks of bighorn and other animals.

Information obtained by observers on foot or horseback that is better than that obtained by observers in airplanes or at water holes includes: (1) days of bighorn use in a given area, (2) preferred habitat types, (3) movement patterns, and (4) ewe-lamb and ewe-yearling ratios.

Pellet-group Counts

Fecal pellets may be used as inert evidence of bighorn presence and distribution that can be subjected to plot sampling and statistical analysis. The application of this technique for determination of big game population trends, census, and distribution has been reviewed by Neff (1968) and is discussed in Chapter 16.

Although the presence of desert bighorn fecal deposits indicates bighorn seasonal distribution patterns, it is unlikely that the pellet groups can be used effectively to determine population trends or to estimate total numbers. In most desert bighorn ranges, fecal deposits are of low density. This results in the least efficient use of the pellet-group sampling technique. A high intensity of sampling is required at the cost of many man-hours of work. The isolated locations of the ranges and the ruggedness of the terrain compound the problem. Another thorny problem in much desert bighorn habitat is the presence of fecal pellets from deer and other ungulates that closely resemble those of bighorn (see Chapter 16).

Line Transects

Experiments were conducted on Desert Wildlife Range to determine whether line transects could be used to estimate numbers of desert bighorn in certain areas (Morgan, 1961). Charles G. Hansen has discovered that a minimum of 100 observations had to be made on at least 380 transects in a short period of time to obtain population data of about 90 percent reliability. He concludes that manpower and fund requirements for such an effort make the technique impractical.

We have no other record of tests conducted on line transects in desert bighorn habitat in other areas. It is our opinion that, in most desert bighorn ranges, line transects would be an inefficient and uneconomical means of estimating numbers or determining population trends.

Lamb Surveys and Similar Techniques

Where small, well defined areas can be selected on which ewes commonly have their lambs, it is possible to systematically hike or ride horseback through sample units from year to year and determine trends in ewe : lamb ratios. Ideally an entire sample unit should be searched in a single day so that movements of ewes and lambs will not make counts unreliable. The routes followed should be the same each year. The observers should also be the same. Here lies a potential source of error since observers usually vary from year to year if they are government employees subject to frequent transfer. With a change in observers comes a change in skill, and results may not be comparable.

A systematic lamb survey will produce valuable lamb production trend figures. However, game managers cannot expect to determine the sex of a lamb (Kelly, 1961).

Lamb surveys were started on Kofa Game Range in 1958. Since then game managers have concentrated their efforts on the lambing grounds in the west end of the Kofa Mountains (Eustis, 1962). At presurvey briefings, the west end of the range is divided into several sample areas. These units are subdivided into areas small enough for one man to search thoroughly in a day. Each surveyor searches up to the boundaries of one or more adjacent units. Each evening the surveyors assemble and compare notes in an effort to eliminate duplicate sightings. Kofa bighorn managers feel that their ewe : lamb ratio estimates are reliable and comparable from year to year.

Lamb surveys also are conducted on Desert Wildlife Range but in a different manner. There, five men survey a 40-km (25-mi) long sample unit in the Sheep Range. This takes 5 days of riding horses and jeeps, and hiking, during the last week of March and again during the last week of April. The data resulting from these two counts are compared, so that observer errors can be identified and corrected (Hansen, 1962). Refuge personnel have found that the March and April ratios agree closely. However, as is also the case on Kofa Game Range, they have not measured their survey errors and therefore are unable to assess objectively the reliability of their lamb count trends.

Sometimes foot travel like that used for lamb surveys is employed as a total count effort. A "ride survey" is described by former Refuge Manager Cecil Kennedy of San Andres Wildlife Refuge as the primary survey method used in that area. At least four men on horseback or foot cover both slopes of a survey unit. On each slope, one man walks near the crest and the other halfway down the slope. They walk along the mountain contour the length of the range. Each man records the number of bighorn he sees, and the location and time of the sightings. At the end of the day all sightings are plotted on a map in an effort to eliminate duplicate sightings. About 6 weeks are required to conduct this survey (John Kiger and Cecil Kennedy).

Goodman (1963) claims that "... nothing quite surpasses actually 'hiking out' a canyon, when possible, with frequent stops for observation of sign and scanning the ridges and slopes for sheep." He supervised a survey involving fifty people, mostly University of Redlands college students, in the massive Santa Rosa Mountains of southern California, designed to determine the number of bighorn occupying several canyons.

A major problem encountered in such a hiking survey is the manpower requirement. Few conservation agencies could obtain 50 trained observers to survey part of a mountain range. Even if they could, the individuals would change from year to year, introducing an unmeasurable observer error. A normal effort would involve perhaps a half-dozen men.

Another problem typical of such surveys showed up when the Redlands students compiled their figures. They saw few bighorn. Goodman concludes that the "most significant thing to be derived from the census was that it was virtually impossible with merely an enthusiastic though untrained group of census takers to observe many sheep in the desert canyons of the Santa Rosa Mountains in mid-winter. It is a personal opinion that had they stayed in for four days instead of four hours their chances of success would have been little better." The animals were widely dispersed in the rugged, brush- and tree-covered mountains. Other such hiking surveys conducted in the Santa Rosa Mountains, as well as in mountain ranges of other states, have been similarly disappointing.

On the other hand, Richard Weaver, California Department of Fish and Game, writes, "I am of the school that contends that there is no substitute for getting over the ground (hiking) to see bighorn, as done by Fred Jones in the Santa Rosa Mountains in 1953, and myself. Anyone can quarrel with our projections from reading the signs, but anyone that quarrels with our actual counts is calling us liars. Hiking 'inventories' (I like that word better than surveys) have a place."

In summation, opinions vary as to whether the "hiking" survey is an economical or effective way to obtain total counts, or even counts of a sample of a population of desert bighorn in a survey area. A minimum of 168 man-days was needed to search the narrow mountain range in San Andres Wildlife Refuge. Fifty man-days were probably many too few in the massive and relatively well vegetated Santa Rosa Mountains. The fewer and more widely dispersed the bighorn, the more man-days must be invested to produce an effective count. The same observers should follow the same routes each year to reduce variability in comparisons of counts between years. And, as in all currently used bighorn survey techniques, the error in each count should be estimated to make comparison meaningful.

18. Capturing, Handling, and Transplanting

Charles G. Hansen, Tommy L. Hailey, and Gerald I. Day

Management and research of desert bighorn often require the capturing and handling of wild animals. This capturing may be done by trapping, tranquilizing, or immobilizing, or in combination.

The reasons for handling the animal usually determine the method of capture. If trapping is for transplanting, then one method may be used, whereas capture for marking may require another. The type of marking used and the method or place of transplanting are determined by the prevailing circumstances. In this chapter each method will be treated separately, although they are often difficult to separate in actual use. Of course, new methods and techniques of the future may prove better than those described here.

BIGHORN CAPTURE

Trapping

Trapping of bighorn in North America for management purposes dates back to 1922 (Yoakum, 1963). California bighorn were trapped in British Columbia in 1954 for transplant to Hart Mountain National Antelope Refuge,

Oregon (Deming, 1961). A corral type trap was used, baited with salt, hay, dairy feed, sheep food pellets, and cabbage. The raw cabbage appeared to be the most enticing. The trap was a wooden arrangement with two gates, and snow-fence wings to guide the bighorn into the trap. Dynamite caps were used to sever a small rope holding the gates open at both ends of the trap.

Bighorn produced from the 1954 release were later caught within the enclosure at Hart Mountain and transplanted to other mountains in southeastern Oregon. This trapping method employed a large number of men to drive the animals into a small fenced enclosure, then up along a chute into a wooden holding and loading cage.

On Desert National Wildlife Range in Nevada, bighorn have been caught in several ways. They were chased and lassoed, taken in steel traps, and enticed into corrals built around water (Aldous, 1958). Some of the captured animals were transported to Corn Creek Field Station on the Range; others were marked and released. The most recent trapping for transplant purposes began in 1967 when a 2-acre steel post and hogwire corral was built around a water hole (Fig. 18.1). The bighorn were forced down a chute into a holding pen similar to that at Hart Mountain and then loaded directly onto a truck. This trap was designed to capture a sizable group at one time.

Desert bighorn were trapped on Kofa Game Range, 97 km (60 mi) northeast of Yuma, Arizona. Tom D. Moore describes the operation:

> During the summer months of 1956 through 1959, a joint effort was made by the Arizona and Texas game and fish departments to trap bighorn for restocking on former ranges within their respective states. (Other agencies that cooperated were the U.S. Fish and Wildlife Service, Wildlife Management Institute, National Park Service, and Boone and Crockett Club.)
>
> The most practical trapping method was the use of corral-type traps around water holes, for efforts to bait the bighorn with various feeds and salt were not successful. In this extremely hot and dry region, bighorn come to natural potholes for water during the summer months. But if the range is lush from early rains, or if the weather is cool, they will go without water. Few bighorn, if any, were observed near the traps until mid-June. Forage succulence by then was at a minimum, and food plants grown in the spring had died. Not until after several consecutive days with temperatures exceeding 46C (115 F) did bighorn regularly enter the traps or seem interested in drinking water. This hot period sometimes lasted for 2 weeks. Soon afterward clouds would form and signal the end of the trapping for that year, as subsequent rains filled all surrounding potholes and the bighorn dispersed.
>
> To make the traps effective, all other available surface water near the traps was siphoned onto the ground. Even so, this method of trapping was slow, for in excess of 2000 man-hours were spent in blinds during 2 summers waiting for bighorn to come to water. Weeks passed before any were seen. During peak periods, nine men manned the blinds throughout the daylight hours. One man would enter the blind before daylight, to be relieved at noon by someone else who stayed until after dark.

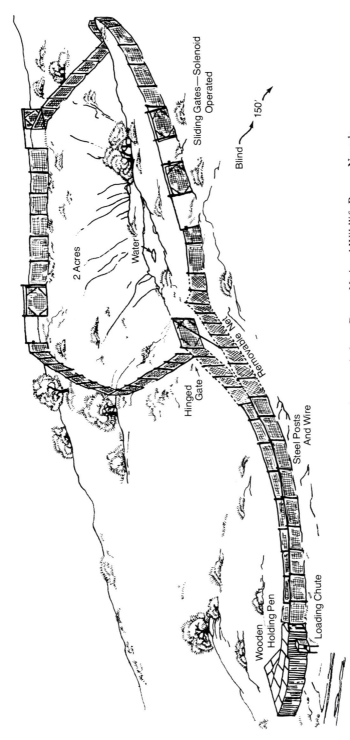

Fig. 18.1. Design of a corral trap built around a water hole on Desert National Wildlife Range, Nevada. The four gates close simultaneously when triggered electrically from a blind.

Simmons (1961) states that sixteen bighorn were caught in a corral trap near Tarryall Reservoir, Colorado, in December 1946, and trucked to Poudre Canyon, Colorado. The trap was similar to the Hart Mountain one, baited with salt and hay. Salt was the preferred bait. The same type of corral trap was used successfully in British Columbia national parks, according to Simmons.

Parry Larsen of New Mexico states that a similar corral trap in the Sandia Mountains was used successfully several times between 1960 and 1968. Permanent, artificial salt grounds were established and enclosed within a corral. However, a similar trap on San Andres National Wildlife Refuge in New Mexico was tried for 2 years with various baits without results.

In 1969 the Nevada Department of Fish and Game successfully trapped eighteen bighorn at the Boulder Beach sewage pond, located within the boundary of Lake Mead National Recreational Area. Water was used as bait and a semiportable drop gate panel trap to capture the animals (Papez and Tsukamoto, 1970).

Trap equipment. The original corral trap on Desert Wildlife Range was about 21 by 21 m (70 by 70 ft) and made of wooden posts and hog-wire fencing. A chute and loading ramp were attached to one side of the trap. The whole arrangement was built around a water trough used by bighorn. There were two gates in the trap, one of which was triggered by a cable from a blind. The other gate was latched shut before trapping operations began for the day.

The subsequent corral traps used to capture bighorn for marking were smaller, about 6.1 by 9.1 m (20 by 30 ft). They were made of pipe posts set in cement and 4.8 cm (2 in.) square cotton fish net. Of the three traps built, two had two overhead sliding gates each, while the third had one. These were built around water and triggered from a blind by a pull cable. One of them used blasting caps to release the two gates simultaneously.

A larger trap, built in 1967, used drill-stem pipe fitted into holes drilled into solid rock. This was more economical because it took only about 15 minutes to drill a hole and set a post, much less time than needed to set posts in cement. The fence was 2.1 m (7 ft) high with five sliding gates, but complications during construction limited the trap to only four gates. The fence and gates were planned to fit the terrain and the pattern of bighorn use of the water hole. The chute and holding pen also followed the terrain and were designed to be as much out of sight of the water hole as possible.

The gates were designed to slide on overhead tracks instead of being placed over the gateways in order to eliminate as much disturbance to animals as possible as they walked into the trap. Each gate was released by an electric solenoid triggered from a blind 46 m (150 ft) away from the nearest gate; the gates then moved by gravity down the tracks. The solenoids were activated by a 12-volt wet-cell automobile battery placed by each gate.

The equipment used on Kofa Game Range is described by Moore:

Five traps were operated, four of which proved successful. Three of these traps were of the corral type, made of nylon netting 2.4 m (8 ft) in height and supported by 4.8 cm (2 in.) steel posts and wire cables. A 2.4 m (8 ft) gate entrance was provided of the same material. A 4.8 cm (2 in.) steel pipe across the bottom of the gate gave additional weight, thus causing the gate to fall straight down when tripped. The gate was held up by a wire stretched to the blind about 30.5 m (100 ft) away, and fell when cut by the observer. Each trap was located so that it encircled the watering site.

The most inaccessible trap site was Tunnel Spring, but it was the most productive as the surrounding granite peaks and canyons contained the highest bighorn population on Kofa Game Range. Furthermore, slick, steep, rocky slopes prevented all other hoofed animals from using it. A small seep, dammed to impound 113 liters (30 gal) of water, was located in a deep fissure or cave high along the canyon wall. As the bighorn entered the cave, a sliding gate of nylon net was closed when pulled by a rope stretched to a blind.

The portable drop gate panel trap used successfully by the Nevada Department of Fish and Game was composed of a minimum of four panels, each measuring 2.1 by 2.4 m (7 by 10 ft). The framework of each panel was constructed of 3.6-cm (1.5-in.) pipe, with a drop-gate mechanism. The drop bar was constructed from 3-cm (1.25-in.) pipe, with four rings attached to the sides of the framework which allowed it to slide freely. Nylon stretch netting (12 cm^2) was attached to the top of the panel, the drop bar, and the rings on the sides. A "Clover" deer trap trip mechanism or an electrical solenoid was used to trigger the drop gate (Papez and Tsukamoto, 1970).

Trap recommendations. The first consideration in trapping desert bighorn is whether or not construction or other activites will disturb the animals unduly. If greatly disturbed, bighorn will abandon the area and possibly deplete the population of more than just those removed by trapping. Consequently, construction should be done during periods of low use. The trap itself should fit into the terrain in order to appear as little as possible like an obstruction to the animals and avoid making them feel they are entering a cul-de-sac.

Wilson, ed. (1975) recommends that water-hole trap size be not less than 61 by 122 m (20 by 40 ft), rectangular in shape and accessible by vehicle. The trap should be constructed of wooden posts instead of metal or pipe to help reduce injuries to animals. Nylon netting is recommended instead of woven wire or wire of any kind. Nylon netting is strong enough to hold bighorns, yet it will stretch when struck by them, thus reducing injuries. The netting should be colored brown or black, with 7.2 by 9.6 cm (3 by 4 in.) size mesh and be a minimum of 2.4 m (8 ft) high.

The less visible an obstruction that the bighorn have to walk under and the wider the entrance, the more readily the bighorn will accept the gateways. Also, the larger the corral, the more acceptable it will be.

Electronic devices for triggering the gates usually are the most satisfactory, because there is no movement to alarm the animals before the gates begin to close, and such devices are relatively quiet.

Intelligent handling of the traps, and of the bighorn after trapping, should minimize loss and the cost of the operation. No traps should be left unattended during the capture operation. Also, adequate manpower should be available to safely handle the captured animals.

Physical restraint of trapped animals. After a desert bighorn has been caught, it should be handled and released as soon as possible. Physical restraint appears to produce an unusual amount of stress. Even bottle-raised lambs handled daily do not accept physical restraint, although they will seek out the handler to be petted. Consequently, when captured bighorns are to be marked and released, the entire process should be accomplished in the shortest possible time.

When handling the animals, several men should be used if needed. Two men will be needed to hold the head and one man to hold the back of a strong adult ram if the animal is to be kept from thrashing around while being treated or marked. The animals should not be tied up if it can be avoided. If it is necessary to tie the animal, a good tranquilizer and the truss described by Thomas et al. (1967) would be the most desirable.

When approaching a captured animal, the handler should move quietly and make no unnecessary noise or quick movements, except perhaps at the last minute. A group of noisy, arm-waving men running up is sure to cause undue fright. If the trap design proves unsuitable, it should be altered appropriately.

Tom D. Moore reports that no great difficulty was encountered in capturing even the largest ram after it was in the trap, for the netting usually became entangled about the animals' horns as they rammed into it. Small sections of netting were sometimes thrown over individuals to effect their capture, but usually this was unnecessary.

Drug Capture

Drugs have been used to safely capture and restrain free-ranging bighorns. The must successful methods of drug capture have been from blinds at water holes and from helicopters. Unfortunately, drug capture is limited to only one animal at a time, but the selection of especially desired animals is possible.

Considerable testing has been done with various drugs and dosages on desert bighorns. Devan (1959) reported poor results with the projectile syringe gun and the drug nicotine alkaloid on captive and free-ranging bighorn. Chew and Goodman (1963) found the drug succinylcholine chloride was relatively safe and effective on domestic sheep and suggested it would probably be effective on bighorn. Logsdon (1969) and Day (1971) tested several different drugs, and both reported M-99 (Etorphine) as the safest and most effective drug for bighorn.

Drugs have been safely injected into bighorn with projectile syringes fired from CO_2 Cap-Chur rifles (Palmer Chemical & Equipment Co.). Projectile syringes fitted with barbed needle points 19 mm (0.75 in.) long reduced bone hits and injuries (Day, 1971). Syringes are easier to remove from animals and cause less damage if half of the barbs are removed. Projectile syringe injections should be made into the large muscle area of the hip or rump to reduce the chances of injuries.

The preferred drug for bighorn capture has been a combination of M-99 and a tranquilizer. Day (1971) reports that the response of free-ranging bighorn to M-99 used alone was extremely variable and unrelated to dosages. He found that the immobilization of bighorn was greatly facilitated by combining M-99 with a neuroleptic tranquilizer.

The following drugs and dosage levels have proven effective for immobilizing free-ranging bighorn (Wilson, ed., 1975). More than forty bighorn have been captured with these recommended drug dosages with no deaths attributed to the drugs.

a. **Ewes and rams**
   2 mg M-99 (Etorphine-DM Pharmaceuticals)
   15 mg Haloanisone (Pitman-Moore) or 20 mg Azaperone
      (Pitman-Moore)
   2.5 mg of Atropine (Merck)

b. **Ewes**
   2.2 mg M-99
   20 mg Azaperone

   Rams
   2.5 mg M-99
   20 mg Azaperone

All of the animals captured with M-99 combinations were given the reversal drug M50-50 (Dyprenorphine, DM Pharmaceuticals) at a rate of two times the M-99 dosages.

Capturing bighorn with drugs at water holes in the summer is not recommended (Day, 1971). Bighorn are forced to utilize water holes during the hot summer months, but the animals then are under severe stress from food and water shortages. Subjecting these animals to additional stress factors with drug capture is risky and only increases the chances of mortality.

The most successful and safest drug capture method has been with a helicopter (Day, 1971; Gates, 1972; Montoya, 1973; Wilson, ed., 1975). The advantages of this method are: (1) Capturing can be done during the cooler months when animals are in better physical condition. (2) Capturing is possible in more remote areas. (3) Individual animals can be more easily selected for capture. (4) Captured animals can be picked up and transported more quickly.

The B-series Bell helicopter with a 260-hp supercharged engine has been the most widely used for bighorn capture (Wilson, ed., 1975). Some of the larger turbojet helicopters can do the job but are more expensive to operate. An experienced pilot is important for safety and for pursuing bighorn. Only the pilot and one gunner should be in the aircraft during pursuit of animals. Loading the helicopter down with a photographer or another observer only increases the weight and risk factors The door on the gunner's side of the helicopter should be removed for better visibility and for shooting purposes.

Once a band of bighorn is located from the helicopter, it is important to try and "cut out" the animal selected for capture. Some hazing of an animal out of dangerous precipitous terrain is usually possible. Continued pursuit of the selected animal is important, so that it will tire out and stop for a close shot. Bighorn may travel 3 to 6 km (2 to 4 mi) and be pursued for 15 to 30 minutes before they can be safely darted.

No attempt should be made to shoot at moving bighorn. Once the animal is tired, it will try to hide under a large shrub, in a cave, or against a cliff. The helicopter can then maneuver the gunner to within 14 to 18 m (15 to 20 yd) for a good rump shot. Normally the downdraft from the helicopter will not affect the flight of the dart. Wilson, ed. (1975) recommends that the gunner shoot at about a 20-degree angle from the helicopter for the best results.

When an animal is hit, the helicopter should back off from the area but the gunner should not lose sight of the animal. The gunner should always remain in the helicopter until the animal is immobilized. Once the gunner is dropped off, the helicopter can go after additional personnel. Ground crew members should wear fluorescent colored vests for better visibility from the air (Montoya, 1973).

Bighorn successfully injected with M-99 will show some signs of ataxia in about 5 minutes. Most of the animals will fall or drop on their briskets in 10 to 15 minutes. A few narcotized animals will remain standing or leaning against some object for support. If no drug reaction is observed in 15 minutes, it usually indicates a bad hit and incomplete drug injection. Such animals can be injected again with a full dosage without danger to them (Wilson, ed., 1975). Immobilized bighorn should always be approached cautiously and quietly on foot with a lariat and the syringe gun available if needed. If possible, approach the drugged animals from the direction most hazardous to the bighorn in case they decide to move.

Captured bighorn should be held by the horns with the head elevated to avoid the aspiration of fluids into the lungs. Check "vital signs" and then blindfold the animal. Montoya (1973) reports that distress signs include rapid breathing, salivation, inverted tail curl, and rapid heart rate. Usually a ¼-reversal dosage of M50-50 will overcome all distress symptoms, except a rapid heart rate. Administer oxygen if the heart rate exceeds 130 beats per minute.

Bighorn to be taken from the capture site should be moved carefully. Gates (1972) found that bighorn captured in rough terrain could be administered ¼ the reversal dose of M50-50 and led to the helicopter. At the helicopter they were loaded into crates which were suspended below the helicopter for the trip to the transportation vehicles. Another method is to secure the legs so that the animals will not injure themselves and pack them out on a canvas stretcher with the animals in a normal lying position. Most helicopters have metal cargo hauling boxes attached to the skid pipes, and the stretcher can be tied into this or the animals placed directly into the boxes. Rubber stretch cords and ropes should be used to secure animals for traveling. Place the drugged animals in a lying position on the gunner's side, so that the gunner can hold the head upright by the horns for the trip out of the mountains.

Once the bighorn are at the transportation vehicles, they should be carefully inspected. All scrapes, minor injuries, and dart wounds should be treated and combiotic injections given to help guard against infections. The reversal drug M50-50 can be administered either intramuscularly or intravenously at a rate of two times the M-99 dosage. Animals should be kept under constant surveillance during the recovery period, which may take over 1 hour.

Transporting Captured Bighorn

The method of transporting desert bighorn will of necessity fit the existing conditions. Tom D. Moore describes how the Arizona and Texas game and fish departments handled animals being moved from trap sites in Kofa Game Range to the Texas Black Gap holding pens:

> Following their capture, bighorn were administered the tranquilizer Diquel (Jansen-Salsbury Lab.) intravenously at a rate of 150 to 250 mg per animal. Then they were loaded on a ton truck with an especially designed hauling compartment, driven 97 km (60 mi) to Yuma, Arizona, and then flown to the Texas release site 966 air-line km (600 miles) away. Upon landing, the bighorn were loaded into a covered trailer for a 4.8 km (3 mi) drive to the 173 ha (427 A) holding pasture in the Black Gap Wildlife Management Area in Brewster County.
>
> The bighorn seemed to adjust to confinement in the darkened hauling compartment of the truck, ate some jojoba, and drank water. This method of transportation for the animals was superior to that by plane; however, since the flights amounted to but 5 hours and only one or two bighorn were caught per week, the use of the truck could not be justified. The airplane used was a single-engine Cessna 180. The rear seat was removed and a slatted crate capable of holding one adult ram and ewe was installed.

The following six points are recommended for transporting bighorn safely (Wilson, ed., 1975):

1. Short hauls can be made with padded wooden crates 1.2 by 0.6 by 0.9 m (4 by 2 by 3 ft).

2. Transporting vehicles for long hauls should afford ample room for the bighorn to lie down in comfort. The vehicle should be well ventilated, have padded sides, and be rendered as dark as possible. When temperatures exceed 30°C (86°F), truck panels should be slotted to allow adequate air circulation.
3. Large rams should be separated from ewes and lambs by a railing or slotted panel.
4. Bedding of good quality, clean prairie or timothy hay, should be distributed approximately 29 to 36 cm (12 to 15 in.) deep on the floor of the vehicle.
5. Bighorn held in a vehicle longer than 48 hours should be fed and watered.
6. When traveling, stop and check the animals every hour.

RESTOCKING FORMER RANGES

Restocking former ranges of the desert bighorn serves several objectives: to preserve a rare game animal, establish a huntable population, and perpetuate a creature of great esthetic and scientific value.

In the southwestern states are many areas where the original native stock is completely gone, and restocking is the only method of bringing the animal back. It is extremely hard to place a price tag on an animal such as the desert bighorn; however, the mere existence of the animal in an area brightens the big game picture and renders the area more natural.

The existence of a huntable population of desert bighorn provides additional benefits to the region. The desert bighorn ram probably is the most sought-after big game trophy in North America and is one of the rarest. Unless a hunter lives in one of the western states that allows hunting of the species, it is extremely hard for him to obtain a hunting permit. Successful restocking of bighorn to its original ranges that are now devoid of them would provide a desirable hunting opportunity.

The esthetic value perhaps is difficult to establish in dollars and to relate to the benefits of a restocking program. The actual observation of a majestic bighorn ram in the wild is a rare sight for the general public. To the wildlife photographer, a series of color slides or movies is an even rarer treasure. That a certain area can boast of having a bighorn population undeniably makes it a valuable public asset.

Methods

Restocking historical ranges with desert bighorn can be expensive. When a program of this type is undertaken, a thorough outline should be developed first. The proposed area should be checked with respect to such relevant factors as the disease problems that might arise, possible predation problems, the public attitude toward such a release, methods to be used, the extent of natural distribution to be expected, and the availability of suitable water and vegetation.

A restoration program should include the following, according to Wilson, ed. (1975): (1) Thoroughly research old documents, reports, journals of early pioneers, and books which might document historic bighorn ranges. (2) Determine the factors responsible for the decline of the former population and take precautions so that the same factors will not reoccur if bighorn are reintroduced. (3) Determine what dominant plant communities existed on historic bighorn ranges and substantiate that the same or similar plant communities exist on reintroduction sites.

Can bighorn be liberated in ranges now utilized by domestic sheep? Although early records discuss the dangers of disease, including parasites, transmitted to wild sheep, one should bear in mind that the science of animal husbandry has greatly improved in recent years and epizootic diseases are no longer common in domestic herds (Yoakum, 1963).

In the Texas transplant herd on the Black Gap Wildlife Management Area, four bighorn died of bluetongue disease. It is unknown how the animals contracted the disease; however, there were domestic sheep on private ranches within 16 km (10 mi) of the enclosure. Although great strides have been made in animal husbandry, the threat of disease striking a herd, and especially an enclosed herd, should be considered when a transplant is proposed. It is also important to ascertain future plans in regard to the utilization of a release site by adjacent landowners, including government agencies.

Certain attempts to establish bighorn populations in the 1930s and 1940s met with failure because of public attitudes (Yoakum, 1963). However, changing relations between range users, land management agencies, sportsmen, and wildlife agencies have altered the picture, and we see more favorable attitudes toward transplants today. A strong factor in favor of restocking animals in an area is the economic importance now attached to big game species.

In most of the first reintroduction attempts, the animals were turned completely loose in their new environment. This practice was not as successful as recent programs in which the animals were retained in large enclosures before being given complete freedom. The method that proved successful in Texas used a brood pasture, 173 ha (427 acres) in size, to maintain a breeding population. The brood pen method allows the animals to adapt to conditions in the area before release. Reproduction was very successful in the enclosure, and, as the breeding herd increased, excess animals were liberated to the surrounding area.

At Red Rock Wildlife Area in New Mexico, bighorn are being propagated in a 259-ha (640-acre) enclosure for future transplants into other mountain ranges (Montoya, 1973, 1976).

Enclosure sites should be selected carefully and with the following criteria in mind: (1) be accessible by vehicle but remote enough to keep public access restricted to limited areas; (2) contain adequate escape terrain such as rocky outcrops, ledges, and vegetative cover; (3) provide year-round water that can be controlled, to reduce disease possibilities; (4) contain year-round forage supplies, although supplemental feeding and salting may be needed (Wilson, ed., 1975).

The size of the enclosure for propagation or restocking bighorn is important to consider. Too small an enclosure may restrict growth and survival of animals by making them more vulnerable to predation and accidents. The larger the fenced area (259 ha per 25 animals is suggested) the better the chances of successfully raising bighorn in captivity that will retain their natural behavioral and wild instincts (Wilson, ed., 1975).

In many historical areas of the Southwest where ranges have been heavily fenced by land owners or surrounded by highways, the natural distribution of bighorn from one mountain range to another would be very slow or nonexistent. The only way to repopulate areas is to restock mountain ranges where the bighorn originally occurred.

Transplant Results and Recent Releases

Of some bighorn reestablishments on native ranges cited by Yoakum (1963), eleven transplants totaling 297 bighorn from Canada took place between 1942 and 1962 in Montana. One herd, released in 1947, increased sufficiently for a harvest in 1955. In South Dakota, twelve bighorn from Alberta were released in 1962 and additional releases for that state are planned.

Oregon, Utah, Washington, and North Dakota also have made releases of bighorn. In 1954, twenty bighorn from British Columbia were transplanted to Hart Mountain Wildlife Refuge in Oregon (Deming, 1961). The animals increased about 200 percent in 5 years, and this herd has been utilized as a nucleus for other transplants. In 1960 and 1961, eleven bighorn from Hart Mountain Refuge were transplanted to Steens Mountains, Oregon, in two separate releases. In 1968, eight more were transplanted to Sheldon National Antelope Range in Nevada. In 1973 twelve bighorn from Lake Mead National Recreation Area were transplanted to Zion National Park, Utah; by 1976 they had increased to twenty-two (McCutchen, 1977).

In 1957, eighteen bighorn from British Columbia were released in the eastern part of Washington. Since their release, two additional transplants have been made in Washington from this nucleus herd of eighteen. Likewise in 1957, North Dakota imported eighteen bighorn from British Columbia. The population more than doubled in the first 4 years.

The trapping of sixteen desert bighorn in 1957 on the Kofa Game Range in Arizona for transplanting to a pasture on Black Gap Management Area in Texas has already been discussed. Heavy losses occurred during transporting and after release within the enclosure. By the 1960 lambing season, only three rams and three ewes remained in the enclosure. However, from 1960 to 1968 the herd increased to forty-four animals. Reproduction was very good and losses low during this 9-year period. A nucleus of brood animals is maintained within the enclosure and the surplus released into the adjacent area. Twenty animals were released from the pasture in January 1971 (Hailey, 1971).

During the joint trapping operation in 1957 in Arizona for the transplant to Texas, eight bighorn were also captured and released into the Horse Camp enclosure located near Klondyke, Arizona, in historical bighorn range. This initial release failed, and only two rams remained by 1965. An additional eight bighorn captured with M-99 were released into the enclosure from 1967 to 1972. This herd by 1972 had expanded to sixteen animals and was released from the enclosure in January 1973 (Weaver, 1973).

In Nevada, ten bighorn were trapped in 1970 on the Lake Mead National Recreation Area and placed within the Dutch Creek Enclosure on Mount Grant (Papez and Tsukamoto, 1970). This transplant in 1970 was used to supplement an earlier release that was reduced to a single animal by a mountain lion or lions.

The New Mexico Game & Fish Department captured five bighorn ewes in mountains southwest of Caborca, in northwestern Sonora, Mexico, for propagation at the Red Rock Wildlife Area enclosure (Gates, 1972). Four lambs from the Mexican ewes were successfully hand-reared at Red Rock. Three rams and five ewes were added to this enclosure in 1972 from the San Andres National Wildlife Refuge. One of these new rams presumably butted and killed three of the ewes from Mexico. The three rams from San Andres all died within a few months. As of early 1977 there were forty-three bighorn in the pasture (Snyder, 1977).

The Mexican government obtained the assistance of the New Mexico Department to jointly capture ten bighorn from the mountains on the mainland of Sonora for transplant to Tiburón Island in the Gulf of California (Montoya and Gates, 1975).

MARKING BIGHORN

Live Animals

Desert bighorn have been marked either after capture or while free without being captured. After an animal has been captured it can be marked in many ways. Regular aluminum livestock ear tags are most often used, although there seems to be quite a loss of tags, and even of part of the ear, after release into the wild. Plastic streamers have been tied into the ears, but these also are often pulled or scratched out. Perhaps the most permanent marking is a heat brand on the horns. This leaves a mark on the animals so permanent that it is

Norman M. Simmons

Fig. 18.2. This 8-year-old ram was marked on its white rump patch with red dye as it entered the water hole under the electrically operated spray nozzle. The nozzle and solenoid trigger can be seen above the ram's head. Eagle Tank, Sierra Pinta, Yuma County, Arizona.

still quite obvious long after death (Aldous and Craighead, 1958). Colored plastic tape (Scotch brand) has been used on bighorn captured in Arizona. The tape and color were clearly visible after 19 months on one ewe, and after 7 months on another. After 29 months a ram killed in the Kofa Mountains, Arizona, still retained blue-colored tape on one horn (Day, 1973).

Day (1973) recommends the aluminum Tamp-R-Pruf seal (LISCO) special sheep tag size for bighorn. The tag should be attached to the upper front edge of the ear as close to the base of the ear as possible. Both ears should be tagged to reduce the chances of losing the identity of an animal should one tag be lost.

Norman Simmons states, "The Canadian Wildlife Service used cattle tags with nylon impregnated polyethylene streamers, and they last at least 5 years if used properly."

There has been some experimentation with freeze branding, but not on desert bighorn. Freeze brands (Farrell et al., 1966) turn the hair white so that numbers or symbols will contrast with the dark coat of the animal. These are permanent marks, like horn brands. Heat brands on the horns and freeze brands on the body may prove an excellent combination.

Free-ranging desert bighorn are sometimes most manageable when they come to a limited area to drink. Accordingly, a method of marking them at water holes with a dye was developed on Desert Wildlife Range (Hansen, 1964a), and improved by Simmons and Phillips (1966) on Cabeza Prieta National Wildlife Refuge. This device was triggered remotely, spraying a blotch of keratin staining-dye on the head or body of the bighorn. This method of marking desert bighorn was used successfully by Simmons (1964) on Cabeza Prieta Refuge, and by Blong (1967) in the Santa Rosa Mountains of California (Fig. 18.2).

See Chapter 16 for additional comments on marking bighorn.

Dead Animals

Desert bighorn remains are often long-lasting under desert conditions. Skulls and horns of old rams, although well weathered, may last for 20 years or more. Skeletal material found in caves may keep much longer. Sometimes it is desirable to mark these remains and leave them in the field because their weight and bulk make it impractical to collect them. On Desert Wildlife Range a system of tagging skulls with numbered metal tags was developed. The tags were wired to the skull usually through one or more of the orifices in order to hold the tag securely. The horn sheaths were marked with India ink to identify them with the tagged skulls.

Various types of paint have been tried on horns, but in most cases paint wears off. Research biologist James K. Baker tried black, white, and yellow highway paint on the horns of dead bighorn at Joshua Tree National Monument. He also worked out a code system for using the three colors.

19. The Impact of Modern Man

Hatch Graham

MODERN MAN IN THE PAST

The Spanish era (1519–1820). The impact of modern man on bighorn began between 1519 and 1521, shortly after the overthrow of the Aztec empire in Mexico by the Spanish. Using Mexico City as a base, the Spaniards of New Spain dramatically pushed the frontier northward for nearly 3 centuries. In 1540, Hernando Alarcón sailed from Mexico north up the Gulf of California to the Colorado Delta, while Francisco Vasquez de Coronado, seeking the Seven Cities of Cíbola, explored northeastern Arizona and the Grand Canyon country and, in 1541, ventured as far as Texas. Around 1600, Don Juan de Oñate, the colonizer of New Mexico, described it as being "plentiful in meat of buffalo, sheep with huge antlers, and native turkey."

By 1699, when Padre Eusebio Kino pioneered El Camino del Diablo along the Mexico border in Sonora and southwestern Arizona, bighorn were encountered at mountain water holes. Padre Junípero Serra, the great colonizer of California, observed bighorn as "great goat-like animals" on his way up Baja California along the inland coast and on top of Sierra San Pedro Mártir.

The Spanish made little effort to occupy Alta California until late in the eighteenth century. The mission period began at San Diego in 1769 and colonization of the California lowlands continued until 1823. In 1774, Juan Bautista de Anza, in seeking to open a land route from Sonora to California, observed that, in extreme southwestern Arizona at what are known as Cabeza Prieta Tanks and other tinajas in the area, certain Papago Indians always camped during the height of the dry season to hunt bighorn that came to drink.

In California, even though the Spanish were aware of the bighorn, they did not threaten the animals. They were a pastoral people who confined their activities to the lowlands, seldom penetrating the mountains. The broad coastal valleys of the San Diego, Santa Marguerita, Santa Ana, Los Angeles, and San Gabriel rivers furnished abundant forage for domestic animals below the mountain chaparral—above which, in the crags and mountain cliffs, was the habitat of the bighorn.

The mountain men (1820-1850). The early mountain men, pushing westward after Lewis and Clark and Zebulon Pike, first met mountain sheep in the Rockies. The later mountain men, explorers, and fur trappers encountered desert bighorn in the Southwest, and California bighorn in the Sierra Nevada of California and as far north as the inland plateau of northeastern California and southeastern Oregon. These men lived off the land, and surely Jedediah Smith, Kit Carson, Cristobal Slover, and others of lesser fame took bighorn when the occasion presented itself. However, the numbers of explorers were so small, and game other than bighorn so abundant, that their impact on bighorn was negligible.

Mineral exploration. The search for mineral riches first brought modern man into real conflict with bighorn. The early Spaniards opened up rich mines in northern Mexico and other parts of the Southwest. The only good source of pure salt needed as flux for ore was the salt flats located adjacent to the northern part of the bighorn range in Texas. For more than 300 years great wagon trains came up from Mexico after the salt. The salt haulers had to camp at the salt lakes for 10 days or longer to let their oxen rest, to load the wagons, and to let the water drain out of the salt. Hunters supplied the big camps with meat. The effect of such meat hunting undoubtedly was detrimental to bighorn (Carson, 1941).

The gold rushes of 1849 and later brought many miners into bighorn country. Gold, silver, lead, and copper were discovered throughout the Southwest. Miners and prospectors utilized wild game for food as an established practice, one that changed little with passing time. Prospectors and miners living in remote areas felt they had an inherent right to take wild game any time, without regard to license, season, or bag limit requirements.

In 1882, a year after the railroads came through Culberson County, Texas, the rich Hazel silver mine was opened, to operate until 1896. Carson (1941) tells us that it was well known that the miners hunted the mountain sheep heavily. The same story was true in California. A Mr. Gordon of Glendora stated that in the winter of 1895 over fifty bighorn were shot on the Prairie Fork of the San Gabriel River to supply fresh meat for miners at the Bighorn Mine (Light et al., 1967). In Arizona, uncontrolled killing became a major problem during the early 1900s and until after World War I during a prospecting boom, which dwindled by the middle 1920s (Russo, 1956).

During the depression years of the 1930s, prospectors again took to the hills in search of minerals. In Death Valley, scarcely a spring could be found without a prospecting camp or millsite within a few feet. In some places the camps were built directly over the water itself. Thus, a water shortage was created of such severity that by 1935 bighorn were believed to be practically extinct in the Death Valley region (Welles and Welles, 1961a). This sort of thing occurred in Arizona and Nevada at this same time.

With the end of the depression years, stabilized gold prices, and a rising standard of living, most of these camps and mills were abandoned, fortunately for the bighorn. But in 1940 uranium was discovered along the Colorado River and its tributaries in southeastern Utah. This brought a tremendous influx of prospectors into the country, from Moab south and west to the Arizona line. A local newspaper estimated that at one time 10,000 prospectors were in San Juan County, Utah, the heart of Utah's bighorn range. Bighorn bones were found in many of the old prospect camps and mines of southeastern Utah during subsequent wildlife surveys.

Wilson (1968) says, "In the 1950s, when many of the uranium mines were in operation, many of the miners were known to have hunted bighorn sheep on their days off for something to do. From all the data available, it appears that illegal hunting was one of the major contributing factors leading to the decline of bighorn in southeastern Utah."

Of all of the activities which have brought man into desert bighorn habitat, mineral exploration and development seem to have been the greatest single adverse factor. Unregulated subsistence hunting and usurpation of bighorn water sources during these periods probably had the greatest total impact on bighorn. The most significant decline in bighorn numbers seems to have coincided with periods of mineral exploration in various parts of the West. But the miner was not the only one to blame.

Ranchers and settlers. After Texas became independent of Mexico, pioneer cattlemen established ranches in the mountains of western Texas. The bighorn had thus far held their own with Spanish hunters, and they still occupied most of the mountain ranges of western Texas. But with the coming of the railroads and permanent civilization, including mining, hunting increased and the bighorn decreased. As Carson (1941) puts it, "The large massive horns carried by the old rams, their many wrinkles and beautiful curling design, was enough to make any early pioneer want to kill a ram just to possess such a pair of horns."

Generally, cattle and sheep ranches established by the early homesteaders were not located directly in bighorn ranges. However, as stock numbers increased, the need for water increased and many mountain springs were diverted to livestock use. There is little doubt that homesteaders and ranchers made use of available wildlife, including bighorn. Nevertheless, in many places the situation that John P. Russo, Arizona Game and Fish Department, describes in Arizona held true: "Because of the isolation of our mountain ranges and the expansive deserts around these mountains, I would imagine that the taking of bighorn was limited."

Anselmo Lewis, Angeles National Forest, says the stockmen had considerable impact in the southern California ranges—especially the migrant sheepmen. But cattlemen also shot the bighorn for fresh meat. An old cattleman from Victorville told Lewis that his father shot bighorn for meat and nailed the heads to the barn. He remembered the barn as being fairly well covered with bighorn heads. Lewis also relates stories of old-time cattlemen shooting bighorn for meat in the Piru and Mutah Meadows country between Santa Barbara and Ventura in Los Padres National Forest. Bighorn were taken for meat and trophies in the San Bernardino Mountains after 1900 (Light et al., 1966).

William H. Rutherford, Colorado Division of Game, Fish, and Parks, believes that the continuity of desert bighorn range extending up the Colorado River from Utah into Colorado was broken by settlement, agriculture, and livestock grazing. A remnant of the former herds persists in the Battlement Mesa area in the western part of White River National Forest, in Mesa and Garfield counties. The fifty to sixty animals that remain are in restricted, isolated ranges that received little, if any, livestock grazing.

The effects of diseases transmitted from domestic sheep may have been great in many parts of the West (Pillmore, 1958). It is believed that much of the loss of California bighorn in the Sierra Nevada was due to epidemics brought in by domestic sheep in the 1870s (Jones, 1950).

Indians again. Although the Indians may have lived in a "balance" with the bighorn before the coming of white man, the introduction of European civilization, domestic livestock grazing, and the modern rifle amplified the Indians' impact on bighorn.

Shortly after 1863, the Indians of the Navajo Indian Reservation in desert bighorn country in southeastern Utah and northern Arizona were overgrazing the range and had to reduce their livestock numbers. Overgrazing made domestic sheep more susceptible to parasites and diseases, which probably spread by epidemic to the neighboring bighorn populations. Since Arizona game laws do not apply to Indians on their own reservations, bighorn may still be hunted there. The same is true in other states. Papago, Havasupai, and Hualapai Indians kill bighorn regardless of age or sex whenever the opportunity presents itself, according to Hal Coss and James A. Blaisdell of the National Park Service. Morongo Indians in California also are known to take bighorn. An instance was related to me by C. E. Wester of Riverside of a visit to the Morongos in the 1950s, at which time he saw a living room furnished with bighorn hides.

MAN TODAY

The rapid growth of civilization in the western United States, the mobility of modern man, the proliferation of cities, transportation arteries, dams and reservoirs, and the desire for recreation developments so that people may "get away from it all" are today bringing mankind more and more into contact with the remaining bighorn and their habitat. What is the impact of man on the bighorn and bighorn range in the late 1970s? This is the question we shall attempt to answer in this chapter. For insight into the general behavior and senses of bighorn, including their reactions to the presence of man, see Chapters 8 and 9.

Bighorn and Non-Hunting Man

Walking alone. The legend that the desert bighorn is the epitome of wildness probably came about because bighorn inhabit wild areas and are seldom seen. The facts demonstrate that man walking alone in bighorn country often causes little disturbance.

Frank W. Groves, former director of the Nevada Game and Fish Commission, writes, "On many occasions, I have been within 'rock throwing distance' of rams, ewes, and lambs, and had them pay very little attention to me as long as they were on a cliff or rock ledge above me."

Anselmo Lewis says, "Hikers have been reporting bighorn on Mount Baldy, East Fork of San Gabriel, etc., for the past 20 years, indicating that the presence of man is tolerated by the bighorn."

Devan (1958) writes, "I rate the animal on the whole as cautious but curious. Being caught in an open valley or at a lower elevation... makes the creature quite nervous, as does approaching the animal in a blind stalk of which the animal is aware. I have approached animals under favorable conditions very closely many times by walking slowly toward them and stopping often."

Francis A. Winter, formerly on the Angeles National Forest: "I've studied the bighorn both in Cattle Canyon and above Highway 2 [in the San Gabriel Mountains of California]. As long as the bighorn were above me (in other words, I wasn't between them and their escape route), they were unconcerned. They would feed, nurse their young, cavort, etc., but they would stop and 'check' on me. As long as I stayed in the open they did not flee; however, if I 'hid' from them, they became nervous."

On the other hand, there have been many occasions when bighorn, surprised by man, have taken off at a full tilt and continued running until they were a kilometer or more away, usually in "safe" terrain.

Welles and Welles (1961a) say:
By nature, the bighorn would appear to be unwary and trusting, but it quickly acquires a wariness from experience which makes it prefer isolation from human activities. However, deliberate attempts on the part of humans to conduct themselves within limits that are clearly acceptable to bighorn sometimes can lead to a gradual but marked reversal from wariness. In such cases the individual personality of the leader of the band, and her previous experiences with humans, become major factors in determining the degree of herd tolerance to human presence.

Herd leadership possibly holds the key to the future of bighorn-human relationships. If the leader (always a ewe) is unafraid of people, an entire band may become tame within days. This matriarchal leadership plays an important part in the overall life history of the naturally gregarious bighorn, controlling to a large extent group movements and reactions to the environment.

Norman M. Simmons, formerly with the U.S. Fish and Wildlife Service in Arizona, believes man's impact varies primarily in relation to the other environmental stresses placed on bighorn by their habitat. Bighorn with adequate water and food and ample escape terrain are quite tolerant of man. Where water and food are scarce and escape terrain limited, bighorn are more disturbed by man's activities. Still, insofar as man is a transient, occasional visitor to bighorn country, he has little lasting impact. Moreover, as bighorn become accustomed to the presence of man, they may tolerate a great variety of activities.

Walking in groups. Occasional groups of people seem to have little more effect on bighorn than single persons. Again, the bighorn's reaction to man varies. Perhaps important is the attitude of the people toward bighorn. Groups of people walking through bighorn territory, engaged in conversation or other activity, seem much less disturbing than groups intent on studying the animals.

Richard Weaver, California Department of Fish and Game, says groups several times have "spooked" bighorn into running, but he has taken Boy Scouts into bighorn country only to have the last boy in the file spot a ram that had quietly watched while the whole group passed by.

Where bighorn are intentionally harassed by groups, they are, of course, disturbed. Blong (1967) tells of teenagers in Magnesia Canyon, California, that have run at bighorn in an attempt to catch them. This aggressive action terrified the bighorn, which fled to safety and eventually abandoned the one water hole in the canyon.

Weaver additionally comments: "In flying helicopters in the Mount Baldy area of the San Gabriel Mountains, we are unable to find bighorn in the vicinity of the trails on the same days that we find hikers on the trails. Admittedly, the displacement is temporary, but as the numbers of hikers increase, the number of days the bighorn can use the area is decreased. My conclusion is that roads and trails in prime bighorn habitat are bad news, because their use *will* increase and bighorn will be displaced."

I believe that many of the observations here exemplify the maximum range of bighorn *tolerance* of man. Evidence of their *intolerance* is not seen, and so it is rarely recorded. Studies by a Forest Service and California Department of Fish and Game team in the San Gabriel Mountains in 1970–71 (Light, 1971; Graham, 1971) measured the effect of human use on bighorn activity and for the first time quantified the amount of human use tolerated by bighorn. The studies showed light to moderate use (up to 500 visitor-days per summer season) had little noticeable effect on use of bighorn home ranges. Heavy use (500–2,000 visitor-days) apparently caused the bighorn to withdraw from their hereditary range.

Walking groups include ever-increasing "backpackers" that invade remote wilderness areas, sometimes the heart of the best desert bighorn habitat. In fact, considerable sentiment against the creation of Wilderness Areas under the Wilderness Act of 1964 is expressed by some people on the grounds that designation of areas as official wilderness serves only to increase the number of people visiting them. Ben Avery, for many years outdoors editor of the Phoenix, Arizona, newspaper the *Arizona Republic* and a well-respected figure in conservation circles, has this to say: "My rationale concerning taking the Cabeza Prieta and Kofa [Game Ranges] into the Wilderness System is quite simple. Every area that is designated as a part of the wilderness system becomes automatically a place that must be visited by hikers and backpackers from all over the nation.

"Unfortunately there are not very many places that are so suitable for fall, winter, and early spring hiking as these two areas, and I feel certain such a designation will cause an even greater invasion of people into desert bighorn range than the threat of new communities and real estate developments ... We are already seeing some of the results of backpacker invasions during the lambing season."

Bighorn tolerate some disturbance. Continued, frequent, and especially new forms of disturbances cause them to avoid an area.

Riding horseback. Bighorn are perhaps fooled by people on horseback. At any rate, their behavior indicates that they consider this strange "man-horse mutant" less of a threat than man himself. John Russo says, "I think the bighorn look upon the horseback rider as they would the horse itself. Bighorn generally remain watching if you are not riding the horse directly toward them. Here, too, is a case of curiosity. I have often used my horses to an advantage, in that once the bighorn were spotted I would tie my horses, and then stalk the bighorn." James A. Blaisdell says bighorn usually are undisturbed by mule trains in the Grand Canyon.

Dr. Loren Lutz, of the Society for the Conservation of Bighorn, says bighorn are not disturbed by men on horseback. The only disturbance occurs when a man gets off. He relates an instance in which a ram came out of the brush 50 feet from a man on horseback, while snorting at a man on foot a half mile away.

On the other hand, Charles G. Hansen, for some years a research biologist with the U.S. Fish and Wildlife Service and the National Park Service, says that where horses are used regularly in bighorn habitat, the bighorn seem to learn that men are associated with them; they then spook as readily from a man on horseback as from a man on foot.

Motor vehicles. Mechanized vehicles—trucks, cars, jeeps, motor boats, motor bikes, ski lifts, tramways—do not appear to disturb the bighorn any more than do other forms of sudden appearance by man.

Richard Weaver says, "Jeeps usually spook bighorn when they happen upon them suddenly, but bighorn continue to bed down on ridges overlooking busy highways year after year with no apparent concern." Terry O. Landrith, Arizona Game and Fish Department, says bighorn are "very tolerant as long as you stay in or immediately around the vehicle." Charles G. Hansen says, "Bighorn are not greatly disturbed by an *occasional* vehicle moving through bighorn habitat at speeds of 10–20 mph. The disturbance comes from the vehicle stopping, the people getting out and milling around, or concentrating their attention on the bighorn." William F. Dengler reports from Joshua Tree National Monument that personnel at the Eagle Mountain Mine have reported observing bighorn along the road which is constantly being used by sixty 120-ton trucks. F. H. Jacot, formerly at Lake Mead National Recreation Area, observed "a ewe with a young lamb feeding less than 7.5 m (25 ft) from U.S. Highway 93-466, 0.4 km (0.25 mi) from Hoover Dam." Bighorn seem little disturbed by heavy weekend traffic on the Palms-to-Pines Highway in the Santa Rosa Mountains, above Palm Desert, and on Highway 2 in the San Gabriel Mountains in the Angeles National Forest.

Pickups rounding a bend on a little-used dirt road in the San Gabriel Mountains of San Bernardino National Forest in the head of the South Fork of Lytle Creek often have startled bighorn out of their beds located on the berm of the road. Here, the bighorn often drop over the side into their escape terrain which is below the road; yet they continue to bed down on the road year after year. Weaver points out that this particular road is receiving much more use by motor vehicles, especially motorcycles, than it used to. There is little doubt, he says, that bighorn use is considerably less in 1970 than a few years ago.

In summary, bighorn seem quite tolerant of steady traffic on through highways or occasional traffic on back roads, but perhaps not so tolerant of patterns of use that result in *unexpected* disturbance.

Motor boats. If bighorn can become tolerant of motor vehicles, this is even more true with respect to motor boats, which are used for desert bighorn censuses along the Colorado River. Several workers along the Colorado River, among them F. H. Jacot, Blayne D. Graves (formerly at Havasu National Wildlife Refuge), Jack Helvie (formerly at Desert National Wildlife Range), and Terry O. Landrith report that bighorn are not concerned about the presence of man as long as he is in his boat. In fact, they seem to be as curious about boaters as most of the boaters are about them. When man lands on the shore, however, bighorn retreat up into the rugged cliffs that surround the areas.

Ski lifts and tramways. Tramways and ski lifts are relatively new intrusions into bighorn territory. Two of these are in southern California, another is in New Mexico.

The Mount Baldy ski lift in California bisects the range of the bighorn, and the animals tend to avoid the area when the lifts are operated. The lift personnel report bighorn traveling through the area during the off periods and when the area is closed down and deserted by people.

However, Richard Weaver states: "As far as I have been able to ascertain the facts, the Mount Baldy ski lift must have considerable lasting effect on bighorn, because bighorn are rarely or infrequently reported traveling through the area, even in periods of nonoperation when it is deserted by people. If we assume that the relatively high use by bighorn of adjacent areas once extended into this area of equally good habitat, then we must conclude that the bighorn have been displaced by the development." The other development in California is the Palm Springs Aerial Tramway. According to William F. Dengler, personnel that operate this tramway occasionally report bighorn on the cliffs watching as the tram car passes.

A major tramway is situated on Cibola National Forest in the Sandia Mountains just east of Albuquerque, New Mexico. William Humphreys of the New Mexico Game and Fish Department says that about seventy transplanted Rocky Mountain bighorn inhabit a 104-sq-km (40-sq-mi) area there. Their chief concentration point is at the upper terminal of the tramway, from which heavy foot travel follows along and just below the crest of the Sandias. Parry Larsen, formerly of the New Mexico Game and Fish Department, reports use has been much reduced in this 3.2-km (2-mi) wide crest area. Cristobal Zamora, Cibola National Forest, states that a problem developed when the tramway operator began to salt the area under the tramway to bait bighorn for the summer sightseers; the salt attracted so many bighorn that the area became trampled out and overgrazed. The Forest Service asked the operator to discontinue the heavy salting.

The "behavior" of man-made objects such as tramway gondola cars, vehicles on regularly used highways, and boats confined always to waterways are predictable. In this sense, they are less disturbing and are tolerated to a greater degree than are the unpredictable actions of people on trails, in recreation areas, or simply abroad on the land. The passage of cars on the tramway has much less effect than people running along nature trails on the mountain top. Insofar as the developments result in bringing more human traffic to an area, they may have a lasting effect, but it appears that people are the problem, not the mechanical objects.

Aircraft. There is little evidence that fixed-wing aircraft seriously disturb bighorn, probably because most fixed-wing aircraft must maintain considerable elevation above the ground. W. Dean Carrier, Los Padres National Forest, and William F. Dengler report little disturbance. Norman M. Simmons says, "Repeated low-level flights cause bighorn to become spooky. Bighorn spooked sooner on a third aerial survey in the same month than they did on the first, for example."

Reactions to helicopters are mixed, but low-flying helicopters often cause considerable panic. W. Dean Carrier believes that larger helicopters, especially the military type, evoke a more violent reaction than the smaller, quieter commercial helicopters. Many observers report fear and agitation among bighorn from low-level helicopter flights. Anselmo Lewis states, "On a number of occasions we've used the helicopter to count or take movies of bighorn. However, we have more or less discontinued this practice, since the bighorn seem to be terrified of the helicopter, especially at low levels. They dart in all directions, bowling over their lambs and in general showing great fear."

Carrier quotes Dave Hipley telling of an old ram that refused to run; he just stood and stared at the helicopters that hovered overhead. His only reaction was to lick his lips constantly. Carrier also relates, "In one instance a ewe and a lamb stood their ground while we hovered nearby. The lamb attempted to keep his head underneath the ewe, and on two occasions seemingly tried to nurse. Bighorn alone or in small bunches are more panicky than larger bands. These larger bands would usually move off at a quick pace but would not become confused and panic. If an object such as a tree, or a large rock, was nearby, they would get behind it and usually wouldn't move from it. One bunch of four sought shelter in a small cave while a fifth, who couldn't fit in, ran over the crest of the hill. Sometimes the bighorn would climb to the highest point on the peak and at other times seemingly 'dive' off into the steep canyons."

The evidence seems to confirm that helicopters do disturb bighorn but that their occasional or transient use would have little serious effect. However, low-level scheduled flights with the establishment of a base heliport in bighorn territory could well cause bighorn to leave.

Noises. Desert bighorn display a great variety of reactions to man-caused noise. The bighorn's location may be all-important. Devan (1958) notes that, "All sheep observed in heavy timber types are quite nervous and will 'stampede' when a horse whinnies or the observer is first seen, which is quite a constrast to the lack of concern they often show when seen in the open." Gerald I. Day, Arizona Game and Fish Department, says, "Jets, sonic booms, and artillery fire practically overhead did not seem to disturb bighorn. They did disturb me and I didn't learn to get used to them in 10 days in one area." John P. Russo says, "I can relate experiences of having seen bighorn become startled with sonic booms. Again [there are] those that pay little or no attention to the boom. I did observe several bighorn go into headlong flight when the scream of rockets [on military reservations] was heard nearby. I might also add that I left the country." Jerry T. Light, U.S. Forest Service, says, "Sonic booms have startled bighorn, causing them to leap into the air and lose their footing while they were being observed in the Santa Rosa Mountains."

Monte Dodson, of the U.S. Fish and Wildlife Service, believes that bighorn continually exposed to sonic booms, as on the Cabeza Prieta Wildlife Refuge in Arizona, may develop severe stress problems that inhibit normal daily living patterns, as well as reproduction.

Other sounds must be equally startling to bighorn. Yet, for the most part, they appear unconcerned. Bonnar Blong tells of house-building activity in Bradley Canyon, Riverside County, California. The activity went on throughout the summer. He says, "The noise of power saws, hammering, and the blasting of airguns that drive nails into concrete did not deter the bighorn from their routine trips to water." Blong points out that these are the same bighorn that had been harassed at Magnesia Canyon Spring, but they tolerated nearby activity in Bradley Canyon when they were allowed to water in peace.

Anselmo Lewis says, "Noises have only a temporary effect on bighorn. When Highway 2 and 39 were constructed through the Kratka Ridge country, which bisects bighorn range, there was considerable blasting and heavy equipment working in this area for over 10 years. The camp located at Cedar Springs was near the summer bachelor ram feeding grounds. Although this activity went on for years, bighorn persisted and appeared to be as numerous as ever. Our experience, in this respect, indicates that bighorn will tolerate the noise of construction work, providing it is not of such extent it does not drive them out of the area on a permanent basis. They will retreat before it and go in behind the disturbance when it passes."

With respect to the human voice, William F. Dengler says, "My experience indicates that even shouting seems to have little effect on bighorn, so long as it is indirect; i.e., it is between people whom the bighorn can see and is not directed toward them at close range. The most nervous bighorn I have seen were observed from a fairly well-concealed blind. In this situation even whispers seem to bother them."

Gerald I. Day says, "Voices definitely seem to alert them, whispers are detected, but not considered too disturbing to bighorn. One year we used walkie-talkie units between blinds. However, they have an echo that appears to run the bighorn off." On the other hand, Terry O. Landrith says, "On one survey I was carrying a portable two-way radio when Phoenix made a transmission. At the time, I was watching two rams on a high mesa. As I was within 46 m (150 ft) of the animals, they heard the message. Instead of running away, they immediately ran toward the noise, stopping when they saw me lying some 15 m (50 ft) away. I then made no pretense of hiding and began turning the squelch button on the radio. This only made them more curious. Finally, I got up and walked away. The rams never moved.''

What about man's imitation of bighorn? Richard Weaver says, "Try bleating; a ewe will come and investigate you." Devan (1958) says, "I've tried calling sheep, too, for lambs, ewes, and rams have very distinct calls at certain times of the year. I have bleated, blatted, and made all sorts of adenoidal sounds but only with the results of an occasional 'cocked eyebrow and tilted head.'" Perhaps Devan has the wrong accent, for Day says, "Another time, I was sitting in a rock blind as a young lamb approached. It was bleating while it walked. I gave out with a similar bleat. The lamb stuck its head into the blind. I was able to stroke the ear and chin of the lamb for about 1 minute. When I attempted to get out of the blind and catch it, it promptly moved away and I never got closer than 14 m (45 ft).''

Blong has found that, during rutting season, he can get rams to run up close to him by pounding rocks together and grunting.

Odors. Devan (1958) reports, ''I believe that our bighorn do have an acute sense of smell. However, they do not employ it as one would expect an animal to, which is in a defensive manner, but appear to rely solely on nimbleness and splendid eyesight. During a period last fall, I was trapping sheep for our tagging program at one of our springs. The ram came right up by my blind, about 7 m (20 ft) away and on the lee side. I had just finished boiling a pot of tea and was smoking a pipe. None of these odors even turned his head; he went to the trap that had odors of sheep blood from slitting an ear to accept a tag on an earlier-caught ram and the odor of kerosene that I spilled from our branding iron heater. These odors were also ignored, but near one corner of the outside of the trap where there were tracks made by several ewes very early that morning, he lowered his head and neck and sniffed out quite a few of the trails in hounddog fashion.''

Richard Weaver says: ''Some of the spookiest bighorn I have ever seen were at Buzzard Spring in the Eagle Mountains, California. Lowell Sumner and I were in the shade of a bluff, high on the canyon wall, and not in view of the bighorn. A ewe group approached down the canyon and apparently caught our scent from the updraft. Even though these were gaunt animals and it was a hot July day, they spooked clear away from the water and never drank.''

For further treatment of this subject, see Chapter 8.

Dogs. Dogs are met with the same mixed reactions as is man. Lewis H. Myers of the Bureau of Land Management describes an occasion when he had a dog with him which seemed to evoke no particular interest on the part of the bighorn. On the other hand, Richard Weaver says, ''Dogs seem to infuriate mountain sheep. I had one experience of a ewe butting the hell out of a Norwegian elkhound and paying no attention to me or the fellow that owned the dog. I've heard stories of dogs that were killed by mountain sheep.'' Frank W. Groves says, ''One evening Pierre [Albert Van S. Pulling] was walking with his Irish setter by his side and had a mountain sheep ram challenge him stiff-legged up to within a few feet of the dog before taking off." Charles G. Hansen has seen desert bighorn in Nevada threaten dogs by lowering their heads as if to butt the enemy.

Man and Habitat

Bighorn habitat requires adequate living space and the rough, rocky slopes, cliffs, and escarpments in which the animals seek escape. It requires the trails and travel routes between these areas. It requires the grazing and forage areas adjacent to and within the bighorn's living space, and the water sources within and nearby. Man's activities bring him in, through, and around such bighorn habitat with variable effects on the bighorn.

Man's activities can be transient or permanent. Occasionally they fall into a category somewhere in between. Most of man's transient activities have only a passing effect and have already been described. *But man's permanent occupancy more often has a lasting effect and constitutes the greatest threat to the future of bighorn.*

Most bighorn workers would probably agree with this summary by Welles and Welles (1961a): "Man and what he has brought with him comprise the only significant threat to the bighorn survival.... Only unchecked human encroachment appears actually to threaten the future status of the bighorn."

Space and terrain. Because of the nature of the bighorn's escape terrain, man seldom occupies the very heart of bighorn habitat. However, permanent developments on the periphery have disturbing effects, which vary with respect to their "psychological" effect on, and continued use of, the area as escape cover for bighorn.

As a rule, bighorn seek escape by climbing uphill away from man. When surprised from above, they take headlong flight down a steep escarpment and continue running until distance provides safety or they climb above the intruder.

Within his living space are other areas essential to the bighorn's well-being, including lambing areas and bedgrounds. Lambing areas generally are on steep slopes, with fairly abundant green grass and forbs in late winter and in spring. In many parts of a bighorn range, lambing areas are limited. Moreover, ewes will seldom lamb in an area disturbed by outsiders. Lambing areas are particularly critical. *Permanent human occupancy near key lambing areas will most certainly cause bighorn to move away.*

Bedding grounds are used daily, in the heat of the summer and at night. A bighorn that is constantly driven from his rest is unlikely to persist in bedding down where the disturbance is regular. On the other hand, certain types of man-made improvements seem to provide additional bedding areas. Dirt roads and trails constructed through bighorn habitat provide convenient ledges where bighorn can scrape out their beds and camp on the edge of a drop-off.

Impact on forage and grazing areas. Man's impact on desert bighorn forage and grazing areas, aside from use by his livestock, appears to have been, for the most part, minimal. The only losses occur where man's developments have, in themselves, occupied grazing lands. This has happened where large mines have been developed within the bighorn home range, and in a few places in southern California and western Arizona where subdivisions have occupied bighorn range. Also to be included are feeding grounds inundated by reservoirs, where, however, additional grass and other plants may grow along the edges of the reservoirs. We are not considering here the direct effects of livestock on bighorn range—which are treated in Chapters 14 and 20.

Sometimes man's impact is favorable. Some land management agencies are attempting to improve forage in bighorn areas. Sometimes man's activities have resulted in accidentally improved forage conditions. In the San Gabriel Mountains, firebreaks—that is, areas of brush cleared to help slow or stop the spread of large watershed fires—have produced increased grass forage. Anselmo Lewis says, "For a number of years an old ewe lambed on the firebreak around the village of Mount Baldy. She seemed undisturbed by the activity in the village, and she and her lamb flourished on the green grass of the firebreak." This is an example of the tolerance of bighorn to man, providing he does not preempt their territory.

There is speculation that protection of areas from fires has tended to support the succession to heavier shrubs and browse in many parts of the Southwest, fire probably being a mechanism for the successional replacement of the climax, or mature, association. Richard Weaver suggests this is a good field for research and points out that bighorn occurred in Ventura County, California, on Los Padres National Forest, until 1916. Forest fire control for watershed protection may have caused changes detrimental to bighorn. Other studies indicate early heavy grazing by livestock decreased grass cover and led to invasion by shrubs and other subclimax plant species. This change favored deer and other subclimax wildlife species but worked against the climax bighorn. On the desert slopes of San Bernardino National Forest, cattle grazed regularly until about 1960. Since grazing ended, deer numbers declined and the heavy pinyon and desert scrub stands have become overmature and decadent, with little or no establishment of new shrub plants, and with eventual restoration of grasses which are favored by bighorn. Where fires have occurred, the resultant grass stands cause the fire scars to remain clearly evident 10 to 20 years later.

Lanny O. Wilson tells of Bureau of Land Management and Utah Department of Fish and Game proposals to eradicate a large portion of the pinyons and junipers from the south and east portions of Wingate Mesa, these areas to be reseeded to provide grass for the bighorn. Nelson reports (1966) that bighorn make use of field crops and orchards when developed within their range. He says, "The herd in the Bill Williams Delta [Arizona] has become accustomed to humanity—in fact, they profit by the products of the Planet Ranch where they feed in the fall and winter every year." John P. Russo indicates that bighorn have been known to utilize agricultural pastures and farm lands and the water from irrigation ditches.

Impact on travel and movement. Bighorn in at least parts of their overall range must move from feeding areas to water, from water to bedgrounds, from winter feed to summer feed to spring feed to fall feed, and back again. Roadways, fences, and canals tend to cross the travel routes of bighorn. Therefore the impact of man on the habitat includes the effect on bighorn travel routes and movements.

John P. Russo observes that, in Arizona, fences probably obstruct travel more than any other type of man-made barrier, aside from cement-lined irrigation ditches. His observations seem to be borne out in other areas as well. Many high-speed highways have rights of way completely fenced with five-strand barbed wire, and bighorn have been caught in such wire more than once. Rams are particularly vulnerable, because they catch their big horns in the wire and in their struggling sometimes cut their throats and bleed to death (Sizer, 1967). Highway departments are cooperating with game and fish departments by using "Page" wire or hog wire to prevent animals from going through the fences. In other places, underpasses and culverts provide crossings for bighorn.

According to Fred L. Jones, Bureau of Outdoor Recreation, and others, the Los Angeles Aqueduct from Owens Valley has been a death trap for bighorn trying to cross between the Sierra Nevada and mountain ranges to the east. This corroborates Russo's observations on irrigation ditches in Arizona. The problem seems to be one of getting out of the water up the steep cement sides of the aqueducts.

Lakes and rivers may interfere with bighorn travel. Crossings of the Colorado River apparently were fairly common in the early days. But, since the advent of dams and reservoirs, such crossings may not be as frequent. Terry O. Landrith believes this to be the case, and Sleznick (1963) reports that bighorn have drowned in the reservoirs. He found a ¾-curl ram floating in Lake Mead in 1962. F. H. Jacot says fishermen report bighorn crossing Lake Mohave in Black Canyon south of Hoover Dam in areas where the lake is less than 0.2 km (⅛ mi) wide.

Highways result in a few bighorn being killed by automobiles. But this would seem to be a minor loss and no barrier to movement over the long run. Some highways through bighorn country have for years been ideal places for observing the animals. For example, Tevis (1959) reports, "Fifty-one years ago, when even the desert floor around Palm Springs was wild, Dr. Joseph Grinnell explored Deep Canyon on land now owned by the University [of California]. He wrote, '... it was the immediate vicinity of Deep Canyon which ... was the metropolis of the sheep. On the steep walls and nearby mesa, a few hundred yards back from the rim, 2,500 to 4,000 feet [760 to 1220 m] altitude, well worn trails, footprints, and feces were plentiful. In places it looked as though a herd of domestic sheep had been over the region.'

"Today the same description—a fresh sign of abundance—fits the scene even though the heavily traveled Palms-to-Pines Highway closely parallels Deep Canyon and cuts right across the very mesa mentioned by Dr. Grinnell."

Anselmo Lewis and Francis A. Winter have both reported that the presence of unfenced highways in the San Gabriel Mountains does not seem to disturb bighorn to any extent. Lewis feels it is important that bighorn have natural terrain to cross, such as over tunnels. He says, "On Highway 2 there are two tunnels which serve as natural crossings for the bighorn. Numerous observations of bighorn crossing these 'bridges' are reported by our personnel. As far as we can determine the bighorn have persisted in this area ...without any apparent reduction in population....However, before generalizing, we must remember that the bridges provided by the tunnel may be the key to their tolerance of the highway barriers."

In southeastern Utah, Lanny O. Wilson reports instances of bighorn crossing Utah State Highway 95. Welles and Welles (1961a) report numerous examples of bighorn feeding beside highways and roads in Death Valley National Monument.

In these instances, the highways did not have a right-of-way fence, such as those along Interstate Highway 10 of which Russo speaks and which resulted in the death of several bighorn.

Bonnar Blong feels a wide highway, of the freeway type, presents a greater barrier. Interstate Highway 8 in San Diego County, California, is wide and bisects bighorn range. This area should be watched closely to determine its effect on bighorn.

In summary then, it appears that fences can well be a barrier to travel unless they are designed in a manner which will permit bighorn to pass through safely and easily. Concrete-lined canals and ditches often are death traps. Unfenced highways, in general, do not deter bighorn from crossing, although as the highways widen and the traffic increases, we can expect the mortality of bighorn to increase. Natural-appearing overpasses or underpasses would seem to be the best solution for bighorn crossings, and game departments would do well to urge these upon the highway commissions when major highways are proposed through bighorn ranges. Lakes and reservoirs probably are barriers to bighorn when they are very wide but often provide additional sources of water and feed along their shores.

Man can move his commerce and dam the rivers without permanently destroying bighorn travel routes, but it will take study and planning and the will to succeed to prevent bighorn habitat loss.

Impact on the air. Man is currently nature's greatest polluter. Air pollution is greatest around cities, but cities are not bighorn habitat. Atomic tests in the southwestern desert have increased the amount of radioactivity to some degree locally. But Ray Brechbill of the former Atomic Energy Commission, now the Energy Research and Development Administration, says levels of strontium 90 measured in hock joints of bighorn browsing on Desert National Wildlife Range in Nevada and elsewhere show that bighorn feeding close to the AEC Nevada test site have less strontium 90 (and less other radionuclides) than bighorn grazing in many other areas. Brechbill adds: "Man-made radioactivity now [1973] is deposited primarily from worldwide fallout other than from testing areas of the Southwest, and rainfall is the principal mode of its distribution."

Air pollution by smog from the Los Angeles basin may be having more serious, though as yet unknown, effects on the bighorn of the San Gabriel Mountains, and perhaps the San Bernardino Mountains. Smog in southern California has reduced the production of oranges by 50 percent in the foothills just below the Cucamonga Wilderness, the heart of the San Gabriel bighorn herd area. It has affected all the ponderosa pine in the San Bernardino Mountains, 10 miles east of the principal San Gabriel–Cucamonga bighorn range (Miller, 1969). And one may only guess as to the effect that it may have had on the production and nutritive value of various plants upon which the bighorn depends. One may wonder also at the smog's effect on the animal itself. We can only suspect that smog can have nothing but an adverse effect on the bighorn and its habitat.

Impact on water. Most of the desert bighorn's range is arid country. Water is a scarce commodity there, and man has put it to his own use, often to the detriment of the bighorn. Many times the only water source for miles has been completely taken over by man. Miners built their cabins over the springs; in many of the higher mountains, water was diverted for the operation of logging machinery; corrals and fences were built around water sources for livestock; and, during Prohibition, bootleggers often located their still at out-of-the-way desert springs. According to Charles G. Hansen, "Between the loggers [in the Charleston Mountains, Nevada], livestockmen, and bootleggers, the wild sheep were hard-pressed to find water. Fortunately there was more water in the early days than now, so the sheep were not completely excluded, but they were forced away from the more suitable habitat."

Sometimes the location of improvements and dwellings, and even the presence of man near a spring, has been tolerated by bighorn. These instances seem to depend on the animal's location in relation to the observer. For instance, Anselmo Lewis says, "Bighorn will water at areas frequented by man, providing man's use is not on a permanent or steady basis. Bighorn will water in the upper portions of San Antonio Falls, while man, a few hundred feet below, is recreating in the falling waters. On a number of occasions, I have observed bighorn 'studying' the sightseers below without any evidence of alarm or fear. Of course, there was a sheer drop of 30 m (100 ft) or more between them.... At Kelly's Camp on Ontario Peak, the bighorn frequented this spring development when use was light. However, as use increased, the lack of interval between camping groups was such that bighorn use of the spring dropped off. Evidence now indicates that the use of the spring by desert bighorn has greatly diminished ... Continued camping at known water holes will definitely drive the animals away.''

Sometimes bighorn will adjust their habits to fit man's visitation. Richard Weaver says that in Chino Canyon, along the road to the foot of the Palm Springs Tramway, bighorn once were seen commonly at different times of day. Now, they only frequent the area in the early hours, before the tram opens for business.

Blong (1967) says,

> There is a limit to the bighorns' tolerance of people invading their summer concentration areas, especially near water. The Magnesia Canyon bighorn could tolerate people at a distance of 60 yards [55 m] from their supply of water, because they have been living near people for many years. However, during water-hole counts in the remote areas of the Santa Rosa Mountains, bighorn have been observed to not tolerate people within 200 yards [185 m] of where the sheep intend to drink. It is difficult to recommend what distance people should stay from water holes during the summer.
>
> Where residential areas have developed gradually over a period of years near bighorn summer concentration areas, the sheep have gradually become conditioned to living close to people. This would indicate that if people would stay ½ mile [0.8 km] from water holes during the summer, bighorn can adapt to living near people. More study is needed to determine bighorn tolerance limits to people in relation to their summer concentration areas. People should be educated to stay away from the water holes during the summer, and to not harass the bighorn in the summer concentration areas.

Emmett Ball, Forest Service mining engineer, reports bighorn watering at the Eagle Mountain Mine in Riverside County, California, within 60 m (200 ft) of an office building where a water pipe was dripping water. The bighorn would drink one at a time in full view of the office building, while others waited in the protection of the rocks.

Welles and Welles (1961a) summarize their observations: "A major danger to the future existence of the bighorn lies in a continued and accelerating usurpation of its ancestral water supplies by man. Bighorn appear to prefer to remain in one home area if conditions allow them to, being born, living, and dying within a radius of 20 miles [32 km] of their home water supply. However, unlike some other ungulates, they will move rather than starve. The exhaustion of non-permanent water supplies probably is the most frequent cause for moving.''

The many observations made of bighorn at water holes seem to bear out these general impressions: bighorn can become accustomed to man and will tolerate some disturbance, particularly where a reasonable amount of height and/or distance adds to their sense of security.

Man's impact on bighorn water falls into four categories — three bad, one good:

1. Interference with bighorn at the water hole.
2. Usurping the bighorn water by capping the spring and piping the water away, by filling the spring with rocks and so making it unavailable, and by lowering the water table by pumping wells within the same groundwater basin.
3. Introducing competition for bighorn water by livestock, exotic wildlife, or feral animals such as wild horses and burros.
4. Development or improvement of water sources available to bighorn for human use.

In 1965, the band of at least twenty-five bighorn which was driven away from its home range in Magnesia Canyon in the Santa Rosa Mountains of California by aggressive teenagers moved its summer concentration area 1.6 km (1 mi) north to Bradley Canyon, where residents had provided a water trough. The bighorn accepted this new source of water the first summer it was available. Blong (1967), who reported on this occurrence (which I have already mentioned in relation to walking in groups and noises), says, "The Bradley Canyon band of bighorn have been living close to a highway and residential area for a long time. Looking down on the highway, homes, and golf courses adjacent to their range, they apparently have become conditioned to living close to people." Blong emphasizes that in this new location they were not harassed. Welles and Welles (1961a) cite several instances in which bighorn readily accepted new or rehabilitated water sources.

Human sewage effluent is accepted by bighorn in at least one area. The sewage disposal unit for the Boulder Beach development of Lake Mead National Recreation Area attracts a number of bighorn to the green grass of the leach fields, as reported by several observers.

Weaver points out that water developments for livestock have had a beneficial effect where uncertain or unreliable water holes have been developed and maintained in good condition.

The latest, and perhaps greatest, threat to bighorn on the desert ranges is the growing human population and its increasing mobility. With more leisure time, man the recreationist is exploring the desert ranges that once were the domain only of the miner, the bootlegger, and the rancher. Many of today's desert explorers come well-equipped with self-contained camp vehicles which include water, stoves, and most of the comforts of home—including the kitchen sink. Still, the desert water hole has a charm which seems to draw men like a magnet. Perhaps it is the comfortable green, riparian growth, the sound of the doves and Gambel quail, or the shade of a cottonwood or mesquite. Whatever the case, one finds man camped at the bighorn's water holes. He may travel out from camp on his trail bike or shoot into the rocks with his .22. His children may throw rocks in the spring to watch the water splash—until the pool is full. He may leave behind his cans, his waste, his bonfire ashes—and a water source from which the bighorn have fled.

But are we justified in pointing a finger of scorn at all men? Surely today, more and more people seek the bighorn range to share his environment and to protect and conserve qualities that support the bighorn's existence. Many of these same people, given the facts, would avoid the water hole in deference to the bighorn.

Habitat protection. In the period before World War II, when public awareness of the need for desert bighorn protection first became evident, four areas were incorporated into the National Wildlife Refuge System specifically for the management of these animals. The four areas are:

Desert National Wildlife Range. 643,000 ha (1,588,459 acres) in Clark and Lincoln counties, southern Nevada. Established 1936.

Kofa Game Range. 267,200 ha (660,000 acres) in Yuma County, Arizona. Established 1939.

Cabeza Prieta Wildlife Refuge. 348,200 ha (860,000 acres) in Yuma County, Arizona. Established 1939.

San Andres National Wildlife Refuge. 23,165 ha (57,216 acres) in Dona Ana County, New Mexico. Established 1941.

It is noted that these areas have different titles. The inconsistency is mainly due to political and bureaucratic pressures that are too involved and complicated to explain here. It is hoped that some day they will all have the same nomenclature, as their main responsibilities are identical.

These national areas were formed chiefly out of the public domain and are administered by the U.S. Fish and Wildlife Service. All have undertaken programs of reduction or elimination of wild burros and domestic cattle numbers, water development, stoppage of poaching that was rife, and predator control, as well as research into many aspects of bighorn life history. Beginning with the Desert Wildlife Range in 1952, bighorn and deer hunting, in both cases rigidly controlled in cooperation with the respective state departments in charge of big game, are permitted on all four areas (with the exception of deer on Cabeza Prieta Refuge).

Public access to Cabeza Prieta and San Andres refuges is severely limited on account of military use overlays. Such military use has not adversely affected bighorn numbers, aside from the possible effect of sonic booms, mentioned earlier. Likely it has had the opposite effect by preventing human entry and consequent harassment. Hunting on the two areas occurs during a period of non-use by the military.

Desert bighorn are also protected and managed as part of the programs on National Park Service areas within the desert bighorn's range, notably on Death Valley National Monument, California, and Grand Canyon National Park, Arizona, and to a lesser extent on Joshua Tree National Monument, California, Lake Mead National Recreation Area, Arizona-Nevada, and Organ Pipe Cactus National Monument, Arizona. There are likewise management programs on Canyonlands and Zion National parks, Utah, and on Glen Canyon National Recreation Area, Utah-Arizona. A number of state parks, and special units administered by the Bureau of Land Management, have bighorn programs, as do national forest units in southern California, Nevada, Arizona, and New Mexico. Very often the degree of desert bighorn management exerted is due to the amount of interest displayed by individual superintendents or supervisors.

THE FUTURE IMPACT OF MAN

There is in this nation a great groundswell of support for improving and restoring the quality of our environment. Many public agencies are improving habitat for bighorn. Enlightened policies regarding land use are protecting key areas of the bighorn habitat. Lay groups such as the Society for the Conservation of the Bighorn Sheep in California and the Arizona Desert Bighorn Society are growing in membership and have a good influence on man's attitude toward the bighorn and his habitat.

But much more needs to be done. Bighorn ranges, if they are not to become filled with ill-advised man-made intrusions, should be in public ownership. Old railroad grants created a checkerboard pattern of private lands throughout a great deal of public lands—both public domain and national forest in southern California and elsewhere. Consolidation of key ranges on these lands through land exchange procedures is possible and should be vigorously supported. Efforts by the states to acquire key water holes on private lands are needed. Private groups like The Nature Conservancy can lend financial assistance toward these goals.

In the management of public lands, bighorn values need to be understood and fully considered by land managers and supported by the public. The National Environmental Policy Act of 1969 provides good direction in this regard. It requires an interdisciplinary analysis of the environment and public involvement in the decision-making process before any major *significant* federal action is to be taken on federal lands.

It is essential to recognize that actions which significantly increase human activity in key portions of bighorn ranges can do great harm. At the same time, wildlife biologists must guard against overprotectiveness and overreaction if their views are to be taken as responsible professional findings.

There are too many examples of bighorn tolerance to man (as described above) to give credence to opinions that bighorn in all cases must have an inviolate sanctuary to survive, although there is no question that this often is the desired goal. More studies which measure the range and limits of bighorn tolerance to human disturbance, such as J. T. Light's work (Graham, 1971; Light, 1971) in the San Gabriel Mountains of California, are required to demonstrate more precisely how much disturbance can and should be permitted. As a corollary, the rate of adjustment to man by bighorn needs to be studied. Along with this, there is the philosophic question of just how "tame" should a wild creature be? Certainly, we have skilled wildlife specialists and ample opportunity to answer these questions.

One cannot prepare a chapter such as this without mixed emotions about the future impact of man. History may be a guide to this future: we have seen a few men exploit our resources in the apparent belief that there was a limitless supply of all game, forage, and other resources. But as the exploitation grew, as the supply dwindled and the demand increased, as our knowledge broadened and as civilization advanced throughout the continent, there was a gradual awakening to the problem. There is a startled public, eager to learn, and a rapid race toward environmental conservation is underway.

We cannot be complacent; but I believe we have the knowledge, the skill, the public support, and the desire to win the race. We must—to insure that our children and their children's children can search for, and see, upon some rugged ledge, that massive curl of horn, that magnificent beast, the desert bighorn.

20. Habitat Protection and Improvement

William Graf

Although the desert bighorn lives in an arid, rocky, poorly vegetated, and poorly watered habitat, it should not be assumed that this is the most favorable habitat. Deming (1962) presents evidence that the ranges of desert bighorn formerly were much wetter and concludes that this subspecies may now be a relict, surviving under conditions more severe than those which favored its original penetration into the Southwest. Even in the past 75 to 100 years, natural conditions over much of the bighorn range were more favorable than they are today. McColm (1963) investigated the history of desert bighorn in central Nevada and produced evidence of a former abundance of grassland where today there is little or no grass, or at best a shrubland succession. Similar inquiries into past conditions of other western range lands indicate that 50 to 100 years ago grasslands were much more prevalent.

The records also show that water was more abundant, even in the recent past. Whether we ascribe this to climatic changes (Deming, 1962) or to changes in range conditions and the consequent lowering of the water table, the result is the same: less favorable forage and less water on most desert bighorn ranges today.

The problem of how to improve and protect the habitat of desert bighorn falls into the following categories:

1. Improve the forage conditions of the range.
2. Decrease the competitive utilization and/or abuse of the range.
3. Improve and increase drinking water for the bighorn.

IMPROVEMENT OF FORAGE CONDITIONS

Natural successional changes in plant communities are familiar to most persons. For example, a fallow field will be invaded by weedy species, which may later be replaced by shrubs, which will in turn be replaced by trees. The so-called climax vegetation of a given area depends largely upon climatic and edaphic (soils) factors. Thus, on soils that will support shrubs and trees, shrubs tend to replace grassland communities, and trees tend to replace shrub communities. One seral stage of vegetation may remain for an indefinite period of time if the conditions that perpetuate it persist. For example, periodic fires are known to perpetuate grassland communities by periodically killing invading species of woody plants. Other factors affecting natural succession are flood action, climatic change, overgrazing by wildlife or livestock, and, of course, activities of man.

Charles G. Hansen found up to 44 percent of all bighorn remains in burned areas on the Sheep Range mountains of Desert National Wildlife Range in Nevada. The burned areas were in various associations of pinyon-juniper woodland. This is a significant percentage when one considers that the burned areas encompass only a small part of the range. The burns, however, are primarily grassland, and the indication is that bighorn prefer these because of the good grazing (also see Chapter 6).

The Sheep Range contains some of the best bighorn habitat in the Southwest. However, the entire eastern slope of this mountain range has been taken over by juniper *(Juniperus* sp.*)* and pinyon *(Pinus* sp.*)* trees. In some areas, the growth is so dense as to preclude the production of even a shrub understory that might provide forage for bighorn. It is in these dense stands that some of the previously mentioned burns are located.

Grasslands can be increased at the expense of trees and shrubs by burning tracts under controlled conditions or by mechanical removal of trees by chaining. Burning would be the least expensive and most expedient method of establishing a grassland succession on the Sheep Range. The bighorn habitat lends itself well to this, in that there are sufficient breaks in the vegetation to control fire and limit it to desired areas. The entire range is removed from areas of potential damage, such as extensive timber tracts.

The biggest problem with burning is not its feasibility but rather a matter of overcoming prejudices. So ingrained has become the belief as to the danger of fire in the forests that proposals for prescribed burning often meet with resistance if not outrage. Stoddard (1931) points out the advantage of fire for management of wildlife habitat. Burning has been used as a management tool in the Southeast for decades, and its use is spreading to other regions. Perkins (1967), among others, summarizes the problems and advantages involved. Wildlife biologists need to consider range improvement in areas where changes in plant succession would be favorable for bighorn and other wildlife. If prescribed burning appears to be warranted, it should be used.

Call (1966) describes a plan for clearing 2,800 to 3,200 hectares (7,000 to 8,000 acres) of bighorn range in Utah having a plant association similar to that on the Sheep Range in Nevada. The plan encompasses "chaining" the area, using large bulldozers to draw a heavy chain between them. This uproots the large trees and can be an effective method where it is feasible. Chaining usually is followed by burning of brush piles and reseeding. In most instances standing, incompletely burned trees are not desirable, but unburned downed trees or brush may provide additional wildlife cover. In addition, downed trees also may protect young seedlings from frost, help to retain moisture, and check erosion. Removal of trees with heavy machinery is effective but costly; it is not feasible on rough terrain.

Herbicides also may be used to kill shrubs or trees, but we do not know the side effects of many of these chemicals; what we have learned has raised grave questions about environmental damage. The possibility of chemical breakdown resulting in secondary, more harmful chemicals (Carson, 1962) always exists. We need a much more careful evaluation of herbicides from the standpoint of primary and secondary toxicity to animals.

Where the rigorous desert climate permits, cleared areas may be seeded with a cover crop to assure ground cover in the shortest time possible. Annual grasses probably provide the quickest means to this end and also permit the native perennials to take over eventually. Native plants usually are present in the form of dormant seeds, or in small communities, and soon will dominate unless soils have been drastically altered. Native species usually are better suited for survival, having developed under the local climatic and edaphic conditions.

All vegetative type conversion projects should be planned in conformance with the basic principles for successful game range restoration identified by Plummer et al. (1968). These procedures have wide application on similar sites in the Southwest. They are referred to as "the ten commandments for success" of game ranges:

1. Alteration of plant cover, by whatever means, must first be determined to be desirable.
2. Terrain and soil types must be suited to the changes selected.

3. Precipitation must be adequate to assure establishment and survival of seeded plants.
4. Competition must be low enough to assure that desired species of plants can be established.
5. Only species and strains of plants adapted to the area should be planted.
6. Mixtures, rather than single species, should be planted.
7. Sufficient seed of acceptable purity and viability should be planted to assure getting a stand.
8. Seed must be covered sufficiently.
9. Planting should be done in the season of optimum conditions for establishment.
10. The planted area must be adequately protected.

Wildlife generally uses the ranges without abusing it, whereas livestock frequently overgrazes the forage. Overuse of the range by livestock deprives bighorn of potential forage; in addition, the highly selective use of plants by livestock may hasten the return of undesirable species previously cleared from the range. Many examples can be found of cattle and other domestic stock suppressing grasses by overgrazing, while undesirable species flourish because they are not grazed or are grazed very lightly. In the interest of the general public, it does not seem unreasonable to insist on the removal of the livestock from that part of the bighorn range that will be adversely affected by its presence.

COMPETITIVE UTILIZATION AND ABUSE

Competition is a major factor in the survival of a species living precariously under marginal conditions. Banko (1963), in discussing the Wynne-Edwards theory in relation to desert sheep, emphasizes that under natural conditions there is a balance between the organism and the environment. Much of the environment in which desert sheep live today is one grossly unbalanced as a result of man's influence. Man, the causative factor of this unbalanced condition, must assume the responsibility to correct the imbalance.

Little can be done to change conditions of the range that has deteriorated because of decreased rainfall. However, when climatic conditions for plant growth are below optimum, we might be able to improve matters by reducing other stresses on the plants, such as grazing pressure. Each species of plant can withstand a limited amount of grazing or browsing, but excessive cropping causes its decline and eventual death. In some cases, a species of plant may be replaced by another species which may be more hardy, or less palatable, or not palatable at all. The replacement species may not be utilized to the same extent as the original plants, if at all, by range animals.

Detrimental human activities include the introduction of exotic animal species. When a foreign species is introduced, there is the possibility of introducing a disease against which the native species has no immunity. The results could be disastrous. The possible consequences have been recognized by Lee (1960) and Graf (1963). The main argument against such an introduction is that there is no justification for introducing a "biological equivalent" into a habitat. The native species, by virtue of its evolutionary history, may not always have the advantage in a contest for resources. As long as the native species exists, every effort should be made to preserve it and to increase the size of its population. Feral grazers and browsers such as burros and goats are not part of the natural ecosystem and should be completely removed from bighorn ranges, or very closely controlled (Gallizioli, 1977). Where the native species is declining, an effort must first be made to determine the cause of the decline, for the same factors will quite likely affect any other species proposed for the same environment. Failure to do this has been the single biggest cause of the waste of millions of dollars in futile attempts to introduce exotics. The loss may be not only in dollars but also in terms of the native species, which must compete with the exotic.

Not all competition is attributable to livestock and other forms of wildlife. Man and his activities also can be harmful to bighorn populations and to the range. Generally, the mere presence of people on a bighorn range, including the normal activities of hikers, wilderness photographers, range survey crews, etc., will cause no great harm to bighorn populations. However, excessive use by people can result in undesirable consequences and become actual harassment. Excessive human use and abuse of privileges on bighorn ranges should be eliminated. Desert ranges having few water sources are especially fragile. Thoughtless campers can monopolize a spring and deprive bighorn of water, which they desperately need for survival. Mining, prospecting, and similar activities of man may deprive bighorn of vital water by diverting it for human use. Such activities in remote areas can also lead to persistent poaching.

Military use of public lands also affects some wild areas. Large tracts of land have been taken over for military training and in some instances are used for bombing practice. The result of these activities is to close off extensive tracts of bighorn range to biologists and others who are responsible for, or are interested in, bighorn sheep. Wildlife ranges and refuges should not be taken over for such purposes because other large tracts of land are available which are not vital to the welfare of endangered species. All such undesirable uses should be stopped entirely or sufficiently curtailed so that their effects are minimized.

Among suggestions to govern the effects of human operations on the value of habitat to desert bighorn are the following:

1. Mineral exploration must be rigidly regulated to minimize habitat destruction and insure adequate rehabilitation. Prior to developing a mine on

desert bighorn habitat, a developmental and operational plan should be filed and approved by the appropriate land management agency. Water sources must be neither disturbed nor usurped by mineral interests.

2. Timber harvests should be done at a time of year when desert bighorn are absent from the range. It should be accomplished in a manner not detrimental to bighorn (Deforge, 1972). Logging roads and skid trails should be reseeded with a proper vegetational mix. Selected logging roads should be closed whenever feasible.

3. Activities such as hiking, camping, rock-climbing, horseback riding, picnicking, sightseeing, and other recreational pursuits (see Chapter 19) can be detrimental to desert bighorn habitat and may require restrictions. Restrictions may include but are not limited to no camping within 400 m (¼ mi) of a water source, no new hiking trails built through desert bighorn habitat, existing trails rerouted away from critical areas, and the number of people allowed in key areas limited either on a seasonal or permanent basis. Key areas should be closed to off-road vehicles, and vehicular traffic should be limited to designated areas.

4. Highways, power lines, repeater sites and their access routes, subdivisions, aerial tramways, trespass use of public land, dams, reservoirs, canals, aqueducts, and associated construction and maintenance activities should be kept to a minimum in desert bighorn habitat.

5. Rights-of-way and livestock fencing should be constructed according to specifications developed by Helvie (1971). Highways, canals, and other arteries should be routed through tunnels to allow safe passage of bighorn over the structure. Undercrossings should be at least 7.5 m (25 ft) wide, 4.6 m (15 ft) high, and 17 m (55 ft) long and should be on known bighorn routes. New roads should not be developed in any occupied desert bighorn habitat area or areas designated for transplant.

6. All lands, together with waters, that are habitat for desert bighorn should be obtained in solid ownership and/or management blocks for the primary purpose of perpetuating desert bighorn. Habitat should be segregated against all forms of land disposal. Land use planning and zoning should be encouraged for desert bighorn habitat and maintenance.

WATER IMPROVEMENTS AND INCREASE

Water is of course one of the most important factors in maintaining our desert bighorn population. No range development will be of any value if water is not present or cannot be made available. Development and protection of the range can succeed only with complementary development and protection of the water. Methods of developing water supplies and various types of water catchments for bighorn are discussed by Weaver (1958), Kennedy (1958), Schadle (1958), Halloran and Deming (1956, 1958), and Duncan (1963), as well as in Chapter 7. Details concerning each type of water development, and methods of improving and/or construction are unnecessary here. We can discuss only in a general way the problems attending some of these facilities and methods improvement, which involve two basic types of water supply, natural and artificial.

Natural water sources include springs, seeps, and natural storage basins or tanks (tinajas). Tanks are natural rock depressions that hold rainwater for part or all of a season. Locating natural water sources in rough country is difficult; Weaver (1958) suggests using a light plane (or helicopter) to speed up this work. From the air it is often easy to detect a spot of green and converging game trails to a water source. Even the sources that do not provide free water without development may be detected in this way. A bit of greenery in an otherwise dry, brown, and gray country often will stand out and indicate water—if not at the surface, then close below.

Weaver also points out that certain plants may be used as water indicators. He cites Department of Interior, U.S. Geological Survey Water Supply Paper No. 577, "Plants as Indicators of Ground Water," by Oscar Edward Meinzer, as well as additional U.S. Geological Survey Papers Nos. 224, 497, 499, and 578.

If the permanent water supply is not too far below the surface, it may be practical to develop it. Here again, the determination must be made by the local workers. Weaver considers 3 m (10 ft) to be about the practical maximum depth for such development. Seeps that may not be producing enough water for wildlife often can be developed by digging out, impounding, or shading to reduce evaporation. Flow also may be increased by reducing the vegetation around a limited water source. Since plants may use all the available water at a small spring, their girdling or removal may increase the flow.

Kennedy (1958) discusses the clogging and filling of natural tanks by sand and rock carried into them during heavy rains. The problem was solved in some cases by building a dam immediately above the tank, so that the fall of the water would clean the tank of debris. In other cases a bypass was constructed which carried the flood water and excess debris away from the tank during floods. Problems of this type will tax the ingenuity of the worker, and each site may call for a different appraisal and solution.

Protecting developed springs from flood damage is a problem; Weaver (1958) suggests burying a short length of perforated "Orangeburg" pipe at the source, covering it with packed gravel, and then piping the water from the collecting point to a basin or trough.

Where water occurs in deep sumps, whether natural or artificial, an access ramp may have to be constructed for wildlife. This is not only for the use of bighorn but also for birds and small mammals that use the water. Failure to do so not only will limit the use of the water but will result in its contamination by animals, including bighorn, that fall into such tanks and drown.

Charles G. Hansen

Fig. 20.1. Bighorn drinking at water collected from a developed spring on Desert National Wildlife Range, Nevada.

Artificial water sources consist of watering devices constructed by man. Examples of these are:

1. Artificial sumps cut into rock or other water-holding surface.
2. Dams to provide storage of water.
3. Artificial collection devices and storage tanks, commonly called "guzzlers" (Fig. 20.1).
4. Wells drilled for water for wildlife. Such wells usually would have to be powered by windmills.
5. Pipelines and canals.

As is the case with natural water sources, evaporation can result in a major loss of a meager water supply in the hot southwestern desert region. Where a tank, natural or artificial, is exposed to the sun, it can profitably be shaded. This may save enough water to carry wildlife through the critical time of the year.

The construction of artificial tanks (sumps) and dams is difficult and costly in the rough terrain occupied by desert bighorn. Duncan (1963) discusses some of the problems, equipment, and costs of the construction of several such catchments in Arizona. Only when the site can be reached by truck or helicopter for the transportation of air compressors and other mechanical equipment do the problem and the work become somewhat easier. Elsewhere, equipment must often be packed in on horses or mules, or even on the workers' backs. The work then must be accomplished the hard way, with hand tools. Under these extreme and difficult conditions, tank construction is a testimonial to the dedication of wildlife managers and of volunteer groups interested in bighorn welfare.

In some areas the "guzzler" type of water collection device has proven highly successful. It is best known as a quail or dove watering catchment. However, such watering devices can be built as large as facilities and finances permit. On the island of Hawaii, giant "guzzlers" having 210-sq-km (52-acre) collecting aprons of asphalt store more than 2 million gallons of water. Here again the primary problem is one of transporting the materials and equipment into the area and in constructing these catchments under adverse conditions.

The efficiency and dependability of modern helicopters make the transportation of such materials, equipment, and men into remote, rough country increasingly practical. Sometimes cooperative arrangements make possible the use of military helicopters to transport materials and equipment into areas where catchments are to be constructed. Another alternative is to drop the equipment and materials by cargo parachute. Such cooperation has been given in the past, with good results. In most cases, the military services have been glad to help with such projects whenever they could. The transportation problems and logistics are not unlike those of military operations and thus provide excellent training opportunities.

A major problem that can result from the establishment of a water supply is the competitive use of forage by livestock that may be attracted to the new supply. If the land is public domain, it should be possible to regulate livestock grazing to prevent overuse of the range. Gerald I. Day reports that "catchments built in Arizona with sportsmen's funds are fenced to keep livestock out. Some springs have also been fenced." However, such fences must be so constructed that bighorn are not kept from the water.

STEPS TO IMPROVE AND PROTECT BIGHORN RANGES

In summary, the following steps should be taken to improve and protect our desert bighorn ranges:

1. Initiate a program of evaluation of the various bighorn ranges to determine the type of range improvement needed and what type of improvement practices are feasible.
2. Where the program of evaluation reveals feasible opportunities for improvement, carry out the improvments vigorously.
3. In the case of certain ranges with characteristics apparently favorable for the experiment, initiate controlled burning as a pilot program to determine its practicality. Other range improvement methods such as mechanical removal of undesirable vegetation might be developed.
4. Undertake a program of water improvement, including development of new water sources and construction of storage catchments.
5. Study the economics of livestock grazing on the marginal desert ranges of the public lands to determine whether continuing this practice is justified under the land-depleting conditions that often are apparent.

The responsibility for developing a program of protection and improvement for the desert bighorn ranges, and, for that matter, all wildlife ranges, is vested with the administrators of the various wildlife and land agencies. The protection and improvement of the land for preservation of the wildlife will in the long run be for our own preservation and survival as well.

21. Habitat Evaluation

Charles G. Hansen

The following habitat evaluation system for desert bighorn was developed at Desert National Wildlife Range, Nevada. The method uses habitat characteristics, as well as values of the land to man; it is designed to assist land managers in determining which parcels of land are necessary for bighorn and which can be used for other wildlife programs. The system draws heavily upon the methods and nomenclature presented by Rose and Morgan (1964).

FACTORS FOR RATING BIGHORN HABITAT

The rating of desert bighorn habitat is based on a series of components, which in combination directly influence the number of bighorn living in an area. Basically, these components are cover, food, and water. Two other factors that influence use of an area by bighorn are weather and the activities of man. In evaluating habitat for bighorn, it is necessary to consider why bighorn use the area. For example, if the use is for watering, the area may be essential because of the limited amount or availability of water in the general area. However, if the use is for cover, the area may not be essential if adjacent areas of rough terrain afford bighorn cover.

An attempt was made to make this system practical through use of standard maps and data that were readily available. For example, preliminary evaluation or classification of habitat can be made by using aerial photographs, a contour map that includes man-made features, standard weather records, and land use data from agencies officially involved with the land. Livestock range surveys by conservation or land management agencies can be used when available. Aerial surveys are of great assistance, but on-the-ground surveys of water holes and vegetation increase the usefulness of the evaluation, especially if grazing or mining activities have been or may be involved. Where a critical evaluation is needed, an extensive study of the area should be made so that all aspects can be taken into account.

The various components or factors of the habitat have been separated into categories called "Tools" (Rose and Morgan, 1964) and assigned appropriate titles and values (points). These are arranged as follows:

I	Natural topography	0 to 20 points
II	Vegetation type	0 to 20 points
III	Precipitation	1 to 5 points
IV	Evaporation	1 to 5 points
V	Water, type and use	2 to 20 and up to 40 points
VI	Bighorn use	2 to 20 points
VII	Human use	0 to 20 points

Each Tool was assigned a numerical value in relation to its influence on bighorn use, based on knowledge of the animals on the part of Wildlife Range personnel and others. The maximum number of points normally available to each section (= 260 ha or 640 acres) of land is 110. However, since water is so vital to desert bighorn in most of their range, a section of land having a water hole in it can bring its total score up to 130 points with 20 points added for the water hole. Areas with abundant water, such as the White Mountains of California and Nevada, will not require the addition of extra points.

The scoring system follows the classification given in Table 21.1. Each Tool is composed of elements that affect the presence of bighorn. These elements have been rated by a point system. Each section of land that is classified is assigned a score, which is obtained by determining the appropriate element for each Tool, then adding up the numbers of points for all the Tools for a total value or score for each section. For example: Section 18, Township 11 South, Range 58 East, in Desert National Wildlife Range was assigned the following points for the seven Tools: I, 16; II, 16; III, 4; IV, 3; V, 10; VI, 8; VII, 20. These seven individual scores total 77 points for this section. These total scores when compared to major categories (listed in Table 21.2) classify this section of land in relation to its importance to bighorn.

TABLE 21.1.

Categories and Values of the Tools Used to Evaluate Desert Bighorn Habitat

Tool I: Natural topography

- 0 Level or undulating
- 4 Level, within 1.6 km (1 mi) of steep and rocky terrain
- 8 Rolling hills
- 12 (a) Steep and rocky (100%)
  - (b) Rolling hills near steep and rocky terrain
  - (c) Mesa-type terrain
- 16 Steep and rocky interspersed with level stretches or rolling hills
- 20 Steep and rocky, broken with dry drainageways

Tool II: Vegetation

- 0 Dry lake beds, blue clay, or slick rock
- 4 Pinyon-juniper woodland or coniferous forest
- 6 Salt bush and/or other low desert shrubs
- 8 Yucca, Joshua tree, creosote bush, and other desert shrubs
- 10 Upper Joshua tree and pinyon-juniper, or blackbrush community
- 16 Grass-desert shrub or grass-coniferous forest
- 18 Grass-pinyon-juniper woodland
- 20 Grass-transition between desert shrub and pinyon-juniper

Tool III: Annual precipitation

- 1 5 to 7.5 cm (2 to 3 in.), thundershowers
- 2 7.5 to 20 cm (3 to 8 in.), thundershowers
- 3 20 to 30 cm (8 to 12 in.), winter mainly
- 4 7.5 to 20 cm (3 to 8 in.), half in winter, half in summer
- 5 20 to 50 cm (8 to 20 in.), half in winter, half in summer

Tool IV: Evaporation per week

- 1 13 or more cm (5 or more in.)
- 2 9 or more cm (4 in.)
- 3 7.5 or more cm (3 in.)
- 4 5 or more cm (2 in.)
- 5 2.5 cm (1 in.) or less

Tool V: Water sources (type and use, restrictions)
Drinking water (amount and permanence)

- 0 Wet seep
- 2 Often dry
- 3 Sometimes dry
- 4 Seldom dry
- 5 Sufficient

Terrain and obstructions (i.e., fences) to bighorn

- 1 Flat, with obstructions
- 2 Rolling hills with obstructions
- 3 Broken terrain with natural obstructions
- 4 Rocky terrain with natural obstructions
- 5 Rough and no obstructions

TABLE 21.1.
(Continued)

Competition with other animals

- 0 Much livestock
- 2 Some livestock
- 5 Other native big game
- 7 Seldom other native big game
- 10 Only bighorn

Tool VI: Desert bighorn use

- 2 Irregular transient
- 4 Rams feeding, or infrequent
- 6 Fall and/or spring transient
- 8 Feeding areas in fall and winter
- 9 Water holes in summer
- 10 Food and cover for ewes and lambs in spring
- 20 Major crossings for water or to feed

Tool VII: Human use

- 0 High density and high economic potential
- 4 Medium density, medium economic potential; unrestricted human use
- 7 Medium density, restricted human use; economic use limited for the sake of general recreation (parks, refuges, etc.)
- 7 High density, restricted human use (wildlife parks, refuges, etc.)
- 10 Medium density, unrestricted human use; and low economic potential
- 10 Human and economic use planned with wildlife in mind
- 15 Medium density and economic potential but restricted human use (bighorn parks, refuges, etc.)
- 15 Low density and low or no economic potential, unrestricted human use
- 20 Relatively no human or economic use
- 20 Planned development for bighorn with human use where and when consistent with primary objective

In the case of the above-mentioned section of land with a score of 77, the classification according to Table 21.2 is that of Periodic Use or Zone of Deficiency, or Area of Potential Economic Value or Occasional Human Use. To determine which of the five classifications this section falls into, it is necessary to look back at the points assigned to each Tool. The Tool for water was scored relatively low; therefore, it is primarily a Zone of Deficiency. Since the deficiency is water, it would be desirable to improve the water situation in or near this section so that it will be more usable for bighorn. If this improvement is made, it will then be possible to reclassify the section as Important or even Vital. Two scores can be applied to each section when changes through management are possible. For example, the Tool score in

TABLE 21.2.

Classification of Total Scores for One-mile-square Sections of Land Evaluated as Desert Bighorn Habitat, Desert National Wildlife Range, Nevada

Total Score	Classification
0 to 50	*Not important to bighorn, or of high value for human use.* Sections that fall within this rating and have a score for Tool I of 8 or less are not important to bighorn. Sections that have a score of 12 or more for Tool I may be in a zone of deficiency because of an inadequate water supply or shortage of food.
51 to 64	*Buffer zone or zone of deficiency for bighorn; or area of potential economic value or of moderate human use.* Sections that fall within this rating are ones which should be retained at least in their current status as buffers against further human encroachment. However, some of these sections may be improved upon for bighorn use in various ways, including habitat manipulation or change in human use.
65 to 79	*Periodic use or zone of deficiency for bighorn; or area of potential economic value or for occasional human use.* Sections that are within this category would be more valuable to bighorn if some segment of the habitat were improved for them or if the economic potential or human use were restricted or eliminated. However, there may be some feature of the terrain or vegetation in this area that is necessary for survival of bighorn, even though it may be used only periodically. Therefore, sections in this category should be critically examined before being removed from potentially important bighorn habitat.
80 to 100	*Important to bighorn.* Sections in this category are important to bighorn. The importance may derive from items essential to the animals, or from lack of human use or economic potential. Generally, sections in this category are in rough, mountainous terrain, or they are areas that are major crossings to summer and winter ranges or to water holes.
101 and above	*Vital to bighorn.* Sections that are vital to bighorn are those that have some feature without which bighorn cannot survive. On Desert National Wildlife Range, the only sections in this category are those with water holes.

Section 18 above was 77; a habitat improvement score may raise it 15 points for potential water development (or, in some cases, for the removal of livestock, controlled human use, etc.), so that the Potential Tool Score will then be 92.

This example illustrates how the system not only classifies the land but also can be used to determine where habitat management can be applied most effectively in order to improve the habitat for the bighorn population.

Explanation of tools. The Tools listed in Table 21.1 were developed, first, by determining the features or components of the habitat that are used by desert bighorn and, second, by determining the importance or preference of each component. When the importance of each component was established, a comparative numerical value was assigned to it. Thus, when an area was analyzed, the value or values assigned became the points scored by each Tool. These are listed as follows, with an explanation of how the value or score was determined:

Tool I: Natural topography

The sources of information for this Tool were aerial photographs and two U.S. Geological Survey topographic maps: the Las Vegas (1959-NS, MR5913) and the Caliente (1959-NS, MR5193) quadrangles.

Value	Description
0	Level or slightly undulating, 100% (example: dry lake beds and their margins, blue clay, or slick rock); more than 1.6 km (1 mi) from steep and rocky terrain.
4	Level or slightly undulating, 100%; within 1.6 km (1 mi) of steep and rocky terrain.
8	Rolling hills, such as alluvial fans, without washes over 4.6 m (15 ft) wide and/or more than 1.6 km (1 mi) from steep and rocky terrain.
12a.	Steep and rocky, 100%; no washes.
b.	Rolling hills broken frequently by broad washes and within 1.6 km (1 mi) of steep and rocky terrain.
c.	Mesa-type terrain.
16	Steep and rocky terrain with washes, 50 to 90%; plus level or rolling hills, 10 to 50%.
20	Steep and rocky terrain, broken frequently by washes of varying widths, with at least one main wash about 15 m (50 ft) wide, and side washes at various angles for protection from the weather and for escape.

Tool II: Vegetation Type

The vegetation types follow Bradley (1964, 1965) and Bradley and Deacon (1965). The percentage of grass in the various vegetation types is important in the classification of the habitat. Grass is specified in the evaluation of the last three vegetation types because the amount is often limited, whereas browse is relatively abundant. The forbs are dependent upon climatic conditions, and so may be abundant one year but absent the next. Since browse is relatively abundant and forbs are not dependable, they are not considered a suitable guide to the requirements of the bighorn on Desert Wildlife Range.

The values for the grass categories are to be reduced according to the amount of grass that is present or available, but the values will not be less than the value of the vegetation type of that section of land (for example: grass in pinyon-juniper woodland cannot be less than 4, the score for pinyon-juniper woodland). For future reference, include the average elevation for each section with each score for this Tool.

Value	Description
0	Dry lake beds or playas, blue clay, or slick rock.
4	Pinyon-juniper woodland; or coniferous forest including ponderosa pine, white fir, and/or bristlecone pine. Less than 1% grass in this vegetation type.
6	Lower desert shrub, including salt bush community.
8	Middle desert shrub, including yucca, Joshua tree, and/or creosote bush with less than 1% grass.
10	Blackbrush community of the desert shrub vegetation type, including some yucca or Joshua trees and/or pinyon-juniper woodlands with less than 1% grass.
16	Grass, 5% of the cover and 100% available, in the lower desert shrub or the coniferous forest.
18	Grass, 10% of the cover and 100% available, in pinyon-juniper woodland; an area of 81 ha (200 acres) or more that has burned over within the last 20 years will receive the full 18-point value.
20	Grass, 20% of the cover and 100% available, in the blackbrush community; an area of 81 ha (200 acres) or more that has burned over within the last 20 years will receive the full 20-point value.

Tool III: Precipitation

The sources of information are U.S. Department of Commerce climatological data and storage gage precipitation data.

	Amount	Type and Time of Year
Value	Annual total cm	
1	5 to 7.5	Primarily showers; light winter showers or thundershowers.
2	7.5 to 20	Winter showers primarily.
3	20 to 30	Winter showers primarily.
4	10 to 20	Showers or general storms, about half in winter and half in summer.
5	20 to 30	Showers or general storms, about half in winter and half in summer.

TABLE 21.3.
Tool IV: Evaporation

Value	Loss in cm* from a Standard 122-cm-diameter Evaporation Pan	Humidity, %	Wind	
			Velocity	Frequency
1	More than 12	10 or less	Moderate	Daily
			Strong	Occasionally
2	10 to 12	11 to 17	Moderate	Daily
			Strong	Frequently
3	Approx. 7.5	Approx. 20	Gentle	Daily
			Moderate	Frequently
			Strong	Seldom
4	Approx. 5	Approx. 23 to 27	Gentle	Daily
			Moderate	Frequently
			Strong	Seldom
5	2.5 or less	30 or more	Any velocity	Occasionally

Source: Records on file at Desert National Wildlife Range; and Climatological Data Annual Summaries by the U.S. Department of Commerce.
*Weekly average during summer months when temperatures are 32°C (90°F) or more.

Tool IV: Evaporation

The application of this Tool is outlined in Table 21.3.

Tool V: Water Sources, Type, and Use

The sources of information for the categories and values of this Tool include visual examination of water holes, and the findings of Pulling (1946), Halloran and Deming (1958), Denniston (1965), McMichael (1964), Crow (1964), Reffalt (1963), Bradley (1963), Weaver (1958), Welles and Welles (1961a), St. John, Jr. (1965), and Hansen (1965b).

The individual values for each of the three parts below are to be added together in order to get a total value for Tool V.

Only land sections within 9.6 km (6 mi) of a water source are to be given the value as listed below. Sections between 9.6 to 19 km (6 to 12 mi) from a water source are to receive only 50 percent of the value assigned to the nearest water source. Any section that is 21 or more km (13 or more mi) from the nearest water source will receive a value of zero for this Tool.

Any section that has an available water source within it will receive twice the total value as scored for this Tool.

When two or more water sources occur with 9.6 km (6 mi) of a section, the maximum points available are 20. These 20 points are assigned only when each water source has a value of 10 or more points. Otherwise the section will receive only the value of the water source with the highest value.

For future reference, the points scored for each of the three parts of this Tool should be listed with the numerical description for each section of land classified.

Value	Amount and Permanence
1	Water present irregularly, mainly in winter.
2	Often dry when needed in summer during dry years.
3	Dry half the time when needed during dry summers.
4	Seldom dry during the summer.
5	Sufficient and always present.

Value	Type of Terrain and Obstructions
1	Flat land; water surrounded by fences or other barriers; or steepsided dam or pothole.
2	Open rolling hills, surrounded by fences or other barriers that are passable by bighorn; or .8 km (.5 mi) or more from steep or rocky terrain.
3	Rolling hills with timber or other natural or minor obstruction to vision.
4	Steep and rocky but with some timber; or natural, or minor obstruction.
5	Open steep and rocky terrain with a clear view for at least 45 m (150 ft).

Value	Competition
0	Frequent livestock use.
2	Some domestic livestock use and some native or feral animal use.
5	More use by deer or other big game than by desert bighorn.
7	Some native big game use other than desert bighorn, but mostly bighorn use.
10	Only desert bighorn use.

Tool VI: Desert Bighorn Use

The sources of information for these categories and values include personal observations, and the findings of Pulling (1946), Hansen (1965a and 1965b), Denniston (1965), Welles and Welles (1961a), and Welles (1961); plus miscellaneous notes and records in the files of Desert Wildlife Range.

Desert bighorn use may be actual or expected, as determined by a study of the habitat and knowledge of the habits or requirements of bighorn. Development of watering facilities or habitat manipulation probably will alter the values originally assigned to a section.

Crossings between mountain ranges or from one area to another are vitally important because they are usually ancestral routes between summer and winter ranges, or paths taken during times when water or food are in short supply. An entire range can be left unused when such routes are destroyed (Geist, 1967); consequently, this latter category has the highest value in the Tool.

Value	Types of Bighorn Use	Time of Year
2	Transient	Irregular
4	Rams' bachelor quarters, or infrequent use by either sex	Winter, spring, and/or early summer
6	Transient	Fall and/or spring
8	Feeding areas	Fall and winter
9	Water sources	Summer
10	Food and cover for ewes, lambs, and yearlings	Spring
20	Major crossing for bighorn between summer and winter ranges, or for food and water during other seasons, or during years of shortages	

Tool VII: Human Use

The sources of information for these categories and the values for this Tool are from Van den Akker (1960), Duncan (1960), Welles and Welles (1961a), Welles (1961), St. John, Jr. (1965), Tevis (1959), Grater (1959), McMichael (1964), Denniston (1965), and Light et al. (1966, 1967), as well as notes and records in Desert National Wildlife Range files.

The "Class" designations were developed by combining the Bureau of Land Management, U.S. Fish and Wildlife Service, and National Park Service classifications for recreation and general land use and values.

Human use includes buildings, roads, recreation, domestic livestock grazing, prospecting, etc.

Economic use or *Economic potential* refers to mining (oil or mineral), industrial or commercial (including urban) development, farming or ranching, etc.

High density human use refers to urban areas, roads or recreation areas used by hundreds of people each week, or concentrated economic development with a constant use by a few people, such as ore trucks moving many times a day.

Medium density human use refers to recreation areas or a roadway which perhaps only about 500 people use each year. Also included would be small-scale mining, grazing, or other commercial uses.

Low density human use refers to recreation areas or a roadway which less than a hundred people use each year, or where occasional prospecting, grazing, etc., may occur.

High, medium, or low economic use or *potential* refers to land values which would support *High, Medium,* or *Low density human use,* respectively.

Relatively no human use and no economic potential (other than from bighorn) refers to human use for only the basic management needs of bighorn, the habitat, or wildlife in general. There may be only 2 or 3 visitor-days a year by people other than wildlife managers. These two or three visits may be for any of the uses mentioned previously. (The term "No economic potential" is a little ambiguous, because two or three visitors may spend as much as $1000 in order to enjoy the beauty of the scenery and animal life on a particular section of land. However, this type of economic value is dependent upon leaving the animals and habitat in a natural state and so is actually beneficial to the bighorn. Therefore, it is not considered an economic potential of the land for the purposes of this classification system.)

Restricted use refers to parks, refuges, or public or private lands where the entry or activities of people are limited by regulations favoring wildlife.

The background data for the Human Use Tool can be acquired from local, county, or state real property offices, and arranged in the above categories. Points can be assigned as follows:

Points	Class	Description of Density and Utilization
0	I	High density human use and/or economic potential.
4	II	Medium to low density human use and/or economic potential, unrestricted.
7	III	Medium density human use and/or economic potential with some restrictions.
7	IV	High density human use restricted, and medium economic potential, all with some emphasis on bighorn.
10	V	Medium density human use restricted, and low or no economic potential.
10	VI	Planned development for wildlife with some unrestricted human use and with some degree of economic potential or value.
15	VII	Medium density human use with restrictions and no economic potential.
15	VIII	Low density human use restricted, and low or no economic potential.
20	IX	Relatively no human use and no economic potential.
20	X	Planned development for bighorn, with human use where and when consistent with primary objective.

On Desert National Wildlife Range, the largest population of bighorn is in the area most frequently studied, and therefore the best known. This portion of the Wildlife Range was analyzed first in order to determine the value of the system, the correctness of the scores, and the practicality of the method. The preliminary analysis, with only minor changes, led to a map which adequately showed the important areas and the areas needed by bighorn. Also included in the preliminary analysis was a land area not frequented by bighorn because some sections were a long way from water, while some contained dry lake beds or rolling hills far from suitable bighorn habitat. In addition, there were areas with varying degrees of human or economic use.

The finished product was a map of thirty-five townships, a portion of which appears as Figure 21.1, showing the final classification of four townships of land in the southern part of the Wildlife Range. The system was tested on 210 square-mile sections (totaling 54,400 ha or 134,400 acres) and was found to provide an accurate evaluation of all types of terrain with various kinds of human use. Figure 21.2 is presented for comparison of actual bighorn use and abundance with the evaluation in Figure 21.1. These two maps show that the qualitative evaluation in Figure 21.1 compares very closely with the actual situation, including provisions for human use.

The written description and the accompanying discussion illustrate how the classification can be used to assign priorities to land areas so that the maximum benefit for desert bighorn can be obtained. This does not preclude multiple use but directs management of each section of land toward its best use for wildlife or humans. Joint occupancy of most areas by bighorn and people is possible when the needs of the bighorn are planned for and provided. For example, in Tool II, the vegetation often can be manipulated to increase the amount and availability of the forage; in Tool V, competition and obstacles around a water source can be decreased, or artificial water sources can be developed where man has taken over the natural sources; and in Tool VI, bighorn use of an area can be increased by elimination of competition from man or other animals, especially in certain important places such as lambing grounds, around bedding grounds, or where water is a limiting factor.

Provisions for joint occupancy by people and desert bighorn in bighorn habitat should be made whenever human use is anticipated. Bighorn can tolerate many types of human activities on a limited scale. Therefore, roads, trails, or lookout points can be provided beyond which such activities are restricted so that bighorn are not continually disturbed on all sides and can learn to expect and accept certain types of human activity.

If primitive conditions are maintained, the number of people or amount of use will be automatically restricted and disturbance of bighorn lessened. Further restrictions can be placed on the modes of travel. In certain

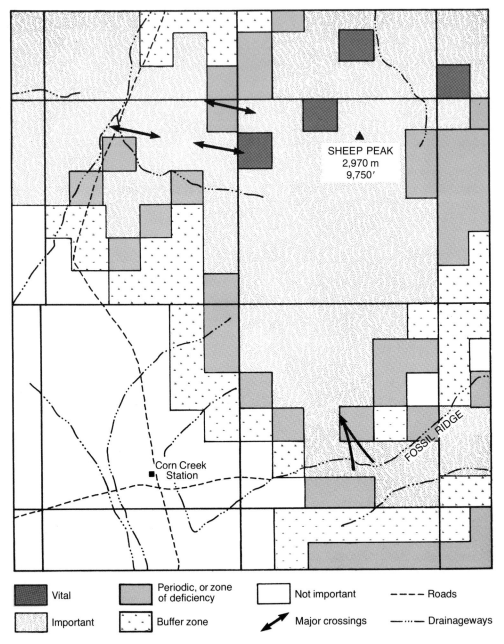

Fig. 21.1. Land sections classified for desert bighorn sheep use (Desert National Wildlife Range).

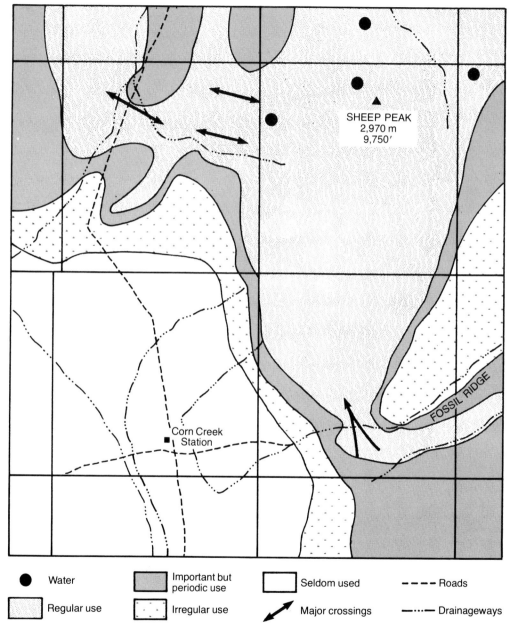

Fig. 21.2. Actual bighorn sheep use of land.

areas, perhaps only foot or horseback travel should be allowed, or some roads maintained in a condition that limits travel only to jeeps or other 4-wheel-drive vehicles. Infrequently used dirt roads are quite acceptable to bighorn, even though the roads traverse areas within their immediate habitat, or terrain used by the animals for crossing between mountain ranges.

If paved or fenced roads are planned, particularly in flat terrain, wide *overpasses* should be provided for bighorn. Parking, picnicking, and even areas developed for economic use should be fenced or otherwise limited in order to keep people within these boundaries. Bighorn are often harassed by low-flying aircraft, unintentionally or not, because of the open terrain where the animals live. Boaters and fishermen along rivers such as the Colorado may also harass bighorn unintentionally, thus keeping them away from their source of drinking water. This could be controlled by the establishment of restricted areas.

The system described provides a classification of land which permits human activity in the areas least important to bighorn but restricts man's activities in areas vital to the animals. Conversely, even artificial water sources or other developments may be provided for bighorn near natural sources in order to attract them away from some human activity that may be in conflict with them. Many areas that are unimportant to bighorn are quite satisfactory for human use; on the other hand, if provisions are not made for bighorn, it is possible to lose a herd from a large area through the destruction or alteration of only one seemingly unimportant part of their habitat. For example, if a water hole or a major crossing is destroyed or lost to them, bighorn may be eliminated from a whole mountain range.

The value of the habitat classification described here is directly related to the amount of time and effort spent in the evaluation of the area. When drastic changes in an area are anticipated, such as recreational or industrial development, a critical examination of the habitat should be made. If it was previously set aside for its wildlife characteristics, action should be taken to provide or maintain the wildlife in the initial plans. Bighorn often will accept changes or artificial situations provided they do not seriously interfere with bighorn behavior and habitat requirements.

This system can also be used to evaluate areas that at present do not have desert bighorn, by changing the title of Tool VI to Anticipated Bighorn Use. Thus, in areas where bighorn transplants are planned, the habitat can be evaluated before any other action is taken.

(Editors' note: The foregoing habitat rating system must be used with some caution, as it may not always be applicable. For instance, as pointed out by McQuivey (1978), a particular mountain range may receive a low rating because it has no water; yet it may receive much use during the winter when bighorn require little, if any, water, and they infiltrate from adjacent mountain areas.)

BIGHORN HABITAT INDEX

Ferrier and Bradley (1970) and Breyen (1971) developed a method of evaluating bighorn habitat. Forage, particularly grass abundance, distance to water, and topography (degree of slope) were evaluated as basic habitat components which were used in formulating a Bighorn Habitat Index (BHI). This index method is an attempt to quantitatively evaluate bighorn habitat. Much valuable information was gathered. However, this system is based on present bighorn use in disturbed areas and assumes that the topography is typical for desert bighorn throughout its range. The idea of using bighorn sign (beds and pellets) to indicate preferred bighorn habitat is questionable, since bighorn usually bed in rough, rocky terrain at night and leave more obvious sign, thus not indicating preferred daytime habitat. Also, the degree of slope as an indicator of bighorn preference is adequate only if the terrain is rocky. Steep slopes of clay or gravel are not particularly good escape cover; therefore, the degree of rockiness needs to be considered.

In the areas studied (Highland and Eldorado ranges, Nevada) bighorn activity was altered by the presence of people, burros, and livestock; therefore, the BHI appears to be a good indicator mainly of present bighorn activity in the particular habitat where the animals were studied.

When evaluating habitat for bighorn it is desirable to compare preferred bighorn habitat with the existing conditions. For example, bighorn prefer (a) steep and rocky terrain broken by broad, relatively level washes or foothills; (b) relatively dense grassland in the transition between open desert scrub and scattered trees; (c) relatively dry climate; (d) small permanent drinking basins with sufficient water in rough, rocky terrain with no obstructions to sight or escape and used primarily by bighorn; and (e) undisturbed travel to forage, escape terrain, and bighorn in adjacent ranges when required. The degree to which the above conditions are met will determine the value of the habitat to bighorn. This may be called an Index or a Classification, but the evaluation should be based on the conditions preferred by bighorn and not on the conditions as they now exist.

22. Hunting

Warren E. Kelly

Early day Indians and first white settlers hunted the desert bighorn for food; however, in recent years bighorn have been pursued as a rare and unique trophy animal. Being a trophy animal has added to the difficulties in developing proper programs for its harvest and management. Each state in the Southwest and Mexico has had its individual problems inaugurating hunting seasons. Controlled desert bighorn hunts have been held in Arizona (since 1953), New Mexico (1954, 1955, and since 1968), Nevada (1952 to 1954 and since 1956), and Mexico (since 1964). Utah conducted its first hunt in 1967 but discontinued hunting in 1972 to resume it in 1975. Bighorn have been protected in California since 1873 and in Texas since 1903.

EARLY HISTORY

Desert bighorn have been harvested by man for many hundreds of years (Chapter 2). The earliest records are found in the Anasazi culture of the plateau country in the Southwest. The Anasazi people, named "the ancient ones" by the Navajos, hunted with darts propelled by atlatls. Petroglyphs in southeastern Utah and southern California portray hunters with atlatls and bighorn. It is unknown what impact, if any, these people with their primitive hunting methods had on bighorn populations. In the sixteenth century the Spanish *conquistadores,* during their exploration of the Southwest and Mexico, came in contact with desert bighorn. The greatest hunting impact on the desert bighorn occurred during the nineteenth century, when the Southwest was settled by the eastern Americans. Common knowledge indicates the peak pressure was between 1850 and 1900. Although illegal hunting still continues to some degree in the Southwest (Jonez, 1959; Welles and Welles, 1961a), it is more prevalent in Mexico (Davila C., 1960; Mendoza, 1976).

Laws to protect wildlife were initiated quite early in the history of some states. Al Jonez writes, "Nevada became a state in 1864, and although some laws to protect game were passed by the territorial government of Nevada, and later by early state legislatures, it would appear they were not effective. Apparently those who wished to hunt did so, and there was little regard as to seasons, limits, or any other legal restrictions. The first relatively comprehensive law for the protection of game, with a penalty for violators, was passed in 1877. This law, however, did not mention big game species. In 1893 a law was passed which did include provisions covering the big game species. However, no limits were imposed and the season was over 3 months long. Very likely each county had options, both as to seasonal length within those prescribed by the state, and to bag limits. In 1895 the season length was changed to September 1 to December 31. In 1917 the season was closed on mountain sheep and it remained closed until modern-day hunts started in 1952. Apparently little attention was paid to this closure, especially in the years prior to 1930. Definite illegal kills were made during these early years and evidence of limited poaching still persists today."

The history of the bighorn in California follows this same pattern. There were serious depredations by the early settlers until the legislature passed laws for total protection of the bighorn. California became a state earlier than the other southwestern states, and an alert legislature recognized the need to preserve this magnificent animal. A law was passed in 1873 giving it complete protection and has remained in effect ever since.

In Arizona and New Mexico, bighorn hunting was illegal from early territorial years, and after both states were admitted to the Union in 1912. In Texas, bighorn hunting was legal in season until 1903. In all these states, however, hunting took place on a considerable scale, legal or illegal, until well into the present century.

Mexico from 1922 to 1964 had, by legislative action, a closed season on its desert bighorn. During this time many quasi-legal permits were issued by local officials. Bighorn also were subjected to additional pressures by both residents and nonresidents who did not go through the formalities of getting permits.

HARVEST OBJECTIVES

Objectives of hunts are twofold and connected. The primary purpose has been and still is to collect biological data on such subjects as disease, parasites, anatomical and physiological conditions, and food habits. Secondly, the hunts make it possible to remove some of the old rams that have reached a maximum age and probably would not live more than a year or two longer, thus making room for a younger increment to the herd. Presumably only these old rams in the last fews years of their lives would be taken. Theoretically this would mean taking rams of 10 to 16 years of age.

The first of these objectives certainly has been realized. Much of this type of data could not be obtained otherwise and has not been available in the past because there has been no legal bighorn hunting since pioneer days. As indicated in other chapters, we have obtained disease and parasite samples, food samples, anatomical data, and data on radiation at or adjacent to atomic test sites; in addition, we have gained considerable knowledge concerning the distribution of the bighorn.

During all seasons of bighorn sheep hunting, the number of hunters has been completely controlled, and we cannot speak of hunter pressure in the same sense in which it applies to other big game species such as deer or elk.

CONSIDERATIONS GOVERNING THE HUNTING SEASONS

Many problems, procedures, and policies had to be decided and evaluated by state game managers and administrators before recommendations could be presented to Fish and Game Commissioners requesting hunting seasons on bighorn. Questions such as time of year to hunt, season length, condition of the animal and the cape, impact of hunting on gravid females, number of nonresident hunters allowed, need for guides, and number of permits needed to be answered. Each state addressed these problems and arrived at solutions that considered local conditions.

REGULATED HUNTING

In Arizona the first two bighorn hunts took place in 1953, one in January, the other in December. The hunts encompassed an area of some 1036 sq km (400 sq mi) in the southwestern portion of the state. Twenty permits were issued for a 10-day season. A legal ram was defined as having at least a three-quarter curl. Except for the first season, all hunts were held in late November or December. Through the first 24 seasons of bighorn hunting, 1,451 hunters harvested 635 rams for a 44 percent hunter success. The number of permits averaged twenty for the first six hunts and seventy for the remainder. The area open to hunting lies chiefly between Phoenix and the Colorado River extending from Lake Mead to the Mexican border.

Nevada's first bighorn hunt was in April of 1952. Portions of Lincoln and Clark counties were opened to hunting. Fifty tags were issued and each hunter was required to have a guide. A legal ram was defined as having a three-quarter curl. With the exception of 1955, there have been bighorn hunts annually. The first 23 bighorn hunting seasons allowed 1,744 hunters to take 573 rams for a hunter success of 33 percent. Most of the hunts were held in late November or early December; however, some seasons were scheduled at other times of the year either to gain access to U.S. Air Force bombing and gunnery ranges or to gain biological data. Areas open to hunting are in the lower one-third of the state, generally south of U.S. Highway 6. Jonez writes of Nevada's hunts, "These seasons have ranged from completely guided hunts to unsupervised hunts. Considerable interest has been generated in Nevada's bighorn sheep populations through these hunts. This intense interest by its sportsmen has undoubtedly been one of the reasons Nevada has attempted to promote the trophy hunting aspects in its recent hunts."

In 1965 the state became concerned about the number of rams under 7 years old that were being taken by hunters. Largely through the efforts of Al Jonez, Charles G. Hansen, and members of the Fraternity of the Desert Bighorn, a new regulation was proposed. The new regulation specified that a legal ram must be 7 years old or score 144 points by Boone and Crockett Club measurements. Hunters were also required to carry a spotting scope with a minimum of 15X magnification and were given an indoctrination session to help familiarize them with the new regulation and the technique of field-judging the size and age of bighorn rams. This first indoctrination session was well received and in future years it became mandatory. See also Chapter 16.

The new regulation and the indoctrination session appear to have helped reduce the young ram harvest but not to the extent hoped for. Tsukamoto (1970) sums it up in his progress report: "After five years of hunting under the 'Trophy Regulation' and with several illegal rams harvested in addition to the unknown number never reported, it has become increasingly apparent that the regulation as a management technique is somewhat less than ideal."

New Mexico's first desert bighorn hunts were in the Big Hatchet Mountains in 1954 and 1955. Hunting was discontinued because of a decline in bighorn numbers. Fish and Game officials believe the decline in bighorn numbers was caused by drought and the fencing of the U.S. border, rather than being related to hunting. In 1968 hunting began on San Andres National Wildlife Refuge and has been continued annually. In the first 10 of these hunting seasons seventy-one hunters harvested fifty-eight rams for a hunter success of 81 percent.

Utah held its first bighorn hunt in 1967 and continued hunting until 1972. In 1972 a telemetric study of this bighorn herd was started. Several adult rams were captured and were radio collared. To insure that none of the collared rams were killed, the hunting season was suspended until the study was completed. Hunting was resumed in 1975. Bighorn hunting in Utah up to 1976 allowed sixty-two hunters to harvest twenty-six rams for a hunter success of 42 percent.

It was believed by some that hunting would have a detrimental effect on bighorn populations. However, this has not been true. The various state wildlife departments involved in bighorn hunting have found no evidence that hunting has had an impact on bighorn populations. It was feared that there would be a change in ram-ewe ratios that would upset the balance of the herds. Russo (1967) maintains there has been no visible change in sex ratios in Arizona and the hunts have had no detrimental effects on the bighorn populations. Data presented by Jonez (1958) and Hansen (1962) indicate the ram-ewe ratios on Desert Wildlife Range in southern Nevada decreased from 83 rams per 100 ewes in 1955 to 35 rams per 100 ewes in 1961. Hansen (1962) states the number of legal rams seen by hunters and the number of rams seen by refuge personnel also decreased during these same years. Cooper (1976) states, "Ram ratios provide the best information as to the effects of hunting on sheep populations. Any significant decrease in individual age class and herd composition ratios would indicate that hunting is having an impact on the population. Average ram ratios in Nevada sheep populations based on aerial surveys since 1969 suggest a total of 60 rams per 100 ewes with no downward trend since 1969." This information is based on a sample of 5,287 bighorn seen in over 547 hours of helicopter surveys since 1969. It appears the ram population has increased since the late 1950s when Jonez and Hansen made their observations.

New Mexico and Utah report that no changes in sex ratios that are hunting related occurred in their respective states.

POACHING

Subsistence hunting, practiced by the early day miners and still followed in some parts of Mexico, by Indians in Arizona, and by scofflaws elsewhere, can serve to eliminate the ewe *nucleus* and thus void a home range of its reproductive capacity. As Charles G. Hansen says, "Bighorn make an easy target for a hunter during the summer months, when the bighorn concentrate around water holes. A man with a high-powered rifle could easily eliminate a herd of bighorn in a few days by shooting the animals that come to drink. Those animals that were not killed would be forced away by the activity of the hunter and might not return if sufficient water was found elsewhere."

A few quotations illustrate man's thoughtlessness as it continues even to the present:

Burandt (1959): "This constant poaching nearly decimated the sheep population in some of California's desert ranges and well could be the cause of extinction of bighorn in their northernmost California range."

Duncan (1960): "Poachers invaded both (Cabeza Prieta and Kofa) Game Ranges. There were two distinct types: the meat hunter and the trophy hunter. One hunter in the former category related that he obtained a year's supply of jerky in 1 day at a water hole in west central Yuma County. He killed four sheep including two ewes. Two other wounded animals were able to get away. The trophy hunter was somewhat more selective, being interested in the large rams, and consequently killed on a smaller scale."

Welles (1967): "In 1963 a team of research scientists found four headless ram carcasses in a remote canyon of Joshua Tree National Monument. In 1960 we found in one forested section of Death Valley Monument in which hunting was so common, hunters had actually cleared a pack trail through the hunting area to facilitate removal of their kill."

Anselmo Lewis: "The bighorn in this area (San Gabriel Mountains, California) have always been subjected to a certain amount of poaching by the white man.... Incidents of shooting bighorn are still common in this area. In 1967, three rams were found near the highway on Kratka Ridge, all within 150 yards of each other. The had evidently been shot and left."

That bighorn poaching is not a historical phenomenon of the Old West but still continues on a considerable scale is attested by the revelation that a ring of poachers, smashed in 1970, had been responsible for the killing of about 150 desert bighorn in California during the previous 2 years (story in the *Los Angeles Times*, Oct. 6, 1970). What was described as the largest bighorn poaching operation in sports hunting history was broken with the arrest in 1971 of a single man who, over a 14-year period, took 200 United States hunters into Baja California on illegal bighorn hunting trips (Associated Press story in the *Arizona Daily Star*, Dec. 17, 1971).

In summary, it can be said that hunting the desert bighorn has produced, first of all, biological data about the bighorn not otherwise obtainable, including a better picture of population dynamics through age class statistics. It has increased and focused public sportsmen's interest on the bighorn. This public interest has moved legislators and administrators into action. Several bighorn studies have been the result, and more adequate funding has been appropriated for bighorn management.

Comment must be made on the theory advanced by Geist (1971) to the effect that the removal of old dominant rams from a herd by hunting may be harmful. In accordance with this concept younger rams then engage in excessive fighting and harassment of the ewes, possibly resulting in lower reproduction as well as danger to the herd by scattering them or by inflicting actual physical injury.

In the case of the desert bighorn, the 3-day-long chases, a form of courtship and sex play recorded by Welles and Welles (1961a), had no such effect so far as could be determined. Although the hair-raising chases back and forth through the herd often disturbed individuals, these activities had no lasting effect on the herd. Such activity is the result of evolution and cannot or will not develop to such a degree that it is harmful to the population.

It is questionable whether an interpretation by Geist of observations on a rather limited segment of a large and widely distributed group of ungulates such as the wild sheep can be universally applied. The adaptability and variability of sheep throughout their widespread distribution in the world would indicate a wide behavioral tolerance. Certainly the desert bighorn can tolerate a high degree of interherd disturbance by its own members and within its own ranks during the breeding season.

Appendix:

Boone and Crocket Club Official Score Sheet

RECORDS OF NORTH AMERICAN BIG GAME COMMITTEE

BOONE AND CROCKETT CLUB

Address Correspondence to:
Mrs. Grancel Fitz, Secretary
5 Tudor City Place, NYC 17, NY

SHEEP

KIND OF SHEEP _____

Measure to a point in line with tip of horn.

SEE OTHER SIDE FOR INSTRUCTIONS	Supplementary Data	Column 1	Column 2	Column 3
		Right Horn	Left Horn	Difference
A. Greatest Spread (Is often Tip to Tip Spread)				
B. Tip to Tip Spread (If Greatest Spread Enter again here)				
C. Length of Horn				/////
D-1. Circumference of Base				
D-2. Circumference at First Quarter				
D-3. Circumference at Second Quarter				
D-4. Circumference at Third Quarter				
TOTALS				

ADD	Column 1		Exact locality where killed
	Column 2		Date killed By whom killed
	TOTAL		Present owner
SUBTRACT Column 3		Address
	FINAL SCORE		Guide's Name and Address
			Remarks: (Mention any abormalities)

I certify that I have measured the above trophy on _____ 19___
at (address) _____ City _____ State _____
and that these measurements and data are, to the best of my knowledge and belief, made in accordance with the instructions given.

Witness: _____   Signature: _____

PN

Bibliography

ABERT, LT. J. W. 1848. Report of Lieut. J. W. Abert, of his examination of New Mexico, in the years 1846–47. 30th Congr., 1st Sess., *Exec. Doc. No. 41*, 520 pp.
ADOLPH, E., AND D. DILL. 1938. Observations on water metabolism in the desert. *Amer. J. Physiol.* 123:369–78.
ALDOUS, M. C. 1957. Seasonal water requirements (group discussion). *Desert Bighorn Council Trans.* 1:51–55.
ALDOUS, M. C. 1958. Trapping and tagging of bighorn sheep. *Desert Bighorn Council Trans.* 2:36–39.
ALDOUS, M. C., AND F. C. CRAIGHEAD, JR. 1958. A marking technique for bighorn sheep. *J. Wildl. Manage.* 22:445–46.
ALDOUS, M. C., F. C. CRAIGHEAD, JR., AND G. A. DEVAN. 1958. Some weights and measurements of desert bighorn sheep *(Ovis canadensis nelsoni)*. *J. Wildl. Manage.* 22:444–45.
ALEXANDER, R., J. HENTGES, J. MCCALL, AND W. ASH. 1962. Comparative digestivity of nutrients in roughages by cattle and sheep. *J. Animal Sci.* 21:373–76.
ALLEN, J. A. 1912. Historical and nomenclatorial notes on North American sheep. *Bull. Amer. Mus. Nat. Hist.* 31(1):1–29.
ALLEN, R. W. 1955. Parasites of mountain sheep in New Mexico with new host records. *J. Parasitol.* 41:583–87.
ALLEN, R. W. 1961. Methods of examining bighorn sheep for parasites. *Desert Bighorn Council Trans.* 5:75–79.

ALLEN, R. W. 1962. Parasitism in bighorn sheep on the Desert Game Range in Nevada. *Desert Bighorn Council Trans.* 6:69–71.
ALLEN, R. W. 1964. Additional notes on parasites of bighorn sheep on the Desert Game Range, Nevada. *Desert Bighorn Council Trans.* 8:5–9.
ALLEN, R. W. 1971. Present status of lungworm and tapeworm infections in desert bighorn sheep. *Desert Bighorn Council Trans.* 15:7–11.
ALLEN, R. W., AND H. G. ERLING. 1964. Parasites of bighorn sheep and mule deer in Arizona with new host records. *J. Parasitol.* 50 (Suppl.):38.
ALLEN, R. W., AND C. A. KENNEDY. 1952. Parasites in a bighorn sheep in New Mexico. *Proc. Helminthol. Soc. Wash.* 19:39.
ALLRED, L. G., AND W. G. BRADLEY. 1965. Necrosis and anomalies of the skull in desert bighorn sheep. *Desert Bighorn Council Trans.* 9:75–81.
ALLRED, L. G., AND W. G. BRADLEY. 1966. Comparative study of necrosis associated with teeth in desert bighorn sheep. *Desert Bighorn Council Trans.* 10:86–97.
ALVAREZ, T. 1976. Status of desert bighorns in Baja California. *Desert Bighorn Council Trans.* 20:18–21.
ANTEVS, E. 1955. Geologic-climate dating in the West. *Amer. Antiquity* 20:317–35.
ARMSTRONG, R. A. 1950. Fetal development of the northern whitetailed deer *(Odocoileus virginianus borealis* Miller). *Amer. Midl. Nat.* 43:650–66.
AUGSBURGER, J. G. 1970. Behavior of Mexican bighorn sheep in the San Andres Mountains, New Mexico. Master's thesis, New Mexico State Univ., Las Cruces. 54 pp.
BAILEY, V. 1931. *Mammals of New Mexico.* N. Amer. Fauna 53. Bureau of Biological Survey, Washington, D.C. 412 pp.
BAKER, L. R. 1967. Variation in the skulls of Nelson bighorn sheep, *Ovis canadensis nelsoni.* Master's thesis. Nevada Southern Univ., Las Vegas. 86 pp.
BAKER, L. R., AND W. G. BRADLEY. 1966. Growth of the skull in *Ovis canadensis nelsoni* from the Desert Game Range. *Desert Bighorn Council Trans.* 10:98–109.
BAKER, R. H. 1960. Mammals of Coahuila, Mexico. *Univ. Kans. Publ., Mus. Nat. Hist.*, 1955–60(9):327–29.
BANDELIER, A. F. 1892. Final report of investigations among the Indians of the southwestern United States, carried on mainly in the years from 1880 to 1885. *Papers of the Archaeol. Inst. of Amer. Series IV*:1, 3. 559 pp.
BANFIELD, A. W., D. R. FLOOK, J. P. KELSALL, AND A. G. LOUGHREY. 1955. An aerial survey technique for northern big game. *Trans. N. Amer. Wildl. Conf.* 20:519–32.
BANKO, W. E. 1963. The Wynne-Edwards theory applied to desert bighorn sheep. *Desert Bighorn Council Trans.* 7:64–71.
BANKS, E. M. 1964. Some aspects of sexual behavior in domestic sheep, *Ovis aries. Behavior* 23:249–79.
BARRETT, R. H. 1964. Seasonal food habits of the bighorn at the Desert Game Range, Nevada. *Desert Bighorn Council Trans.* 8:85–93.
BARRETT, R. H. 1965. A history of the bighorn in California and Nevada. *Desert Bighorn Council Trans.* 9:40–48.
BECKLUND, W. W., AND C. M. SENGER. 1967. Parasites of *Ovis canadensis* in Montana with a checklist of the internal and external parasites of the Rocky Mountain sheep in North America. *J. Parasitol.* 53:157–65.
BENDT, R. H. 1957. Status of bighorn sheep in Grand Canyon National Park and Monument. *Desert Bighorn Council Trans.* 1:16–19.
BENSON, S. B. 1943. Occurrence of upper canines in mountain sheep *Ovis canadensis. Amer. Midl. Nat.* 30:786–89.

BISHOPP, F. C., AND H. L. TREMBLEY. 1945. Distribution of hosts of certain North American ticks. *J. Parasitol.* 31:1–54.

BLAISDELL, J. A. 1961. Bighorn-cougar relationships. *Desert Bighorn Council Trans.* 5:42–46a.

BLAXTER, K., McN. GRAHAM, AND F. WAINMAN. 1959. Environmental temperature, energy metabolism, and heat regulation in sheep: III, The metabolism and thermal exchange of sheep with fleeces. *J. Agric. Sci.* 52:41–49.

BLIGHT, J. 1959. The receptors concerned in the thermal stimulus to panting in sheep. *J. Physiol.* 146:142–51.

BLONG, B. 1967. Desert bighorn and people in the Santa Rosa Mountains. *California-Nevada Sec. of the Wildl. Soc. Trans.* 1:66–70.

BLONG, B., AND W. POLLARD. 1968. Summer water requirements of desert bighorn in the Santa Rosa Mountains, California, in 1965. *Calif. Fish and Game* 54:289–96.

BOLTON, H. E. 1930. *Anza's California Expeditions*, II. Univ. Calif. Press, Berkeley. 473 pp.

BOLTON, H. E. 1936. *Rim of Christendom*. MacMillan, New York. 644 pp.

BOLTON, H. E. 1950. *Pageant in the Wilderness: The Story of the Escalante Expedition to the Interior Basin.* Utah State Hist. Soc., Salt Lake City. 265 pp.

BRADLEY, W. G. 1963. Water metabolism in desert mammals with special reference to the desert bighorn. *Desert Bighorn Council Trans.* 7:26–59.

BRADLEY, W. G. 1964. The vegetation of the Desert Game Range with special reference to the desert bighorn. *Desert Bighorn Council Trans.* 8:43–67.

BRADLEY, W. G. 1965. A study of the blackbrush plant community on the Desert Game Range: 1. *Desert Bighorn Council Trans.* 9:56–61.

BRADLEY, W. G., AND L. G. ALLRED. 1966. A comparative study of dental anomalies in desert bighorn sheep. *Desert Bighorn Council Trans.* 10:78–85.

BRADLEY, W. G., AND D. P. BAKER. 1967. Life tables for Nelson bighorn sheep on the Desert Game Range. *Desert Bighorn Council Trans.* 11:142–70.

BRADLEY, W. G., AND J. E. DEACON. 1965. The biotic communities of southern Nevada. *Desert Research Inst., Reprint Ser.* No. 9. Nevada Southern Univ., Las Vegas. 86 pp.

BREYEN, L. J. 1971. Desert bighorn habitat evaluation in the Eldorado Mountains of southern Nevada. Master's thesis, Nevada Southern Univ., Las Vegas.

BRINTON, E. P., D. E. BECK, AND D. M. ALLRED. 1965. Identification of the adults, nymphs, and larvae of ticks of the genus *Dermacentor* Kock (Ixodidae) in the western United States. *Brigham Young Univ., Sci. Bull., Biol. Ser.*, Vol. V, No. 4.

BROADBENT, R. V. 1969. Nevada's 1968 transplant disappointment. *Desert Bighorn Council Trans.* 13:43–47.

BRODIE, FAWN. 1971. The Brimhall saga. *The American West* 7:4–5, 8:4–9 and 18–23.

BRODY, S. 1945. *Bioenergetics and Growth*. Reinhold, New York. 1023 pp.

BROOK, A. H., AND B. F. SHORT. 1960a. Regulation of body temperature of sheep in a hot environment. *Australian J. Agric. Res.* 11:402–7.

BROOK, A. H., AND B. F. SHORT. 1960b. Sweating in sheep. *Australian J. Agric. Res.* 11:557–69.

BROWN, D. E. 1972. The status of desert bighorn sheep on the Papago Indian Reservation. *Desert Bighorn Council Trans.* 16:30–35.

BROWN, K. W., D. D. SMITH, AND R. P. McQUIVEY. 1977. Food habits of desert bighorn sheep in Nevada, 1956–1976. *Desert Bighorn Council Trans.* 21:32–61.

BROWNING, B., AND H. LEACH. 1959. Predator scat analysis (coyote). *Calif. Dept. Fish and Game, PR Quarterly Report,* Sacramento.

BUECHNER, H. K. 1960. *The Bighorn Sheep in the United States, Its Past, Present, and Future.* Wildl. Monogr. No. 4, The Wildlife Society. 174 pp.

BUNCH, T. D., S. R. PAUL, and H. MCCUTCHEN. 1978. Chronic sinusitis in the desert bighorn *(Ovis canadensis nelsoni). Desert Bighorn Council Trans.* 22:16–20.

BURANDT, V. 1959. Bighorn sheep patrol and protection problems in California. *Desert Bighorn Council Trans.* 3:37–40.

BURANDT, V. 1970. Status of bighorn sheep in Inyo Mountains. *Desert Bighorn Council Trans.* 14:1–8.

BURROUGHS, R. D. (ed.). 1961. *The Natural History of the Lewis and Clark Expedition.* Mich. State Univ. Press, East Lansing. 340 pp.

CAHALANE, V. H. 1939. Mammals of the Chiricahua Mountains, Cochise County, Arizona. *J. Mamm.* 20:418–40.

CAIN, S. A., J. A. KADLEC, D. L. ALLEN, R. A. COOLEY, M. G. HORNOCKER, A. S. LEOPOLD, AND F. H. WAGNER. 1972. Predator control–1971. Report to Council on Environmental Quality and the Department of the Interior by the Advisory Committee on Predator Control.

CALL, M. W. 1966. A proposed desert bighorn sheep range development project. *Desert Bighorn Council Trans.* 10:53–55.

CAPELLE, K. J. 1966. The occurrence of *Oestrus ovis* L. (Diptera: Oestridae) in the bighorn sheep from Wyoming and Montana. *J. Parasitol.* 52:618–21.

CAROTHERS, S. W., M. E. STITT, AND R. R. JOHNSON. 1976. Feral asses on public lands: An analysis of biotic impact, legal considerations and management alternatives. *Trans. 41st N. Amer. Wildl. and Nat. Resources Conf.*:396–406.

CARSON, B. 1941. Man, the greatest enemy of the desert bighorn mountain sheep. *Tex. Game, Fish and Oyster Comm. Bull.* 21.

CARSON, RACHEL. 1962. *Silent Spring.* Houghton Mifflin Co., Boston. 368 pp.

CATER, B. 1968. Scabies in desert bighorn sheep. *Desert Bighorn Council Trans.* 12:76–77.

CHEW, D. W., AND J. D. GOODMAN. 1963. Studies on the effects of the drug succinylcholine chloride on domestic sheep related to its use as an aid in capture of desert bighorn sheep. *Desert Bighorn Council Trans.* 7:139–44.

CLARK, J. L. 1964. *The Great Arc of the Wild Sheep.* Univ. Okla. Press, Norman. 247 pp.

CLARK, SUSAN R. 1966. A tabular summary of plant and animal resources of the Glen Canyon area. *Univ. Utah Anthropol. Paper No. 80, Glen Canyon Ser.* 30:107.

COOPER, J. 1976. In *Big Game Investigations and Hunting Season Recommendations.* G. Tsukamoto, ed. Nev. Dept. Fish and Game.

COSGROVE, H. S., AND C. B. COSGROVE. 1933. The Swarts Ruin. *Papers of the Peabody Mus. Amer. Archaeol. and Ethn.* 15(9):3–5, 56, 57, 75, 92.

COTTAM, C., AND C. S. WILLIAMS. 1943. Speeds of some wild animals. *J. Mamm.* 24:262–63.

COUEY, F. M. 1950. Rocky Mountain sheep of Montana. *Mont. Fish and Game Comm. Bull.* 2:19–66.

COWAN, I. McT. 1940. Distribution and variation in the native sheep of North America. *Amer. Midl. Nat.* 24:505–80.

COZZENS, S. W. 1967. *The Marvellous Country; or, Three Years in Arizona and New Mexico, the Apaches' Home, 1858–1860.* Ross & Haines, Inc., Minneapolis. 532 pp.

CRONEMILLER, F. G. 1948. Mountain lion preys on bighorn. *J. Mamm.* 29:68.

CROW, L. Z. 1964. A field survey of water requirements of desert bighorn sheep. *Desert Bighorn Council Trans.* 8:77–83.

CUSHING, F. H. 1882–83. My adventure in Zuni. *The Century Illustrated Monthly Magazine* 25(n.s. 3):501.

DALQUEST, W. E., AND D. F. HOFFMEISTER. 1948. Mountain sheep from the state of Washington in the collection of the University of Kansas. *Trans. Kans. Acad. Sci.* 51:224–34.

DASMANN, R. F. 1964. *Wildlife Biology.* John Wiley & Sons, Inc., New York. 231 pp.

DAVILA C., J. A. 1960. Borregos y berrendos en Mexico (in Spanish with English translation). *Desert Bighorn Council Trans.* 4:101–6.

DAVIS, W. A. 1966. Theoretical problems in western prehistory. In *The Great Basin Anthropological Conference on the Current Status of Anthropological Research, Reno, 1964.* Desert Research Institute, Social Sciences and Humanities Publ. 1:147–65.

DAVIS, W. B. 1961. Vanished. *Tex. Game and Fish* 19(12):15–22.

DAVIS, W. B., AND W. P. TAYLOR. 1939. The bighorn sheep of Texas. *J. Mamm.* 20:440–55.

DAY, G. I. 1971. Remove injection of drugs. Ariz. Game and Fish Dept. Proj. No. W-78-R-15-WP-1-J4:5–40.

DAY, G. I. 1973. Marking devices for big game animals. Ariz. Game and Fish Dept., Abstract No. 8 (Report W-78-R). 7 pp.

DEAN, H. C., AND J. J. SPILLETT. 1976. Bighorn in Canyonlands National Park. *Desert Bighorn Council Trans.* 20:15–17.

DEEVEY, E. S., JR. 1947. Life tables for natural populations of animals. *Quart. Rev. Biol.* 22:283–314.

DEFORGE, J. R. 1972. Man's invasion into the bighorn's habitat. *Desert Bighorn Council Trans.* 16:112–15.

DEFORGE, J. R., C. W. JENNER, A. J. PLECHNER, AND G. W. SUDMEIER. 1979. *Decline of Bighorn Sheep (Ovis canadensis): the Genetic Implications.* The Society for the Conservation of Bighorn Sheep, Alhambra, Calif. 14 pp.

DEMING, O. V. 1952. Tooth development of the Nelson bighorn sheep. *Calif. Fish and Game* 38:523–29.

DEMING, O. V. 1953. Lambs of the Nelson bighorn sheep in Nevada. Unpublished, in files of Desert National Wildlife Range, Las Vegas, Nev. 143 pp.

DEMING, O. V. 1955. Rearing bighorn lambs in captivity. *Calif. Fish and Game* 41:131–43.

DEMING, O. V. 1961. Bighorn sheep transplants at the Hart Mountain National Antelope Refuge. *Desert Bighorn Council Trans.* 5:56–57.

DEMING, O. V. 1962. Is the desert bighorn a relict species? *Desert Bighorn Council Trans.* 6:93–113.

DEMING, O. V. 1963. Bighorn breeding. *Desert Bighorn Council Trans.* 7:92–111.

DEMING, O. V. 1964. Some bighorn foods on the Desert Game Range. *Desert Bighorn Council Trans.* 8:137–43.

DENNISTON, A. 1965. Status of bighorn in the River Mountains of Lake Mead National Recreation Area. *Desert Bighorn Council Trans.* 9:27–34.

DEVAN, G. A. 1958. Daily movement and activity of the bighorn. *Desert Bighorn Council Trans.* 2:67–72.

DEVAN, G. A. 1959. The use of the CO_2 Cap-Chur gun at the Desert Game Range. *Desert Bighorn Council Trans.* 3:50–52.

DILL, D. 1938. *Life, Heat, and Altitude.* Harvard Univ. Press, Cambridge. 211 pp.

DIPESO, C. C. 1956. *The Upper Pima of San Cayetano del Tumacacori.* The Amerind Foundation (No. 7), Dragoon, Arizona. 589 pp.

Dixon, J. S., and E. L. Sumner, Jr. 1939. A survey of desert bighorn in Death Valley National Monument, summer 1938. *Calif. Fish and Game* 25:72–95.

Dowling, D. 1958. The significance of sweating in heat tolerance of cattle. *Australian J. Agric. Res.* 9:579–86.

Dukes, R. H. 1955. *The Physiology of Domestic Animals*. Comstock Publ. Assn., Ithaca, N.Y. 1020 pp.

Dunaway, D. J. 1970. Status of bighorn sheep populations and habitat studies on the Inyo National Forest. *Desert Bighorn Council Trans.* 14:127–46.

Duncan, G. E. 1960. Human encroachment on bighorn habitat. *Desert Bighorn Council Trans.* 4:35–37.

Duncan, G. E. 1963. Progress report from the Kofa Game Range. *Desert Bighorn Council Trans.* 7:151–55.

Edwards, R. Y. 1954. Comparison of an aerial and ground census of moose. *J. Wildl. Manage.* 18:403–4.

English, P. 1966a. A study of water and electrolyte metabolism in sheep: I, External balances of H_2O, NA^+, K^+, and Cl^-. *Res. Vet. Sci.* 7:233–57.

English, P. 1966b. A study of water and electrolyte metabolism in sheep: II, Volumes of distribution of antipyrine, thiosulfate and T-1824, and values for certain extracellular fluid constituents. *Res. Vet. Sci.* 7:258–75.

Errington, P. L. 1963. The phenomenon of predation. *Amer. Sci.* 51:188–92.

Etkin, W. (ed.). 1964. *Social Behavior and Organization Among Vertebrates*. Univ. Chicago Press. 307 pp.

Euler, R. C. 1966. Southern Ute ethnohistory. *Univ. Utah Anthropol. Paper,* 7:25, 114–15.

Eustis, G. P. 1962. Winter lamb surveys on the Kofa Game Range. *Desert Bighorn Council Trans.* 6:83–86.

Eyal, E. 1954. The body temperature and cardio-respiratory activities of shorn and unshorn Awassi sheep. *Bull. Res. Council Israel, Sec. B.* 4:307–11.

Farrell, R. K., L. M. Koger, and L. D. Winward. 1966. Freeze branding of cattle, dogs and cats for identification. *Amer. Vet. Med. Assoc.* 149:745–52.

Ferrier, G. J., and W. G. Bradley. 1970. Bighorn habitat evaluation in the Highland Range of southern Nevada. *Desert Bighorn Council Trans.* 14:66–93.

Fewkes, J. W. 1903. Hopi katchinas. 21st Ann. Report Bureau of Amer. Ethn. 126 pp.

Fewkes, J. W. 1922. Idols in Hopi worship. *Ann. Report Smithsonian Inst.* 385 pp.

Fontana, B. L. 1962. An archeological survey of the Cabeza Prieta Game Range, Arizona. Manuscript, Ariz. State Museum, Tucson.

Forrester, D. J., and R. S. Hoffman. 1963. Growth and behavior of a captive bighorn lamb. *J. Mamm.* 44:116–18.

Gabrielson, I. N. 1941. *Wildlife Conservation*. The MacMillan Co., New York. 250 pp.

Gallizioli, S. 1977. Overgrazing on desert bighorn ranges. *Desert Bighorn Council Trans.* 21:21–23.

Gates, G. H. 1972. Capture of free-ranging desert sheep in Sonora, Mexico. *Desert Bighorn Council Trans.* 16:97–101.

Geist, V. 1966. The evolutionary significance of mountain sheep horns. *Evolution* 20:558–66.

Geist, V. 1967. A consequence of togetherness. *Nat. Hist.* 76(4):24–31.

Geist, V. 1968a. On delayed social and physical maturation in mountain sheep. *Canad. J. Zool.* 46:899–904.

Geist, V. 1968b. On the interrelation of external appearance, social behavior and social structure in mountain sheep. *Zeitsch. für Tierpsychologie* 25:199–215.

GEIST, V. 1971. *Mountain Sheep: A Study in Behavior and Evolution.* Univ. Chicago Press, Chicago and London. 383 pp.
GILBERT, P. F., AND J. R. GRIEB. 1957. Comparison of air and ground deer counts in Colorado. *J. Wildl. Manage.* 21:33–37.
GILES, R. H., JR. (ed.). 1969. *Wildlife Management Techniques.* 3rd Ed. The Wildlife Society, Washington, D.C. 623 pp.
GLADWIN, H. S. 1957. *A History of the Ancient Southwest.* Wheelwright, Portland, Maine. 383 pp.
GLADWIN, H. S., E. W. HAURY, E. B. SAYLES, AND NORA GLADWIN. 1937. Excavations at Snaketown—material culture. *Medallion Papers No. 25.* Gila Peublo, Globe, Ariz. 305 pp.
GOLDMAN, L. C. 1961. Summary—bighorn predators. *Desert Bighorn Council Trans.* 5:113.
GOODMAN, J. D. 1962. Annual migration of desert bighorn. *Desert Bighorn Council Trans.* 6:43–51.
GOODMAN, J. D. 1963. A report on the first mid-winter and spring bighorn sheep census in the Santa Rosa Mountains, 1962–1963. *Desert Bighorn Council Trans.* 7:135–38.
GORDON, S. P. 1956. Mexican border big game management survey. Pittman-Robertson Proj. W-68-R-3. Comp. Report, New Mexico Game and Fish Dept.
GORDON, S. P. 1957. The status of bighorn sheep in New Mexico. *Desert Bighorn Council Trans.* 1:3–4.
GRAF, W. 1963. Exotics and their implications. *Desert Bighorn Council Trans.* 7:42–50.
GRAHAM, A., AND R. BELL. 1969. Factors influencing the countability of animals. *In* W. G. Swank, R. M. Watson, G. H. Freeman, and T. Jones (eds.). Proc. of the workshop on the use of light aircraft in wildlife management in East Africa. *East African Agric. and Forestry J.* 34 (Spec. Issue):38–43.
GRAHAM, H. 1971. Environmental analysis procedures for bighorn in the San Gabriel Mountains. *Desert Bighorn Council Trans.* 15:38–45.
GRANT, C. 1968. Desert bighorn rock drawings at the Coso Range, Inyo County, California. *Desert Bighorn Council Trans.* 12:40–49.
GRATER, R. K. 1959. Recreational values of bighorn other than hunting. *Desert Bighorn Council Trans.* 3:53–57.
GRINNELL, G. B. 1928. Mountain sheep. *J. Mamm.* 9:1–9.
GROSS, J. E. 1960. History, present and future status of the desert bighorn sheep in the Guadalupe Mountains of southeastern New Mexico and northwestern Texas. *Desert Bighorn Council Trans.* 4:66–71.
GROSSCUP, G. L. 1960. *The Culture History of Lovelock Cave, Nevada.* Univ. Calif. Archaeol. Surv. 52. Berkeley. 91 pp.
GROVES, F. 1957. (In discussion.) *Desert Bighorn Council Trans.* 1:50.
HABERLAND, W. 1964. *The Art of North America.* Crown, New York. 251 pp.
HAILEY, T. L. 1962. Status of transplanted bighorns in Texas. *Desert Bighorn Council Trans.* 6:129–30.
HAILEY, T. L. 1964. Status of transplanted bighorns in Texas. *Desert Bighorn Council Trans.* 8:113–16.
HAILEY, T. L. 1966. Status of transplanted bighorns in Texas in 1966. *Desert Bighorn Council Trans.* 10:59–61.
HAILEY, T. L. 1971. Reproduction and release of Texas transplanted desert bighorn sheep. *Desert Bighorn Council Trans.* 15:97–100.
HALL, E. R. 1946. *Mammals of Nevada.* Univ. Calif. Press, Berkeley. 710 pp.

HALLORAN, A. F. 1949. Desert bighorn management. *N. Amer. Wildl. Conf. Trans.* 14:527–37.
HALLORAN, A. F., AND W. E. BLANCHARD. 1950. Bighorn ewe associates with cattle on Kofa Game Range, Arizona. *J. Mamm.* 31:463–64.
HALLORAN, A. F., AND H. B. CRANDELL. 1953. Notes on bighorn food in the Sonoran Zone. *J. Wildl. Manage.* 17:318–20.
HALLORAN, A. F., AND O. V. DEMING. 1956. *Water Development for Desert Bighorn Sheep.* Fish and Wildlife Serv. Wildl. Manage. Series. 12 pp.
HALLORAN, A. F., AND O. V. DEMING. 1958. Water development for desert bighorn sheep. *J. Wildl. Manage.* 22:1–9.
HALLORAN, A. F., AND C. A. KENNEDY. 1949. Bighorn-deer food relationships in southern New Mexico. *J. Wildl. Manage.* 13:417–19.
HAMMOND, C. P. 1940. *Coronado's Seven Cities.* U.S. Coronado Exposition Comm., Albuquerque. 48 pp.
HANSEN, C. G. 1960. Lamb survival on the Desert Game Range. *Desert Bighorn Council Trans.* 4:60–61.
HANSEN, C. G. 1961. Significance of bighorn mortality records. *Desert Bighorn Council Trans.* 5:22–26.
HANSEN, C. G. 1962. Progress report from the Desert Game Range. *Desert Bighorn Council Trans.* 6:73–82.
HANSEN, C. G. 1964a. A dye-spraying device for marking desert bighorn sheep. *J. Wildl. Manage.* 28:584–87.
HANSEN, C. G. 1964b. Bighorn sheep. *In* Desert Game Range Narrative Rept. Unpublished report in Desert National Wildlife Range files, Las Vegas, Nevada.
HANSEN, C. G. 1964c. Progress report from the Desert Game Range, Nevada. *Desert Bighorn Council Trans.* 8:69–76.
HANSEN, C. G. 1965a. Growth and development of desert bighorn sheep. *J. Wildl. Manage.* 29:387–91.
HANSEN, C. G. 1965b. Management units and bighorn sheep herds on the Desert Game Range, Nevada. *Desert Bighorn Council Trans.* 9:11–14.
HANSEN, C. G. 1965c. Summary of distinctive bighorn sheep observed on the Desert Game Range, Nevada. *Desert Bighorn Council Trans.* 9:6–10.
HANSEN, C. G. 1965d. White spotting in bighorn sheep from the Desert Game Range, Nevada. *J. Mamm.* 46:352–53.
HANSEN, C. G. 1967a. Bighorn sheep populations of the Desert Game Range. *J. Wildl. Manage.* 31:693–706.
HANSEN, C. G. 1967b. Classifying bighorn habitat on the Desert National Wildlife Range. Mimeo report, files of Desert National Wildlife Range, Las Vegas, Nev. 20 pp.
HANSEN, C. G. 1970. Tongue color in desert bighorn. *Desert Bighorn Council Trans.* 14:14–22.
HANSEN, C. G. 1971. Overpopulation as a factor in reducing desert bighorn populations. *Desert Bighorn Council Trans.* 15:46–52.
HANSEN, PATRICIA A. 1964. Tag-along's first year. *Desert Bighorn Council Trans.* 8:145–52.
HANSEN, R. M. 1971. Estimating plant composition of wild sheep diets. *Trans. 1st N. Amer. Wild Sheep Conf.*:108–13.
HARRINGTON, M. R. 1957. A Pinto site at Little Lake, California. *Southwest Mus. Papers* 17:1–91.
HAURY, E. W., K. BRYAN, E. H. COLBERT, N. E. GABEL, CLARA LEE TANNER, AND T. E. BUEHRER. 1950. *The Stratigraphy and Archeology of Ventana Cave, Arizona.* Univ. Ariz. Press, Tucson. 590 pp.

HECKER, J., O. BUDTZ-OLSEN, AND D. OSTWALD. 1964. The rumen as a water store in sheep. *Australian J. Agric. Res.* 15:961–68.

HEIZER, R. F. 1951. The sickle in aboriginal western North America. *Amer. Antiquity* 16:247–52.

HEIZER, R. F., AND M. BAUMHOFF. 1962. *Prehistoric Rock Art of Nevada and Eastern California*. Univ. Calif. Press, Berkeley. 412 pp.

HEIZER, R. F., M. BAUMHOFF, AND C. W. CLEWLAW, JR. 1968. Archeology of South Fork Shelter, Elko County, Nevada. *Univ. Calif. Archeol. Papers 71*. Berkeley. 58 pp.

HELVIE, J. B. 1971. Bighorns and fences. *Desert Bighorn Council Trans.* 15:53–62.

HELVIE, J. B. 1972. Census of desert bighorn sheep with time-lapse photography. *Desert Bighorn Council Trans.* 16:3–8.

HELVIE, J. B., AND D. D. SMITH. 1970. Summary of necropsy findings in desert bighorn sheep. *Desert Bighorn Council Trans.* 14:47–52.

HEMMING, J. E. 1969. Cemental deposition, tooth succession, and horn development as criteria of age in Dall sheep. *J. Wildl. Manage.* 33:552–58.

HONESS, R. F., AND N. M. FROST. 1942. A Wyoming bighorn sheep study. *Wyo. Game and Fish Dept. Rept. No. 1*. 127 pp.

HONESS, R. F., AND K. B. WINTER. 1956. Diseases of wildlife in Wyoming. *Wyo. Game and Fish Dept. Bull. No. 9.*

HORNADAY, W. T. 1908. *Camp-fires on Desert and Lava*. Charles Scribner and Sons, New York. 366 pp.

HUNT, ALICE. 1960. Archeology of the Death Valley salt pan, California. *Univ. Utah Archaeol. Papers 47*. 313 pp.

HUNTER, R. F. 1964. Home range behavior in hill sheep. *In* D. J. Crisp, *Grazing in Terrestrial and Marine Environments*. Blackwell Sci. Publ., Oxford:pp. 155–71.

IVES, R. L. 1962. Kiss tanks. *Weather* 17(6):194–96.

IRVINE, C. A. 1969a. Factors affecting the desert bighorn sheep in southeastern Utah. *Desert Bighorn Council Trans.* 13:6–13.

IRVINE, C. A. 1969b. The desert bighorn sheep of southeastern Utah. *Utah Div. Fish and Game Publ. 69–12*. 99 pp.

JACKSON, A. T. 1938. Picture writing of Texas Indians. *Univ. Texas Publ.* 3809:402–3.

JAEGER, E. C. 1957. *The North American Deserts*. Stanford Univ. Press, Stanford, California. 308 pp.

JENNINGS, J. D. 1957. Danger Cave. Memoirs of the Soc. Amer. Archaeol. 14. *Amer. Antiquity* 23:223–24.

JEWELL, P. A. 1966. The concept of home range in mammals. *Symp. Zool. Soc. London* 18:85–109.

JOHN, R. T. 1968. Utah's desert bighorn. *Utah Fish and Game Publ. 68-9*. 19 pp.

JOHNSON, E. L. 1957. Disease and mechanical injury in desert bighorn sheep. *Desert Bighorn Council Trans.* 1:38–42.

JOLLY, G. M. 1969a. Sampling methods for aerial censuses of wildlife populations. *In* W. G. Swank, R. M. Watson, G. H. Freeman, and T. Jones (eds.). Proceedings of the workshop on the use of light aircraft in wildlife management in East Africa. *East African Agric. and Forestry J.* 34 (Spec. Issue):46–49.

JOLLY, G. M. 1969b. The treatment of errors in aerial counts of wildlife populations. *In* W. G. Swank, R. M. Watson, G. H. Freeman, and T. Jones (eds.). Proceedings of the workshop on the use of light aircraft in wildlife management in East Africa. *East African Agric. and Forestry J.* 34 (Spec. Issue):50–55.

JONES, F. L. 1950. A survey of the Sierra Nevada bighorn. *Sierra Club Bull.* 35:29–76.

Jones, F. L. 1954. The Inyo-Sierra deer herds, a final study report. Calif. Dept. Fish and Game, mimeo. 84 pp.

Jones, F. L. 1955. Bighorn management problems in California. *Proc. Ann. Conf. Western Assn. State Game and Fish Comm.* 35:177–81.

Jones, F. L., and O. V. Deming. 1953. Report of a survey of bighorn habitat in the Twenty-nine Palms Marine Corps Artillery Training Center. Calif. Dept. Fish and Game, mimeo. 4 pp.

Jones, F. L., G. Flittner, and R. Gard. 1957. Report on a survey of bighorn sheep in the Santa Rosa Mountains, Riverside County. *Calif. Fish and Game* 43:179–91.

Jonez, A. R. 1958. Hunting the desert bighorn sheep in Nevada. *Desert Bighorn Council Trans.* 2:1–5.

Jonez, A. R. 1959. 1958 bighorn hunt highlights—Nevada. *Desert Bighorn Council Trans.* 3:20–23.

Jonez, A. R. 1960. The bighorn as a multiple use animal. *Desert Bighorn Council Trans.* 4:45–46.

Jonez, A. R. 1961. Hunting results in Nevada, 1960. *Desert Bighorn Council Trans.* 5:87–89.

Jonez, A. R. 1966. Trophy hunting for bighorn sheep. *Proc. Ann. Conf. Western Assn. State Game and Fish Comm.* 46:72–75.

Jonez, A. R., and G. Monson. 1957. (In discussion.) *Desert Bighorn Council Trans.* 1:90.

Jorgensen, M. C., and R. E. Turner. 1972. A survey of the desert bighorn sheep *(Ovis canadensis)* in the Anza-Borrego Desert State Park. Calif. Dept. Parks and Recreation. 27 pp.

Judd, N. M. 1954. The material culture of Pueblo Bonito. *Smithsonian Inst. Misc. Coll.* 124:64, 140, 199.

Judd, N. M. 1959. Pueblo del Arroyo, Chaco Canyon, New Mexico. *Smithsonian Inst. Misc. Coll.* 138:139.

Kelly, W. E. 1961. Lamb and yearling counts. *Desert Bighorn Council Trans.* 5:31a–31d.

Kemper, H. E. 1947. The spinose ear tick. USDA Farmers Bull. No. 980.

Kennedy, C. A. 1948. Golden eagle kills bighorn lamb. *J. Mamm.* 29:68–69.

Kennedy, C. E. 1957. (In discussion.) *Desert Bighorn Council Trans.* 1:50.

Kennedy, C. E. 1958. Water development on the Kofa and Cabeza Prieta Game Ranges. *Desert Bighorn Council Trans.* 2:28–31.

Kennedy, C. E. 1963. Bighorn sheep of the Angeles National Forest. *Desert Bighorn Council Trans.* 7:126–34.

Kidder, A. V., and S. J. Guernsey. 1919. Archaeological explorations in northeastern Arizona. *Bur. Amer. Ethn. Bull.* 65:128–29, 156, 194–95.

Kiger, J. H. 1970. Helicopter observations of bighorn sheep on the San Andres National Wildlife Refuge. *Desert Bighorn Council Trans.* 14:23–27.

Kirkland, F., and W. W. Newcomb, Jr. 1967. *The Rock Art of Texas Indians.* Univ. Tex. Press, Austin. 239 pp.

Koplin, J. R. 1960. Information on tagging on the Desert Game Range. *Desert Bighorn Council Trans.* 4:49–53.

Lamb, S. M. 1958. Linguistic prehistory in the Great Basin. *Int. J. Amer. Linguistics* 24:95–100.

Lange, R. E., A. V. Sandoval, and W. P. Meleney. 1980. Psoroptic scabies in bighorn sheep *(Ovis canadensis mexicana)* in New Mexico. *J. Wildl. Diseases* 16(1):77–82.

LARSEN, P. A. 1962a. (No title given) Pittman-Robertson Proj. W-100-R-4, Job 12, Comp. Report, N. Mex. Game and Fish Dept., pp. 13–14.
LARSEN, P. A. 1962b. Progress of Mexican bighorn sheep population and management investigations in the San Andres-Big Hatchet Mountains of New Mexico. *Desert Bighorn Council Trans.* 6:126–28.
LAWRENCE, BARBARA. 1951. Mammals found at the Awatovi Site. *Papers of the Peabody Mus. Amer. Archaeol. and Ethn.* 35(3):1–6.
LEE, D. 1950. Studies of heat regulation in the sheep with special reference to the merino. *Australian J. Agric. Res.* 2:200–16.
LEE, D. 1968. Human adaptations to arid environments. *In* Desert Biology. G. W. Brown (ed.). Academic Press, New York. 635 pp.
LEE, D., AND R. PHILLIPS. 1948. Assessment of the adaptability of livestock to climatic stress. *J. Animal Sci.* 7:391.
LEE, D., AND K. ROBINSON. 1941. Reaction of the sheep to hot atmosphere. *Proc. Royal Soc. Queensland* 53:189–200.
LEE, L. 1960. The possible impact of Barbary sheep in New Mexico. *Desert Bighorn Council Trans.* 4:15–16.
LEOPOLD, A. 1933. *Game Management.* Charles Scribner and Sons, New York. 481 pp.
LEWIS, A. 1960. Desert bighorn status on the Mt. Baldy District of Angeles National Forest. *Desert Bighorn Council Trans.* 4:72–75.
LIGHT, J. T., JR. 1971. An ecological view of bighorn habitat on Mt. San Antonio. *Trans. 1st N. Amer. Wild Sheep Conf.*:150–57.
LIGHT, J. T., JR., AND F. A. WINTER. 1967. San Gabriel bighorn management plan, Angeles and San Bernardino National Forests. Unpublished report, U.S. Forest Service. 112 pp.
LIGHT, J. T., JR., F. A. WINTER, AND H. GRAHAM. 1967. San Gabriel bighorn habitat management plan. U.S. Forest Service, Angeles and San Bernardino National Forests, San Bernardino, California. 32 pp.
LIGHT, J. T., JR., T. R. ZRELAK, AND H. GRAHAM. 1966. San Gorgonio bighorn habitat management plan, San Bernardino National Forest. Unpublished report, U.S. Forest Service. 67 pp.
LOGSDON, H. S. 1966. Summary of wildlife management study on determination of safe methods of immobilizing and tranquilizing desert bighorn sheep for the purpose of capture. Mimeo., in files of Desert National Wildlife Range, Las Vegas, Nevada. 17 pp.
LOGSDON, H. S. 1967. Domestic sheep drug trials, conducted at Fort Collins, Colorado. Mimeo., in files of Desert National Wildlife Range, Las Vegas, Nevada. 44 pp.
LOGSDON, H. S. 1969. Experimental use of drugs for capturing desert bighorn. Ph.D. thesis, Colorado State Univ., Fort Collins. 173 pp.
LONG, P. V., JR. 1966. Archaeological excavations in lower Glen Canyon, Utah. *Mus. Northern Ariz. Bull. 42*. 80 pp.
LOWE, C. H. (ed.). 1964. *The Vertebrates of Arizona.* Univ. Ariz. Press, Tucson. 270 pp.
MCCANN, L. J. 1956. Ecology of the mountain sheep. *Amer. Midl. Nat.* 56:297–324.
MCCOLM, M. A. 1963. A history of bighorn sheep in central Nevada. *Desert Bighorn Council Trans.* 7:1–11.
MCCULLOUGH, D. R., AND E. R. SCHNEEGAS. 1966. Winter observations on the Sierra Nevada bighorn sheep. *Calif. Fish and Game* 52:68–84.
MCCUTCHEN, H. E. 1975. Desert bighorn restoration at Zion National Park, Utah. *Desert Bighorn Council Trans.* 19:19–27.

McCutchen, H. E. 1977. The Zion bighorn restoration project, 1976. *Desert Bighorn Council Trans.* 21:9–11.

Macfarlane, W. V. 1964. Terrestrial animals in dry heat: ungulates. In *Handbook of Physiology. Sec. IV: Adaptation to the Environment.* D. B. Hill (ed.). Amer. Physiol. Soc., Washington, D.C. pp. 509–39.

Macfarlane, W. V., R. J. H. Morris, and Beth Howard. 1956. Water economy of tropical merino sheep. *Nature* 178:304–5.

Macfarlane, W. V., R. J. H. Morris, and Beth Howard. 1958. Heat and water in tropical merino sheep. *Australian J. Agric. Res.* 9:217–28.

Macfarlane, W. V., R. J. H. Morris, and Beth Howard. 1962. Water metabolism of merino sheep and camels. *Australian J. Sci.* 25:112.

Macfarlane, W. V., R. J. H. Morris, Beth Howard, and O. E. Budtz-Olsen. 1959. Extracellular fluid distribution in tropical merino sheep. *Australian J. Agric. Res.* 10:269–86.

Macfarlane, W. V., R. J. H. Morris, Beth Howard, Janet McDonald, and O. E. Budtz-Olsen. 1961. Water and electrolyte changes in tropical merino sheep exposed to dehydration during summer. *Australian J. Agric. Res.* 12:889–912.

McKnight, T. L. 1958. The feral burro in the United States: distribution and problems. *J. Wildl. Manage.* 22:163–79.

McLean, D. D. 1930. Desert mountain sheep of Inyo Mountains. *Calif. Fish and Game* 16:79–82.

McMichael, T. J. 1964. Relationships between desert bighorn and feral burros in the Black Mountains of Mohave County. *Desert Bighorn Council Trans.* 8:29–35.

McQuivey, R. P. 1977. Bighorn sheep status report from Nevada. *Desert Bighorn Council Trans.* 21:5.

McQuivey, R. P. 1978. The desert bighorn sheep of Nevada. *Biol. Bull. No. 6.* Nevada Dept. Fish and Game. 81 pp.

Mearns, E. A. 1907. Mammals of the Mexican boundary of the United States. *U.S. Natl. Mus. Bull.* 56, Part I. 530 pp.

Meighan, C. W. 1969. *Indian Art and History: The Testimony of Prehistoric Rock Paintings in Baja California.* Dawson's Book Shop, Los Angeles. 79 pp.

Mence, A. L. 1969. Psychological problems of conducting aerial censuses from light aircraft. In W. G. Swank., R. M. Watson, G. H. Freeman, and T. Jones (eds.). Proceedings of the workshop on the use of light aircraft in wildlife management in East Africa. *East African Agric. and Forestry J.* 34 (Spec. Issue):44–45.

Mendoza, J. 1976. Status of the desert bighorn in Sonora. *Desert Bighorn Council Trans.* 20:25–26.

Mensch, J. L. 1969. Desert bighorn *(Ovis canadensis nelsoni)* losses in a natural trap tank. *Calif. Fish and Game* 55:237–38.

Mensch, J. L. 1970. Survey of bighorn sheep in California. *Desert Bighorn Council Trans.* 14:125–26.

Merriam, C. H. 1890. Results of a biological survey of the San Francisco Mountain region and desert of the Little Colorado, Arizona. *N. Amer. Fauna No. 3.* Bureau of Biological Survey, Washington, D.C. 136 pp.

Miller, P. R. 1969. Air pollution and the forests of California. *Calif. Air Environment* 1:(4). Statewide Air Pollution Research Center, Univ. Calif., Riverside.

Monson, G. 1958. Water requirements. *Desert Bighorn Council Trans.* 2:64–66.

Monson, G. 1960. Effects of climate on desert bighorn numbers. *Desert Bighorn Council Trans.* 4:12–14.

Monson, G. 1963. Some desert bighorn reflections. *Desert Bighorn Council Trans.* 7:61–63.

Monson, G. 1964. Long distance and nighttime movements of desert bighorn sheep. *Desert Bighorn Council Trans.* 8:11–17.

Monson, G. 1965. Group mortality in the desert bighorn sheep. *Desert Bighorn Council Trans.* 9:55.

Montoya, B. 1973. Bighorn sheep capture techniques. *Desert Bighorn Council Trans.* 17:155–63.

Montoya, B. 1976. The future outlook for desert bighorn sheep in New Mexico. *Desert Bighorn Council Trans.* 20:49.

Montoya, B., and G. Gates. 1975. Bighorn capture and transplant in Mexico. *Desert Bighorn Council Trans.* 19:28–32.

Montoya, B., and E. Munoz. 1976. Bighorn today: New Mexico's desert bighorn program. *Desert Bighorn Council Trans.* 20:4.

Moore, T. D. 1958. Transplanting and observations of transplanted bighorn sheep. *Desert Bighorn Council Trans.* 2:43–46.

Moore, T. D. 1961. The Texas bighorn sheep transplant. *Desert Bighorn Council Trans.* 5:53–55.

Morgan, N. B. 1961. Censusing by transect. *Desert Bighorn Council Trans.* 5:30–31.

Morrison, F. B. 1951. *Feeds and Feeding: a Handbook for the Student and Stockman.* Morrison Publ. Co., Ithaca, N.Y. 207 pp.

Morrison, P. 1966. Insulative flexibility in the guanaco. *J. Mamm.* 47:18–23.

Moser, G. A. 1962. The bighorn sheep of Colorado. *Colo. Dept. Game and Fish Tech. Publ. No. 10.* 49 pp.

Muir, J. 1901. *The Mountains of California.* The Century Co., New York. 381 pp.

Murie, A. 1944. The wolves of Mount McKinley. *U.S. National Park Service Fauna Ser. No. 5.* Washington, D.C.

Musgrave, E. W. 1926. A unique flock of hybrids. *Sci. Amer.* 135:434.

Myers, L. H. 1970. The role of public lands and the Bureau of Land Management in bighorn habitat management in Nevada. *Desert Bighorn Council Trans.* 14:94–106.

Neff, D. J. 1968. The pellet-group count technique for big game trend, census, and distribution: a review. *J. Wildl. Manage.* 32:597–614.

Nelson, M. C. 1966. Problems of recreational use of game ranges. *Desert Bighorn Council Trans.* 10:13–20.

Nichol, A. A. 1937. Desert bighorn sheep study—1937. Unpublished report, Ariz. Game and Fish Dept. file, Phoenix.

Nichol, A. A. 1940. The desert bighorn sheep in Arizona. Unpublished report, Ariz. Game and Fish Dept. file, Phoenix.

Nicholson, A. J. 1953. An outline of the dynamics of animal populations. *Australian J. Zool.* 2(1):9–65.

Ober, E. H. 1931. The mountain sheep in California. *Calif. Fish and Game* 17:27–39.

Odum, E. P. 1959. *Fundamentals of Ecology.* 2nd Ed. W. B. Saunders Co., Philadelphia. 546 pp.

Ogren, H. A. 1954. A population study of the Rocky Mountain bighorn sheep *(Ovis canadensis)* on Wildhorse Island. Master's thesis, Montana State Univ., Missoula. 77 pp.

Osgood, W. H. 1913. The name of the Rocky Mountain sheep. *Proc. Biol. Soc. Wash.* 26:57–62.

Osgood, W. H. 1914. Dates for *Ovis canadensis, Ovis cervina,* and *Ovis montana. Proc. Biol. Soc. Wash.* 27:1–4.

PACKARD, F. M. 1946. An ecological study of the bighorn sheep in Rocky Mountain National Park, Colorado. *J. Mamm.* 27:3–28.

PANARETTO, B. 1963. Body composition, III, The composition of living ruminants and its relationship to tritium water space. *Australian J. Agric. Res.* 14:944–52.

PAPEZ, N. J., AND G. K. TSUKAMOTO. 1970. The 1969 sheep trapping and transplant program in Nevada. *Desert Bighorn Council Trans.* 14:43–50.

PERKINS, C. J. 1967. Prescribed burning on International Paper Company lands. *Proc. 6th Ann. Tall Timbers Fire Ecology Conf.* Tall Timbers Res. Sta., Tallahassee, Florida.

PILLMORE, R. E. 1958. Problems of lungworm infection in wild sheep. *Desert Bighorn Council Trans.* 2:57–61.

PLUMMER, A. P., D. R. CHRISTIANSON, AND S. B. MONSEN. 1968. Restoring big game range in Utah. *Utah Div. Fish and Game Publ. 68-3.* 183 pp.

POURNELLE, G. H. 1964. Desert bighorn sheep at the San Diego Zoological Garden. *Desert Bighorn Council Trans.* 8:1–3.

POWELL, J. W. 1874. Report of J. W. Powell (of the exploration of the Colorado River) to the Smithsonian Institution. 291 pp.

PULLING, A. V. S. 1945. Porcupine damage to bighorn sheep. *J. Wildl. Manage.* 9:329.

PULLING, A. V. S. 1946. Monthly summary of biological accomplishments. Report in files of Desert National Wildlife Range, Las Vegas, Nevada.

RANSOM, B. H. 1911. The nematodes parasitic in the alimentary tract of cattle, sheep, and other ruminants. *Bull. 127, Bureau Animal Industry,* U.S. Dept. Agric., Washington, D.C. 132 pp.

REED, J. J. 1960. Highlights of the 1959 Arizona bighorn sheep hunt. *Desert Bighorn Council Trans.* 4:81–84.

REFFALT, W. C. 1963. Some watering characteristics of two penned bighorn sheep on the Desert Game Range, Nevada. *Desert Bighorn Council Trans.* 7:156–67.

REICHARD, G. A. 1950. *Navaho Religion—A Study of Symbolism.* Bollingen Series XVIII, Princeton Univ. Press, Princeton, New Jersey. 804 pp.

RICHERT, R. 1964. Excavation of a portion of the East Ruin, Aztec Ruin National Monument, New Mexico. *Southwestern Monuments Assn. Tech. Series* 4:21–30. Globe, Arizona.

ROBINETTE, W. L. 1966. Mule deer home range and dispersal in Utah. *J. Wildl. Manage.* 30:335–49.

ROBINSON, C. S., AND F. P. CRONEMILLER. 1954. Notes on the habitat of the desert bighorn in the San Gabriel Mountains of California. *Calif. Fish and Game* 40:267–71.

ROBINSON, R. M., T. L. HAILEY, C. W. LIVINGSTON, AND J. W. THOMAS. 1967. Bluetongue in the desert bighorn sheep. *J. Wildl. Manage.* 31:165–68.

ROSE, B. J., AND H. R. MORGAN. 1964. A priority system for Canadian wetlands preservation. *N. Amer. Wildl. Conf. Trans.* 29:249–58.

ROSENMANN, M., AND P. MORRISON. 1963. The physiological response to heat and dehydration in the guanaco, *Lama guanicoe. Physiol. Zool.* 36:45–51.

RUNNELLS, R. A. 1954. *Animal Pathology.* Iowa State Univ. Press, Ames, Iowa. 718 pp.

RUSSI, T. L., AND R. E. MONROE. 1976. Parasitism of bighorn in the Anza-Borrego Desert State Park, California. *Desert Bighorn Council Trans.* 20:36–39.

RUSSO, J. P. 1956. The desert bighorn sheep in Arizona. *Ariz. Game and Fish Dept. Wildl. Bull. No. 1.* 153 pp.

Russo, J. P. 1957. (In discussion.) *Desert Bighorn Council Trans.* 1:50.
Russo, J. P. 1967. Fifteen years of bighorn sheep hunting in Arizona. *Desert Bighorn Council Trans.* 11:86–93.
Sanchez D., Rebecca. 1976. Analysis of stomach contents of bighorn sheep in Baja California. *Desert Bighorn Council Trans.* 20:21–22.
Sanderson, G. C. 1966. The study of mammal movements—a review. *J. Wildl. Manage.* 30:215–35.
Schaafsma, Polly. 1966. A survey of Tsegi Canyon rock art. Unpublished manuscript. National Park Service, Santa Fe, New Mexico. 48 pp.
Schadle, D. 1958. Arizona's catchment then and now. *Desert Bighorn Council Trans.* 2:32–35.
Schmidt-Nielsen, K. 1964. *Desert Animals: Physiological Problems of Heat and Water.* Oxford Univ. Press, London. 277 pp.
Schmidt-Nielsen, K., E. Crawford, A. Newsome, K. Rawson, and H. Hammel. 1967. Metabolic rate of camels: effect of body temperature and dehydration. *Amer. J. Physiol.* 212:341–46.
Schmidt-Nielsen, K., Bodil Schmidt-Nielsen, S. Jarnum, and T. Houpt. 1956. Water balance of the camel. *Amer. J. Physiol.* 185:185–94.
Schmidt-Nielsen, K., Bodil Schmidt-Nielsen, S. Jarnum, and T. Houpt. 1957. Body temperature of the camel and its relation to water economy. *Amer. J. Physiol.* 188:103–12.
Scott, J. P. 1963. *Animal Behavior.* Doubleday and Co., Garden City, New York. 331 pp.
Seton, E. T. 1929. *Lives of Game Animals.* Doubleday, Doran and Co., New York. 4 vols.
Shaw, G. 1804. *Naturalist's Miscellany.* London. Vol. 15 (text to plate 610).
Shreve, F. 1942. The desert vegetation of North America. *Botanical Rev.* 8:195–46.
Shutler, R., Jr. 1961. Lost city; Pueblo Grande de Nevada. *Nev. State Mus. Archaeol. Papers 5.* Carson City, Nevada. 169 pp.
Simmons, N. M. 1961. Daily and seasonal movements of Poudre River bighorn sheep. Master's thesis, Colo. State Univ., Fort Collins. 180 pp.
Simmons, N. M. 1963. A desert bighorn study, part one. *Desert Bighorn Council Trans.* 7:72–86.
Simmons, N. M. 1964. A desert bighorn study: part two. *Desert Bighorn Council Trans.* 8:103–12.
Simmons, N. M. 1969a. Heat stress and bighorn behavior in the Cabeza Prieta Game Range, Arizona. *Desert Bighorn Council Trans.* 13:55–63.
Simmons, N. M. 1969b. The social organization, behavior and environment of the desert bighorn sheep on the Cabeza Prieta Game Range, Arizona. Ph.D. thesis, Univ. Ariz., Tucson. 145 pp.
Simmons, N. M., S. Levy, and J. Levy. 1963. Observations of desert bighorn sheep lambing, Kofa Game Range, Arizona. *J. Mamm.* 44:433.
Simmons, N. M., and J. L. Phillips. 1966. Modifications of a dye spraying device for marking desert bighorn sheep. *J. Wildl. Manage.* 30:208–9.
Sizer, B. 1967. Bighorn crosswalk. *Wildl. Views* (Ariz. Game and Fish Dept.) 14(5):10.
Sleznick, J., Jr. 1963. The bighorn sheep of Lake Mead National Recreation Area. *Desert Bighorn Council Trans.* 7:58–60.
Smith, A. E. 1969. Burro problems in the Southwest. *Desert Bighorn Council Trans.* 13:91–97.

SMITH, D. R. 1954. The bighorn sheep in Idaho. *Idaho Dept. Fish and Game Wildl. Bull. No. 1*. 154 pp.

SMITH, E. D. 1890. New Mexico game galore. *Forest and Stream* 35:309.

SMITH, E. L., J. H. WITHAM, AND W. S. GAUD. 1978. Studies of desert bighorn sheep *(Ovis canadensis mexicanus)* in western Arizona. Report on findings—year I. E. Linwood Smith & Associates, Tucson, Arizona. 149 pp.

SMITH, R. A. 1966. Records of the San Andres Refuge deer hunts. *Desert Bighorn Council Trans.* 10:36–46.

SMITH, W. 1952. Kiva mural decorations at Awatovi and Kawaika-a. With a survey of other wall paintings in the Pueblo Southwest. *Papers of the Peabody Mus. Amer. Archaeol. and Ethn.* 37. 483 pp.

SMITHSON, CARMA L., AND R. C. EULER. 1964. Havasupai religion and mythology. *Univ. Utah Anthropol. Papers No. 64*. 132 pp.

SNYDER, W. A. 1977. New Mexico's bighorn sheep reintroduction program. *Desert Bighorn Council Trans.* 21:3.

SPARKS, D. R., AND J. C. MALECHEK. 1968. Estimating percentage dry weight in diets using a microscopic technique. *J. Range Manage.* 21:264–65.

SPENCER, C. C. 1943. Notes on the life history of the Rocky Mountain bighorn sheep in the Tarryall Mountains of Colorado. *J. Mamm.* 24:1–11.

STEEN, C. R., L. M. PIERSON, AND KATE P. KENT. 1962. Archaeological studies in the Tonto National Monument, Arizona. *Southwestern Monuments Assn. Tech. Series 2*. Globe, Arizona. 126 pp.

ST. JOHN, K. P., JR. 1965. Competition between desert bighorn sheep and feral burros for forage in the Death Valley National Monument. *Desert Bighorn Council Trans.* 9:89–92.

STOCK, A. D., AND W. L. STOKES. 1969. A re-evaluation of Pleistocene bighorn sheep from the Great Basin and their relationship to living members of the genus *Ovis*. *J. Mamm.* 50:805–7.

STODDARD, H. L. 1931. *The Bobwhite Quail, Its Habits, Preservation and Increase*. Charles Scribner and Sons, New York. 559 pp.

STRONG, E. 1969. *Stone Age in the Great Basin*. Biniforts and Mort, Portland, Oregon. 274 pp.

SUGDEN, L. G. 1961. *The California Bighorn in British Columbia*. Dept. Recreation and Conservation, Victoria. A. Sutton, Queen's Printer, Ottawa. 58 pp.

SUMNER, E. L. 1948. An air census of Dall sheep in Mount McKinley National Park. *J. Wildl. Manage.* 12:302–4.

SUMNER, E. L. 1957. Report of Joshua Tree bighorn survey, July 1957. Unpublished report, National Park Service files, Joshua Tree National Monument, California.

SUMNER, E. L. 1959. Effects of wild burros on bighorn in Death Valley National Monument. *Desert Bighorn Council Trans.* 3:4–8.

SUSHKIN, P. P. 1925. The wild sheep of the world and their distribution. *J. Mamm.* 6:145–57.

SUTTON, ANN, AND M. SUTTON. 1966. *The Life of the Desert*. McGraw-Hill Book Co., New York. 232 pp.

TABER, R. D., AND R. F. DASMANN. 1958. The black-tailed deer of the chaparral, its life history and management in the north coast range of California. Calif. Dept. Fish and Game, *Game Bull.* 8. 163 pp.

TAYLOR, C. R. 1968. Hygroscopic food: a source of water for desert antelopes. *Nature* 219:181–82.

TAYLOR, C. R. 1969a. The eland and the oryx. *Sci. Amer.* 220:88–95.

TAYLOR, C. R. 1969b. Metabolism, respiratory changes, and water balance of an antelope, the eland. *Amer. J. Physiol.* 217:317–20.
TAYLOR, C. R., AND C. P. LYMAN. 1967. A comparative study of the environmental physiology of an East African antelope, the eland, and the Hereford steer. *Physiol. Zool.* 40:280–95.
TAYLOR, R. E. L. 1973. Disease losses in Nevada bighorn. *Desert Bighorn Council Trans.* 17:47–52.
TAYLOR, R. E. L. 1976. Mortality of Nevada bighorn sheep from pneumonia. *Desert Bighorn Council Trans.* 20:51–52.
TEVIS, L., JR. 1959. Man's effect on bighorn in the San Jacinto–Santa Rosa Mountains. *Desert Bighorn Council Trans.* 3:69–75.
THOMAS, J. W., R. M. ROBINSON, AND R. G. MARBURGER. 1967. A rope truss for restraining deer. *J. Wildl. Manage.* 31:359–61.
THWAITES, R. G. (ed.). 1905. Pattie's personal narrative, 1824–1830, Vol. XVIII of *Early Western Travels 1748–1846*. The Arthur H. Clark Co., Cleveland, Ohio. 379 pp.
TILL, A., AND A. DOWNES. 1962. The measurement of total body water in sheep (tritiated H_2O). *Australian J. Agric. Res.* 13:335–42.
TINLEY, K. 1969. Dik-dik, *Madoqua kirki,* in southwest Africa: notes on distribution, ecology, and behavior. *Madoqua* 1:7–33.
TRAINER, D. O., AND M. M. JOCHIM. 1969. Serological evidence of bluetongue in wild ruminants of North America. *Amer. J. Vet. Res.* 30:2007–11.
TSUKAMOTO, G. K. 1970. Nevada's 1969 bighorn sheep hunt. *Desert Bighorn Council Trans.* 14:54–62.
TURNER, J. C. 1970. Water consumption of desert bighorn sheep. *Desert Bighorn Council Trans.* 14:189–96.
TURNER, J. C. 1973. Water energy and electrolytic balance in the desert bighorn sheep *(Ovis canadensis cremnobates* Elliot). Ph.D. thesis, Univ. Calif. Riverside. 150 pp.
VAN DEN AKKER, J. B. 1960. Human encroachment on bighorn habitat. *Desert Bighorn Council Trans.* 4:38–40.
VOSDINGH, R. A., D. O. TRAINER, AND B. C. EASTERDAY. 1968. Experimental bluetongue disease in white-tailed deer. *Canad. J. Comp. Med. Vet. Sci.* 32:382–87.
WALLACE, W. J. (n.d.). Archaeological explorations in the northern section of Death Valley National Monument. Unpublished manuscript, Death Valley Natl. Monument, California. 109 pp.
WATKINS, R. A., AND W. W. REPP. 1964. Influence of location and season on the composition of New Mexico range grasses. *Bull. 486,* Agric. Exp. Sta., New Mexico State Univ., Las Cruces.
WATSON, R. M. 1969. Aerial photographic methods in censuses of animals. *In* R. G. Swank, R. M. Watson, G. H. Freeman, and T. Jones (eds.). Proceedings of the workshop on the use of light aircraft in wildlife management in East Africa. *East African Agric. and Forestry J.* 34 (Spec. Issue):32–37.
WEAVER, R. A. 1957. (In discussion.) *Desert Bighorn Council Trans.* 1:70–76.
WEAVER, R. A. 1958. Game water development on the desert. *Desert Bighorn Council Trans.* 2:21–27.
WEAVER, R. A. 1959. Effects of burro on desert water supplies. *Desert Bighorn Council Trans.* 3:1–3.
WEAVER, R. A. 1961. Bighorn and coyotes. *Desert Bighorn Council Trans.* 5:34–37.

WEAVER, R. A. 1968. A survey of the California desert bighorn *(Ovis canadensis)* in San Diego County. Calif. Dept. Fish and Game. Mimeo. 26 pp.

WEAVER, R. A. 1972a. Conclusion of the bighorn investigation in California. *Desert Bighorn Council Trans.* 16:56–65.

WEAVER, R. A. 1972b. Desert bighorn sheep in Death Valley National Monument and adjacent areas. Calif. Dept. Fish and Game. Mimeo. 23 pp.

WEAVER, R. A. 1972c. Feral burro survey. Calif. Dept. Fish and Game. PR Proj. W-51-R-17. Completion Report. July 1, 1968–June 30, 1972. Mimeo. 14 pp.

WEAVER, R. A., AND J. M. HALL. 1971. Desert bighorn sheep in southeastern San Bernardino County. Calif. Dept. Fish and Game. Mimeo. 28 pp.

WEAVER, R. A., AND J. M. HALL. 1972. Bighorn sheep in the Clark, Kingston and Nopah mountain ranges (San Bernardino and Inyo Counties). Calif. Dept. Fish and Game. Mimeo. 17 pp.

WEAVER, R. A., AND J. L. MENSCH. 1969. A report on desert bighorn sheep in eastern Imperial Co. Calif. Dept. Fish and Game. PR Proj. W-51-R-14B. Mimeo. 17 pp.

WEAVER, R. A., AND J. L. MENSCH. 1970a. Bighorn sheep in southern Riverside County. Calif. Dept. Fish and Game Wildl. Manage. Admin. Rept. No. 70-5. Mimeo. 36 pp.

WEAVER, R. A., AND J. L. MENSCH. 1970b. Bighorn sheep in northwestern San Bernardino and southwestern Inyo Counties. Calif. Dept. Fish and Game. Mimeo. 17 pp.

WEAVER, R. A., J. L. MENSCH, AND W. V. FAIT. 1968. A survey of the California desert bighorn *(Ovis canadensis nelsoni)* in San Diego County. Calif. Dept. Fish and Game. PR Proj. W-51-R-13B. Mimeo. 21 pp.

WEAVER, R. A., J. L. MENSCH, W. TIMMERMAN, AND J. M. HALL. 1972. Bighorn sheep in the San Gabriel and San Bernardino mountains. Calif. Dept. Fish and Game. Mimeo. 38 pp.

WEAVER, R. K. 1973. Progress at Aravaipa. *Desert Bighorn Council Trans.* 17: 117–22.

WELCH, R. D. 1969. Behavioral patterns of desert bighorn sheep in south-central New Mexico. *Desert Bighorn Council Trans.* 13:114–29.

WELLES, R. E. 1957. Status of the bighorn sheep in Death Valley. *Desert Bighorn Council Trans.* 1:22–25.

WELLES, R. E. 1960. Progress report on current Death Valley burro survey. *Desert Bighorn Council Trans.* 4:85–87.

WELLES, R. E. 1961. How much room does a bighorn need? *Desert Bighorn Council Trans.* 5:99–102.

WELLES, R. E. 1962. What makes a valid observation? *Desert Bighorn Council Trans.* 6:29–40.

WELLES, R. E. 1965. Progress report on Joshua Tree National Monument bighorn research. *Desert Bighorn Council Trans.* 9:49–52.

WELLES, R. E. 1967. Status of the bighorn in Death Valley and Joshua Tree National Monuments. *Desert Bighorn Council Trans.* 11:59–63.

WELLES, R. E. 1968. Unanswered questions. *Desert Bighorn Council Trans.* 12: 70–75.

WELLES, R. E., AND FLORENCE B. WELLES. 1961a. The bighorn of Death Valley. U.S. National Park Service Fauna Ser. No. 6. Washington, D.C. 242 pp.

WELLES, R. E., AND FLORENCE B. WELLES. 1961b. The feral burro in Death Valley. *Desert Bighorn Council Trans.* 5:32–33.

WELLES, R. E., AND FLORENCE B. WELLES. 1966. The water book. Unpublished report, National Park Service files, Joshua Tree National Monument, California.

WELSH, G. 1964. Boat surveys as a technique in bighorn sheep classification counts on Lakes Mead and Mojave in northwestern Arizona. *Desert Bighorn Council Trans.* 8:37–42.

WETHERILL, M. A. 1954. A Paiute trap corral on Skeleton Mesa, Arizona. *Plateau* 26:109–38.

WILLIAMS, O. B. 1969. An improved technique for identification of plant fragments in herbivore feces. *J. Range Manage.* 22:51–52.

WILSON, L. O. 1966. Research and future rehabilitation of the bighorn sheep in southeastern Utah. *Desert Bighorn Council Trans.* 10:56–58.

WILSON, L. O. 1968. Distribution and ecology of the desert bighorn sheep in southeastern Utah. *Utah Div. Fish and Game Publ. 68-5.* 220 pp.

WILSON, L. O. 1970. Whitey, the unique desert bighorn. *Desert Bighorn Council Trans.* 14:116–22.

WILSON, L. O. 1971. The effect of free water on desert bighorn home range. *Desert Bighorn Council Trans.* 15:82–89.

WILSON, L. O. (ed.). 1975. Guidelines for re-establishing and capturing desert bighorn. In *The Wild Sheep in Modern North America,* J. B. Trefethen (ed.). Boone and Crockett Club, New York. 302 pp.

WINKLER, C. K. 1977. Status of the Texas desert bighorn program. *Desert Bighorn Council Trans.* 21:4.

WINSHIP, G. P. 1896. The Coronado expedition, 1540–1542. Bur. Amer. Ethn., 14th Ann. Rept. (1892–93), Part 1. Washington, D.C. pp. 329–613.

WOODBURY, A. M. 1965. Notes on the human ecology of Glen Canyon. *Univ. Utah Anthropol. Papers 74.* 62 pp.

WOODGERD, W. R. 1964. Population dynamics of bighorn sheep on Wildrose Island. *J. Wildl. Manage.* 28:381–91.

WOODGERD, W. R., AND D. J. FORRESTER. 1962. Observability of colored earmarkers on Rocky Mountain bighorn lamb *(Ovis c. canadensis). Desert Bighorn Council Trans.* 6:65–67.

WRIGHT, F. E. 1968. Bighorn sheep transplant program—interim report. Nev. Fish and Game Dept. Spec. Rept. 68:3–15.

YOAKUM, J. D. 1963. Re-establishing native bighorn ranges. *Desert Bighorn Council Trans.* 7:122–25.

YOAKUM, J. D. 1964. Bighorn food habits-range relationships in the Silver Peak Range, Nevada. *Desert Bighorn Council Trans.* 8:95–102.

YOAKUM, J. D. 1966. Comparison of mule deer and desert bighorn seasonal food habits. *Desert Bighorn Council Trans.* 10:65–70.

Index

Abundance, 6, 40–51
Acalypha, 93
Accidents, 184–85
Agastache, 87
Agave, 90, 247; desert *(deserti)*, 70, 88; lechuguilla *(lechuguilla)*, 74; maguey *(shawii)*, 72; Palmer *(palmeri)*, 74; Utah *(utahensis)*, 86
Age classes, 218–20
Age determination, 170; by horns, 166–70; by size and shape, 171; by teeth, 171
Aggression, 137
Aircraft: in bighorn capture, 279–81; in bighorn transport, 281; disturbances by, 120, 297; in materials transport, 318; surveys with, 260–65
Allthorn *(Koeberlinia)*, 74
Amaranthus, 93
Ambrosia, 93
Ancestral areas, 2
Ancestral patterns, 76
Aoudad (Barbary sheep), 2
Apricot, desert *(Prunus)*, 88
Archaeological evidence, 7–8, 39
Arizona Desert Bighorn Society, 308
Art, rock, 10–28, 32
Artifacts: basketry, 38; bone, 37; hide, 38; horn, 36–37
Argali, 2
Arrival of sheep: in North America, 3; in Southwest, 7
Ash, singleleaf *(Fraxinus)*, 83
Atlatls, 8, 13, 19–22, 37
Atrichoseris, 238
Audubon (badlands) bighorn, 5

Bacterial diseases, 172
Badlands (Audubon) bighorn, 5
Baja California Mountain Desert, 72–73; food plants, 93

Bands (groups), 126–27, 221–22
Barbary sheep (aoudad), 2
Bassia, fivehook *(Bassia)*, 83
Beardgrass *(Andropogon)*, 85
Beargrass *(Nolina)*, 70, 95, 182, 185
Beds, 78, 137–40, 141–43, 244–46, 300
Bees, 197, 214
Bee sage *(Hyptis)*, 88
Behavior, 124–44, 292–93; aggression, 137; atypical, 257; changes in group composition, 126–27, 221; changes in group size, 126–27; communication, 133–35; daily activities, 133–37, 245, 259; dispersals, 130; and environment, 141–43; of ewes and lambs, 125–26; nighttime movements, 143, 267; play, 140–41; of rams, 127; rut, 125, 137, 146, 166; seasonal movements, 25, 127–30; sleep, 140; social dominance and leadership, 135–37; and topography, 144
Bentgrass *(Agrostis)*, 66
Bering land bridge, 2, 3
Bharal, 2
Bibliography, 345–63
Birds, 214
Bitterbrush *(Purshia)*, 68
Blackbrush *(Coleogyne)*, 68, 69, 83, 86, 96
Black sheep, 5
Bladderworm, 176, 179
Blinds: hunting, 8; water-hole, 268–69
Bluegrass *(Poa)*, 66, 82, 85, 94
Blue sheep, 2
Bluestem *(Andropogon)*, 69
Bluetongue, 174, 283
Bobcat, 11, 186, 190
Body conformation, 52–54, 156–61, 170
Boojum *(Idria)*, 72
Boone and Crockett Club, 57, 253–54, 339, 343
Bovidae (family), 2

[365]

Breeding. *See* Reproduction
Brittle bush *(Encelia)*, 70, 71, 72, 73, 90, 91, 93, 204
Brome *(Bromus)*, 66, 69, 83
Brooming of horns, 165, 166
Browse. *See* Food
Brucellosis, 174
Buckwheat, wild *(Eriogonum)*, 70, 85, 86, 88
Bud sage *(Artemisia)*, 68
Burns (habitat), 311
Burro bush: *Ambrosia*, 69; *Hymenoclea*, 90
Burro grass *(Scleropogon)*, 74
Burros, wild, 11; competition from, 107, 202–7, 314
Bur sage *(Ambrosia* or *Franseria)*, 73, 203, 205
Butting, 62, 91, 95, 118, 163–64, 165, 247, 257

Cactus; 90, 91; barrel *(Echinocactus)*, 70, 72, 74, 93, 102, 274; cardón *(Pachycereus)*, 72; cholla *(Opuntia)*, 91; hedgehog *(Echinocereus)*, 102; *Machaerocereus*, 72; pencil cholla *(Opuntia)*, 102; pincushion *(Mammillaria)*, 92; prickly pear *(Opuntia)*, 72, 74, 95; saguaro *(Carnegiea)*, 70, 91, 247; as water source, 91, 102
California bighorn, 5, 6, 44, 45, 149, 291
Candelilla *(Euphorbia)*, 74
Capture: by drug, 278–81; transporting the captured, 281–82; by trapping, 273–78
Carvings, 33, 34, 37
Catclaw *(Acacia)*, 72, 74, 88, 89, 90, 204
Cattle. *See* Competition; Management
Censusing, 41, 260–72
Ceramic figures, 34
Ceremonial use, 28–30, 34
Cherry *(Prunus)*, 86
Chihuahuan Desert, 73–74; food plants, 94–95, 98
Cholla *(Opuntia)*, 91
Chuckwalla, 91
Chukar, 215
Classification of bighorns. *See* Taxonomy
Cliff rose *(Cowania)*, 68, 85
Climate (habitat): changes in, 310; drought, 79, 183; effect on lambs, 151; effect on survival, 89, 100–1, 183–84; evaporation, 79; precipitation, 78, 326–27; temperature, 78–79
Coccidia, 176, 179, 180
Coffeeberry *(Rhamnus)*, 86
Color, of hair and skin, 54, 153–54, 253–57
Colorado Desert, 71–72; food plants, 88–89
Communication, 133–35
Competition, 196, 197–216, 306, 313, 328; within bighorn herds, 215–16; from birds and bees, 214; from cattle, 74, 207–8, 209–13; from deer, 183, 197–202; from domestic sheep, 6, 74, 207, 208–9, 291; from elk, 214; from exotic species, 314; from goats, 74, 213; from horses, 207, 213; from jackrabbits and rodents, 214; from javelina, 214; from man, 288–309, 314–15
Coral-vine *(Antigonon)*, 93
Cottontail, desert, 11
Cottontop *(Tricachne)*, 94
Counts. *See* Population surveys
Coyote: in Coso Range, 11; as predator, 186, 187–90, 195; as tapeworm host, 179
Cream bush *(Holodiscus)*, 70, 86
Creosote bush *(Larrea)*, 11, 68, 70, 71, 73, 74, 203
Crowding, 74, 217
Curiosity, 122–23, 294
Cyprian (red) sheep, 2

Daily activities, 133–37, 245
Daily movements, 137–41, 259
Dall sheep, 5, 8, 124, 149, 161
Dams, 10
Death. *See* Mortality
Death trap, 107, 184
Deer, 6, 136; competition from, 183, 197–202; in Coso Range, 11
Definition of desert bighorn, 6
Dehydration, 101, 112
Desert iguana, 91
Deserts: Baja California Mountain, 72–73; Chihuahuan, 73–74; Colorado, 71–72; definition of, 64; extent of, 64; Great Basin, 66–67; Mohave, 68–70; Painted, 67–68; Sonoran, 70–71
Desert thorn *(Lycium)*, 90
Desert trumpet *(Eriogonum)*, 86
Dew, 78
Dietary deficiencies, 181
Disappearance of bighorn, 6, 25–26, 45, 198, 205, 211
Diseases: bacterial, 172–74; effects on numbers, 291; fungal, 174
Dispersals, 130
Ditches as barriers, 302, 303, 315
Dogs, 9, 14, 299–300
Dominance, social, 106, 135–37, 215–16
Douglas fir *(Pseudotsuga)*, 66, 74
Dove, white-winged, 214
Downy chess *(Bromus)*, 82, 85
Drawings, rock art: color of bighorn, 8, 10–28, 29, 32, 41; with atlatls, 13, 14, 20; with bows, 14, 22; ceremonial, 19; of dogs attacking bighorn, 14; use as location markers, 19
Drinking. *See* Water
Droppings. *See* Pellets
Dropseed *(Sporobolus)*, 69
Drought, 79
Drug capture, 278–81

Eagles, 24, 186, 193–94
Ears: compared with goats', 2; plugged, 114
Ecologic entity, desert bighorn as an, 6
Effigies, 32, 33, 4
Elephant tree: *Bursera*, 70; *Pachycormus*, 73

Elk, 214
Enclosures, 47, 48, 192–93, 283–85
Enterotoxemia, 173
Environment: and behavior, 141–43; and growth and development, 171
Ephemeral plants, 87, 89, 92, 98
Estrus period, 146
Eugenics, 195
Euphorbia, 93, 95, 238; *platysperma*, 72
Evaporation, 109–11, 317, 327
Evening primrose *(Oenethera)*, 87
Evolution, 3
Eyes, 58

Fairy duster *(Calliandra)*, 90, 93
Family relationships, 125–26, 137
Fat (insulation), 56
Fear, 119–20
Fences, 302–3, 315
Fendler bush *(Fendlera)*, 95
Fern, 95
Fescue *(Festuca)*, 66, 85
Fetal development, 152–53
Field identification of individual bighorn, 249–59; actions, 257; artificial markings, 257–59; behavioral characteristics, 257; body markings, 253–55; family characteristics, 256–57; hair, 253–55; horns, 166–70, 250–53; injuries, 255–56; lambs, 158–60; miscellaneous, 170, 256, 264; sex identification, 158–60, 249
Filaree *(Erodium)*, 86, 92, 93
Finch, house, 214
Fires, 301, 311
First mention of bighorn, 3, 41
First published picture of bighorn, 3
Fleece. *See* Hair
Fluffgrass *(Tridens)*, 69, 85, 90, 95
Food: amount required, 76; availability and distribution, 76, 98–99, 217; browse, 82, 85–86, 88, 90, 93, 95, 97; competition for, 200–1, 203–5, 209–1, 213; forbs, 82, 86–87, 88–89, 92, 93, 95, 97; found in canyons, 98; grasses, 81–82, 84, 85, 88, 89, 90, 93, 94–95, 97; habits, 80–99, 247; improvement in forage, 311–13; response to new growth by bighorn, 92, 95; shortage, 183–84; of young bighorn, 96
Food, use of bighorn as, 39, 291, 336, 341
Forbs. *See* Food
Fox, gray, 186, 189–90
Fraternity of the Desert Bighorn, 339
Fungal disease, 174

Gaits, 60–61
Galleta *(Hilaria)*, 66, 69, 70, 72, 81, 83, 85, 90, 96
Game laws, 291
Genetics, 171, 185, 195
Glands: feet, 2, 60; lachrimal, 2, 60, 247; sweat, 60, 62; tail (lack of), 2

Goats, 2, 74, 213
Gourd, wild *(Cucurbita)*, 14
Grama *(Bouteloua)*, 67, 70, 74, 85, 90, 94
Grasses. *See* Food
Grazing. *See* Competition; Man, impact of; Management
Grease bush, spiny *(Glossopetalon)*, 86
Greasewood *(Sarcobatus)*, 66
Great Basin Desert, 66–67
Green molly *(Kochia)*, 68, 69
Group mortality, 185
Growth and development, 152–71; factors influencing, 171; fetal development, 152–53; general, 156–63; horns, 165–70; lambs, 153–56; skull, 163–64; teeth, 171
Guayule *(Parthenium)*, 74
Guzzlers, 317, 318

Habitat, 6, 64–79, 98; burns, 311; carrying capacity, 106, 195; classification of, 320–35; clearing, 312; effects of man on, 300–27; evaluation of, 319, 320–35; evaporation and, 327; factors of, 320–35; index, 335; and precipitation, 326–27; protection and improvement of, 310–19; requirements, 6, 74–76; and topography, 77, 78, 325; water sources, 316–18, 319, 327–28
Hair, 56, 248, 253–55
Hairworms, 176, 180
Harassment: by man, 25–26, 43, 75, 293, 306, 314; by other bighorn, 75
Harvest. *See* Hunting
Headdresses, Indian, 28, 32
Hearing, 114
Herbicides, 312
Highstrung bighorn, 75
Highways, as barriers, 302–3, 315
Holly, desert *(Atriplex)*, 69, 85
Home range, 107, 130–33
Honeysweet *(Tidestromia)*, 86
Hop sage *(Grayia)*, 66, 68, 69, 85
Horns: age determination by, 166–70; and dominance, 135–36; in field identification, 250–53; goat, 2; growth of, 165; piled by Indians, 30, 31; record heads, 57; rings, 166; sheep, domestic, 2; size, 56–57; uses by bighorn, 58
Horsebrush *(Tetradymia)*, 68
Horsemint *(Monardella)*, 86
Horses, 11, 213, 247
Human (modern man) impact. *See* Man, impact of
Hunting, 289–92, 336–42; by aborigines, 8–10, 13, 19, 22–26, 28, 39, 265, 336–37; communal, 9, 10; excessive, 6, 44, 341–42; harvest objectives, 338; by Indians generally, 291–92; in Mexico, 337; by Navajo Indians, 10, 291; by Papago Indians, 6, 10, 30–31, 291; poaching, 337; on refuges, 307; regulations, 338–40; results, 338–40; seasons, 338; by trapping, 9–10

Hunt shamans, 24–25
Hybrid with domestic sheep, 209

Identification, field. *See* Field identification
Immobilization. *See* Drug capture
Impact of man. *See* Man
Inbreeding, 185
Indian wheat *(Plantago)*, 92
Injuries, 255–56
Insulation from heat, 56
Intelligence, 115–23; curiosity, 122–23; fear, 119–20; learning, 115–18; memory and recognition, 118–19; response to noise, 121–22; wariness, 120–21
Introductions, 192–93, 196, 218, 231
Ironwood *(Olneya)*, 70, 72, 89, 90, 91, 93

Jackrabbit, 11, 91, 214, 247
Jaguar, 186, 193
Javelina, 214
Jimson weed *(Datura)*, 182
Jojoba *(Simmondsia)*, 70, 72, 88, 89, 90, 93, 209–10, 211
Joshua tree *(Yucca)*, 11, 68, 85
Juniper *(Juniperus)*, 11, 66, 69, 86, 311

Kachina, mountain sheep, 30
Kenai sheep, 5, 8
Kidneys, 111–12

Lambing, 78, 128–29, 148, 221
Lambing areas, 127–28, 154, 246, 300
Lambs: birth, 148, 153; bummer lambs, 156; color, 153–54; distinguishing features, 158–61; and ewes, 125–26, 154–56; freezing, 184; raising artificially, 156; range, 131; surveys, 271–72; survival, 106, 183–84; survival rates, 156; voice, 134, 154; weaning, 155–56
Lava beds (California) bighorn, 5
Leadership, 135–37, 293
Learning, 115–18
Lechuguilla *(Agave)*, 74
Leptospirosis, 174
Life tables, 222–30
Limber bush *(Jatropha)*, 73
Limiting factors, 59, 77, 100, 106, 195, 201, 204, 211
Livestock. *See* Competition; Management
Long-distance movements, 133, 259
Lung diseases, 172–73, 177, 179–80
Lungworm, 177, 179–80

Mallow *(Sphaeralcea)*: big, 95; globe, 92
Man, impact of, 288–309, 329–30; aircraft disturbances, 297; air pollution, 304; barriers, 302–3, 315; dogs, 299–300; future impact, 308–9; on habitat, 299–300; human odors, 299; hunting. *See* Hunting; livestock, 44, 301–2, 313; mineral exploration, 289–90; motor boats, 295–96; motor vehicles, 295; mountain men, 289; noises, 297–99; ranchers and settlers, 290–91; recreationists, 44, 306–7, 314–15; residence by man, 44, 304–5; riding horseback, 294–95; ski lifts and tramways, 296, 305; Spanish era, 288–89; tolerance by bighorn, 293, 294, 304, 305; use of water, 304–7, 314; walking, 292–94
Management: burning, 301, 311–12, 319; clearing, 312, 319; of food supplies, 97–99, 311–13; general, 185, 196, 265, 331–32; herbicides, 312; land, 301–2, 307–9, 310–19; land planning, 314–15; livestock and bighorn, 312–13; 319; public attitudes, 282, 283, 308–9; reseeding, 312; of water supplies, 107–8, 319
Maple *(Acer)*, 74, 93
Marco Polo sheep, 2
Mariposa lily *(Calochortus)*, 83
Markings: artificial, 132, 249, 257–59, 267, 285–87; body, 253–55; natural, 249, 250–53, 267–68; on remains, 287
Maturity: physical, 161, 163; sexual, 145
Measurements: chest, 52; ear, 161–62; girth, 161–62; height, 52, 161–62; horns, 56–57; lambs, 153, 159–61; legs, 52, 161–62; length, 161–62; skull, 164; tail, 161–62; weight, 52, 54, 161–62
Memory and recognition, 118–20
Mesquite *(Prosopis)* 70, 73, 74, 88, 89, 90
Metabolism, 62, 101
Mexican bighorn, 5, 6
Mexican tea *(Ephedra)*, 74
Midges, biting, 174
Milk-aster *(Stephanomeria)*, 85
Milkweed *(Asclepias)*, 14
Mineral deficiencies, 181
Mineral requirements, 96–97, 150, 181
Mining, 83, 289–90, 314–15
Mistletoe, desert *(Phoradendron)*, 88, 91
Mock orange *(Philadelphus)*, 74
Mohave Desert, 68–70; food plants, 84–88, 98
Molt, 54–55, 248
Molt pattern, 54–55
Mormon tea *(Ephedra)*, 11, 66, 68, 83, 85, 86, 88, 90, 91, 95, 203
Mortality, 112, 217, 220, 228–30, 234–35; group, 185; lamb, 224
Motor vehicles and boats, impact of, 295–96
Mouflon, 2
Mountain lion, 186, 190–93, 243
Mountain mahogany *(Cercocarpus)*, 66, 68, 70, 74, 85, 95; desert, 86
Mountain men, 289
Movements. *See* Behavior
Muhly *(Muhlenbergia)*, 69, 90, 93, 95
Mythology, Indian, 30–2

Needlegrass *(Stipa)*, 67, 72, 82, 85, 94
Nelson bighorn, 5, 6
Nematodes (stomach worms), 174, 176
Nighttime movements, 143, 267
Noise, response to, 121–22, 297–99

Nose bots, 178
Numbers: in Arizona, 41–44; in Baja California 49–50; in California, 44–45; in Chihuahua, 50; in Coahuila, 50; in Colorado-Wyoming, 49, 291; grand totals, 51; in Nevada, 45–46; in New Mexico, 46–47; prehistoric, 6, 9; in Sonora, 51; in Texas, 47; in Utah, 48
Nursery system, 155, 221
Nutrition: and horn growth, 166; in lamb production, 150–51

Oak *(Quercus):* live, 73, 74; scrub, 85, 88
Ocelot, 186, 193
Ocotillo *(Fouquieria),* 70, 72, 73, 74, 90, 203–4
Origin and relationships of bighorn, 1–6

Painted Desert, 67–68; food plants, 81–83, 98
Paintings: rock, 8, 10, 14, 27–28; sand, 18, 24; wall, 30, 32–33
Paloverde *(Cercidium* and *Parkinsonia),* 70, 71, 73, 89, 90, 83
Panicum (grass), 93
Panting, 60, 110, 141
Parasites, 174–80
Parsley, wild (Umbelliferae), 87
Parturition, 119, 125, 148
Patination, rock, 13, 14, 17–18
Pelage. *See* Hair
Pellets: analyses, 80; at beds, 240; contents, 240; counting, 239, 270; like deer's, 238; lambs', 240; relationship to sex and age, 240; weathering, 238–39
Peninsular bighorn, 5, 6, 49
Penstemon *(Penstemon),* 87
Peppergrass *(Lepidium),* 93
Petroglyphs. *See* Drawingss, rock art
Physalis, 238
Physical characteristics, 52–63
Pictures, rock. *See* Drawings, rock art
Pincushion, pebble *(Chaenactis),* 87
Pine *(Pinus),* 66, 69, 74
Pinworm, 177, 179
Pinyon *(Pinus),* 11, 66, 68, 69, 311
Play, 140–41
Pleistocene epoch (Ice Age), 2, 3, 13, 26
Pneumonia, 172, 173, 180
Poaching, 48, 337, 340–41
Poisonous plants, 181–82, 185
Poppy, desert *(Arctomecon),* 87
Population(s): composition, 218–19; decline, 233–35; dynamics, 217–35; effect of hunting, 340; fluctuation, 320–31, 340; group composition, 221–22, 223; growth, 231; increase, 218; life tables, 222; survey methods. *See* Population surveys; turnover, 235
Population surveys: aerial, 260–65; boat, 269–70; foot and horseback, 270–72; lamb, 271; transects, 204, 270–71; water-hole, 54, 221, 265–69
Porcupine damage to bighorn, 182
Precipitation, 40, 78, 326–27

Predators, 108, 184, 186–96
Prickly pear *(Opuntia),* 72, 74, 95
Protozoan blood parasites, 178–79
Pulse, 63

Quail, Gambel, 214

Rabbitbrush *(Chrysothamnus),* 11, 66, 68
Races (subspecies), 5–6
Ratany *(Krameria),* 88, 90
Ratios: lamb-ewe, 149, 219, 220; sex, 149–50, 218, 220, 340
Redberry *(Rhamnus),* 70, 86
Red (Cyprian) sheep, 2
Refuges, 307
Relationships, taxonomic, 1–6
Relationships with other animals. *See* Competition
Releases. *See* Transplants
Reproduction, 145–51; gestation, 146; lambing, 148; parturition, 119, 125, 148; percentage of productive ewes, 149; puberty, 145–46; relation to nutrition, 150–51; reproductive periods, 125, 146; reproductive potential, 148–49; rut, 125, 137, 146, 166; sex ratios, 149–50, 220
Reproductive potential, 148–49
Respiration, 63, 110–11, 141
Restocking, 282–85
Ricegrass, Indian *(Oryzopsis),* 67, 68, 82, 83, 85
Rimrock (California) bighorn, 5
Rock art, 10–28
Rock moss *(Selaginella),* 92
Rocky Mountain bighorn, 5, 8, 25, 45, 47, 49, 124, 131, 161
Roundworms (general), 179
Rubber weed *(Hymenoxys),* 83
Russian thistle *(Salsola),* 83
Rutting season, 125, 137, 146, 166
Ryegrass *(Elymus),* 68, 82

Sacahuista. *See* Beargrass
Sacaton *(Sporobolus),* 70, 85
Sagebrush *(Artemisia),* 11, 66, 68, 85, 203
Saguaro *(Carnegiea),* 70, 91, 247
Salt bush *(Atriplex),* 68, 69, 85, 93, 205
Salt cedar *(Tamarix),* 107, 118
Saltgrass *(Distichlis),* 69, 85
Salvia, 93
Sarcocyst, 176
Scabies, 6, 177, 178, 209
Screw worms, 178
Seasonal movements and use, 127–30, 328–29
Sego lily *(Calochortus),* 83
Senses, 113–15; hearing, 114; smell, 114; vision, 114
Sex identification, 158–60, 165, 167–70, 249
Shadscale *(Atriplex),* 66, 68, 85
Shedding. *See* Molt
Sheep, domestic, 152. *See* Competition; Man, impact of; Management

Sign reading, 236–49; beds, 244–46; feeding, 247; odor, 248; pellets, 237–40; remains, 247–48; sign interpretation, 248–49; signposts, 247; tracks, 241–43; trails, 243–44; urine, 240–41
Silktassel *(Garrya)*, 86, 95
Silver bush *(Ditaxis)*, 90, 91
Sinusitis, 178
Size. *See* Measurements
Skin color, 56
Skull: abnormalities, 59–60, 173; growth of, 163–64
Skunk bush *(Rhus)*, 74, 85
Sleep, 140
Smell (sense), 114–15
Smoke tree *(Dalea)*, 71
Snakeweed *(Gutierrezia)*, 66, 68, 70, 71
Snowberry *(Symphoricarpos)*, 83, 86
Social dominance, 135–37
Social relationships: with burros, 207; with cattle, 212–13; with deer, 201–2; with goats, 213; with other bighorn. *See* Behavior
Society for the Conservation of Bighorn Sheep, 308
Sonic booms, 122, 298, 307
Sonoran Desert, 70–71; food plants, 89–92, 98
Sotol *(Dasylirion)*, 74, 75
Space, need for, 74–76, 217, 300
Speeds, 62
Spiderling *(Boerhaavia)*, 93
Spineflower *(Chorizanthe)*, 89
Squirreltail *(Sitanion)*, 66
Starvation, 183
Stickweed *(Stephanomeria)*, 69
Stomach analyses, 80, 81, 84, 89–90, 92, 93, 94, 97, 98
Stone sheep, 5, 8
Streptococcus, 172
Subspecies (races), 5–6
Sumac, laurel-leaf *(Rhus)*, 88
Survival rates, 219, 222–30; lambs, 156
Sweating, 62–63, 111
Sweet bush *(Bebbia)*, 69, 72, 85, 205
Swimming, 302

Tag-along (lamb), 114, 115–17, 119
Tanglehead *(Heteropogon)*, 95
Tapeworms, 176, 179
Taxonomy, 5–6
Teeth: abnormalities, 59–60, 171, 174; age determination by, 171; dental formula, 59; wear, 59–60, 171
Telemetry, 133, 339
Temperature: body, 62, 63; climatic, 78–79, 245; regulation, 62–63
Tertiary period, 2
Thinhorn (Dall) sheep, 5
Thistle, Mojave *(Cirsium)*, 86
Three-awn *(Aristida)*, 90, 93
Ticks, 114, 174, 175, 178, 180
Timber harvest, 315
Tinajas (water holes), 28, 30, 71, 72

Tongue color, 56, 75
Topography, 198, 325; and behavior, 139, 144, 300
Tracks, 241–43; deer, 241–42; use in sign reading, 243; weathering, 242–43
Trails, 144, 243–44, 329
Transplants, 44, 47, 48, 49, 51, 192–93, 196, 218, 231–32, 274, 276, 282–85
Transporting, 281–82
Trapping, 9–10, 273–78
Tridens, slim *(Tridens)*, 90, 95
Trophies, 252–53, 336, 341
Tumors, 181
Turpentine broom *(Thamnosma)*, 85
Twinning, 148, 218
Type specimen, 3, 5

Urial (shapu), 2
Urine, 111–12, 135, 240–41

Varnish, desert, 17–18
Varnish bush *(Flourensia)*, 74
Vegetative classes, 80–81, 84, 93, 94, 200
Vegetative seasons, 87, 98
Vegetative types, 66–74, 325–26
Viral diseases, 174
Vision, 114
Voice, 134, 154, 257

Wariness, 120–21, 292–93
Water, 100–12; availability, 77, 89, 100, 208, 212, 316; balance, 100–1, 104, 108–12, and behavior, 107–8, 120, 121, 129–30, 136, 138; competition for, 201, 205–7, 208, 211, 212, 213, 216, 318; conservation (retention), 62–63, consumption, 87, 101–6; and daily movements, 138; dependence on, 105, 108, 129; developments, 107–8, 315–19; and disease, 106; lakes, 269; and lambs, 102–3, 106; nursing ewes, 132; plasma volume, 63, 112; seasonal use, 104–7, 129, 136; sewage ponds, 206; shortages, 106–7, 183; sources, 101, 102, 106, 107, 315–18, 319, 327–28; time of watering, 101–2, 266–67; unsurpation by man, 304–7
Water-food relationships, 89, 102, 107, 111
Water holes, 29–30, 72, 77, 104, 108, 129, 265–69
Weems bighorn, 5, 6, 49
Weights. *See* Measurements
Wells, 317
Wheatgrass *(Agropyron)*, 66, 69
Whipworm, 176–77
Whitethorn; *(Acacia)*, 70, 74, 90; chaparral *(Ceanothus)*, 70, 86; mountain *(Ceanothus)*, 70
Wild Horse and Burro Act, 203
Willow, desert *(Chilopsis)*, 71
Winter fat *(Eurotia)*, 66, 68, 69, 85, 95
Wolf, 186, 187
Wool. *See* Hair

Yucca *(Yucca)*, 68, 72, 74, 85–86, 95